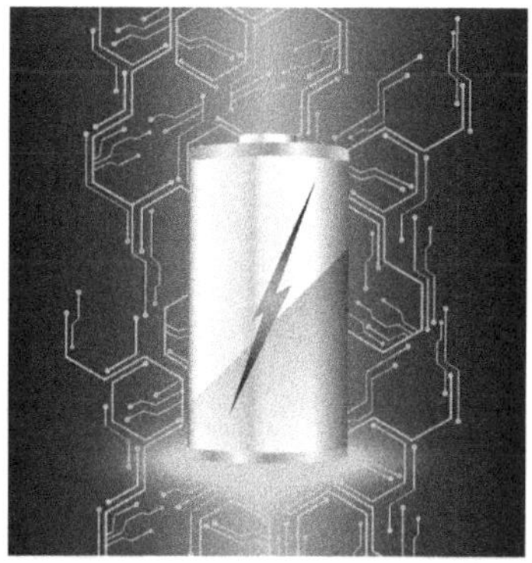

HANDBOOK ON SMART BATTERY CELL MANUFACTURING

The Power of Digitalization

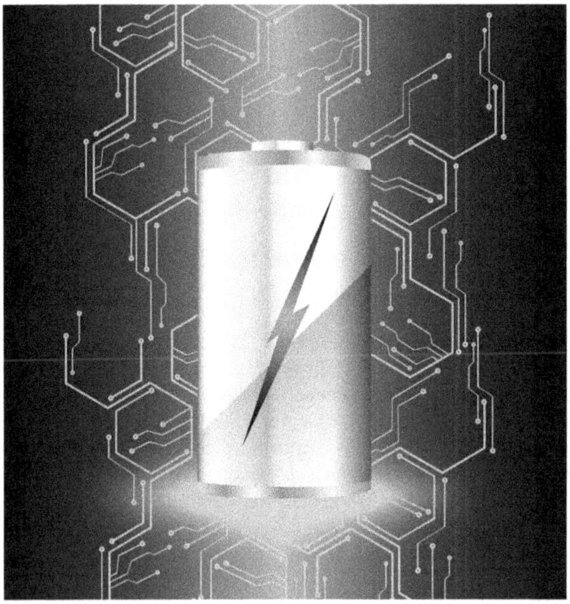

HANDBOOK ON SMART BATTERY CELL MANUFACTURING

The Power of Digitalization

Editors

Kai Peter Birke
University of Stuttgart, Germany
Fraunhofer Institute for Manufacturing Engineering and Automation (IPA), Germany

Max Weeber
Fraunhofer Institute for Manufacturing Engineering and Automation (IPA), Germany

Michael Oberle
Fraunhofer Institute for Manufacturing Engineering and Automation (IPA), Germany

World Scientific

NEW JERSEY • LONDON • SINGAPORE • BEIJING • SHANGHAI • HONG KONG • TAIPEI • CHENNAI • TOKYO

Published by

World Scientific Publishing Co. Pte. Ltd.

5 Toh Tuck Link, Singapore 596224

USA office: 27 Warren Street, Suite 401-402, Hackensack, NJ 07601

UK office: 57 Shelton Street, Covent Garden, London WC2H 9HE

Library of Congress Cataloging-in-Publication Data

Names: Birke, Kai Peter, editor. | Weeber, Max, editor. | Oberle, Michael, editor.
Title: Handbook on smart battery cell manufacturing : the power of digitalization / editors Kai Peter Birke, University of Stuttgart, Germany, Fraunhofer Institute for Manufacturing Engineering and Automation (IPA), Germany, Max Weeber, Fraunhofer Institute for Manufacturing Engineering and Automation (IPA), Germany, Michael Oberle, Fraunhofer Institute for Manufacturing Engineering and Automation (IPA), Germany.
Description: Singapore ; Hackensack, NJ, USA : World Scientific Publishing Co. Pte. Ltd, [2022] | Includes bibliographical references and index.
Identifiers: LCCN 2021038192 | ISBN 9789811245619 (hardcover) | ISBN 9789811245626 (ebook for institutions) | ISBN 9789811245633 (ebook for individuals)
Subjects: LCSH: Electric batteries--Design and construction.
Classification: LCC TK2896 .H36 2022 | DDC 621.31/242--dc23/eng/20211022
LC record available at https://lccn.loc.gov/2021038192

British Library Cataloguing-in-Publication Data

A catalogue record for this book is available from the British Library.

First published 2022 (hardcover)
Reprinted 2026 (in paperback edition)
ISBN 9789819833337 (pbk)

For any available supplementary material, please visit
https://www.worldscientific.com/worldscibooks/10.1142/12511#t=suppl

Desk Editors: Christina Ramalingam/Amanda Yun

Typeset by Stallion Press
Email: enquiries@stallionpress.com

https://doi.org/10.1142/9789811245626_fmatter

About the Editors

Kai Peter Birke is a physicist, materials scientist and a full Professor at the University of Stuttgart, Germany, covering the field of Electrical Energy Storage Systems, including new energy storage cell materials and technologies, advanced Li-Ion batteries and Power-to-X. He obtained his Ph.D. in Materials Science (ion conducting ceramics) from the University of Kiel, Germany, in 1998. In 1999, he joined the Fraunhofer Institute for Silicon Technology, Itzehoe, Germany, to work on the development of proprietary Li-Ion laminated hybrid solid cells with novel functional ceramic separators. He also co-founded two spin-offs to put this technology into production.

After being involved in the development and production of pouch-type laminated Li-Ion cells (PoLiFlex) in leading positions at VARTA AG for five years (2000–2005), Prof. Birke joined Continental AG, Business Unit Hybrid Electric Vehicle, in Berlin in 2005, as a Senior Expert and Project Leader in Energy Storage systems. He subsequently became Senior Technical Expert in Battery Technology and Team Leader in Cell Technology, and in 2010 he was appointed Head of Battery Modules and Electromechanics. In 2013, he joined the JV SK-Continental e-motion as Head of Advanced Development. He was one of the most relevant pioneers for this JV. In 2015, Prof. Birke became a full professor at the University of Stuttgart, Institute for Photovoltaics, ipv, as head of Electrical Energy Storage Systems. Since 2018, he has also led the Center for Battery Cell Manufacturing at the Fraunhofer Institute for Manufacturing Engineering and Automation (IPA), Germany, with an extended focus on battery cell production and digitalization.

Max Weeber specializes in manufacturing engineering and renewable energy systems, and received his engineering degree from the Technical University of Munich (TUM), Germany. In 2013, he started working as a researcher at the Fraunhofer Institute for Manufacturing Engineering and Automation (IPA), Germany. Since then, he has developed and coordinated several research projects together with partners from industry and science. His project engagements cover a broad spectrum of research domains including resource efficiency, digitalization, battery manufacturing and recycling. His research focus is on improving manufacturing sustainability using the tools of digitalization, simulation and optimization. He is author and co-author of more the 30 peer-reviewed publications. In 2018, he was entrusted to establish the Center for Battery Cell Manufacturing at the Fraunhofer IPA together with Professor Kai Peter Birke. Today, Mr. Weeber leads a group of researchers which focus on advancing simulation and optimization methods and their application within battery design and manufacturing.

Michael Oberle is a computer scientist working in the field of smart manufacturing. He obtained his Master's degree from the University of Auckland, New Zealand, in 2012 and joined the Fraunhofer Institute for Manufacturing Engineering and Automation (IPA), Germany, working as a researcher in the Competence Center for digital tools in manufacturing. Initially focusing on manufacturing control in the semiconductor industry, he began leading projects with a focus on the digitalization of battery manufacturing since 2014. One of his main achievements is the fully connected cloud-controlled battery manufacturing pilot line at the Fraunhofer IPA center for battery cell manufacturing. Today, he leads the group for Data and Application Services for Digital Production with a focus on data-driven and event-driven production control services.

https://doi.org/10.1142/9789811245626_fmatter

About the Contributors

Andreas Aichele is a research associate at the Fraunhofer IPA. In addition to researching new approaches to quality management methods for current problems in industry, he supports companies by training and moderating these methods. In interdisciplinary projects, Mr. Aichele supports and works with the respective process experts on the industrialization and optimization of production processes for battery cell manufacturing.

Sabri Baazouzi received his Master's degree in Mechanical Engineering from the University of Stuttgart in 2019. In 2020, he joined the Fraunhofer Institute for Manufacturing Engineering and Automation IPA as a scientific researcher. He is currently working on several research projects dealing with circular economies for battery systems and battery cell production and pursuing his Ph.D. on battery cell design with a focus on tab systems.

Thomas Bauernhansl has been Director of the Fraunhofer Institute for Manufacturing Engineering and Automation IPA in Stuttgart, Germany, since September 2011. He is also Director of the University of Stuttgart's Institute of Industrial Manufacturing and Management (IFF). His scientific focus is on production organization, factory planning, energy-efficient production, surface engineering, automation and process engineering, and digital and biological transformation. Bauernhansl studied mechanical engineering at RWTH

Aachen University. After completing his Ph.D. with honors in 2002, he embarked on his career with Freudenberg in 2003, advancing to become Director of the Technology Center at Freudenberg Sealing Technologies. Bauernhansl was responsible for the regions Europe, North and South America, optimizing the "global footprint" of the production network made up of over 50 production sites.

David Brandt studied computer science at the Offenburg University for Applied Sciences and got his Bachelor's Degree in Applied Computer Science. Since 2019, he has been employed at the Fraunhofer IPA and supports the team with his expertise in database and web development projects for the IT systems landscapes for digital battery manufacturing.

Anna Buss is a research associate at the Fraunhofer Institute for Manufacturing Engineering and Automation in Stuttgart, Germany. There, she develops concepts for flexible manufacturing systems and is engaged in the assessment of different technology solutions for specific challenges of industrial clients. Prior to this, she studied mechanical engineering at HS Osnabruck and the University of Stuttgart.

Joachim Doppler is a Management Consultant at Boston Consulting Group in Stuttgart, with a focus on the Automotive, Industrial and Energy sector. Prior to this, he studied Mechanical and Process Engineering at TU Darmstadt and concentrated on Automotive and Production technology. In 2019, he wrote his Master's thesis at Fraunhofer IPA, in Stuttgart, on activity recognition in assembly processes. During this time, he supported the Fraunhofer as a research assistant in the field of digital battery production.

Ira Effenberger has a diploma in mathematics and conducted her Ph.D. thesis in the field of automated 3D data evaluation for multi-sensor metrology at the University of Stuttgart. Since 2008, Dr. Effenberger has been group manager at the department of "Machine Vision and Signal Processing" at the Fraunhofer IPA in Stuttgart. The research activities of her group are focused on the intelligent processing and evaluation of sensor data in the field of quality assurance as well as object detection and digital models for production environments where the digital shadow plays an important role. In addition, Dr. Effenberger is deputy head of the Center for Cyber Cognitive Intelligence (CCI) at the Fraunhofer IPA, which deals with artificial intelligence and machine learning in industrial applications.

David Ensling received his Ph.D. in materials science from TU Darmstadt. He worked at Ångström Advanced Battery Centre (ÅABC) at Uppsala University before joining VARTA Microbattery in 2009, where he is now working as R&D Manager in business unit V4Drive.

Sascha Gärtner, born in 1993, is a research associate at the Fraunhofer Institute for Manufacturing Engineering and Automation (IPA) in Stuttgart, Germany. As part of his work in the Competence Center for Digital Tools in Production, he researches different topics in the context of digitalization and industry 4.0, particularly focusing on the practical application of artificial intelligence methods in the production environment. He earned his bachelor's degree in mechanical engineering and master's degree in product and process development (2020) at Karlsruhe University of Applied Sciences. Prior to his studies, he completed an apprenticeship as an industrial mechanic in the field of manufacturing technology and gained practical experience in production through his subsequent work in setting up and operating various CNC machines.

Simon Glöser-Chahoud graduated from TU Berlin with a diploma in Industrial Engineering and Management in 2010 and he holds a doctorate degree from TU Clausthal. Before joining the Institute for Industrial Production (IIP) at the Karlsruhe Institute of Technology (KIT), where he currently holds a position as the head of the "Sustainable Value Chains" research group, he worked as a project manager and research associate at the Fraunhofer ISI for more than six years. Simon Glöser-Chahoud is coordinating the accompanying research in the DeMoBat project.

Julian Grimm is research associate at the Fraunhofer Institute for Manufacturing Engineering and Automation IPA in Stuttgart, Germany. There, he is engaged in planning and developing concepts for the planning of industrial energy systems. Furthermore, he is working on concepts for digital twins of machines. Between 2017 and 2019, he worked as a consultant for manufacturing engineering and logistics. Prior to this, he studied Industrial Engineering and finished his Master's degree (M.Sc.) at the University of Kassel (Universität Kassel), Germany.

Peter Grüninger received his Ph.D. in Physical Chemistry from the University of Tübingen, Germany, under the supervision of Prof. Dr. Thomas Chassé and Prof. Dr. Holger F. Bettinger in 2019. In the same year, he joined the Fraunhofer Institute for Manufacturing Engineering and Automation IPA in Stuttgart, where his current research focuses on functional materials.

Christoph Haar, M.Sc., studied industrial engineering at the University of Stuttgart and the Technical University of Ingolstadt. Since 2017, he has been working as a scientist at the Fraunhofer IPA in the department of Factory Planning and Production Management. He is engaged in research in the field of data-based process optimization of different manufacturing processes as well as the automation of process planning tasks.

Silke Hartleif received her Master's degree in industrial engineering from the Karlsruhe Institute of Technology in 2015. In the same year, she joined the Fraunhofer Institute for Manufacturing Engineering and Automation IPA, where she is currently a research associate. Her research interests include the lean and demand-oriented digitalization of battery cell production.

Caroline Hellmich is currently completing her Master's degree in Industrial Engineering at the University of Bayreuth, Germany. In 2019, she completed her bachelor thesis about the *Feasibility of Preparation of Low-Temperature Fuel Cells Concerning Remanufacturing* at the Chair of Manufacturing and Remanufacturing Technology. In addition to her studies, she is working as a research assistant at the Fraunhofer Institute for Manufacturing Engineering and Automation IPA, Bayreuth, Germany. In her work, she investigates solutions for sustainable e-mobility with a focus on electric vehicles and bicycles.

Thomas Hilzbrich is Co-Founder of the Fraunhofer Spin-off plus10, which is developing and offering self-learning optimization software for fully automated and complex manufacturing lines. Between 2016 and January 2020, he worked as a research associate and project leader at the Fraunhofer Institute for Manufacturing Engineering and Automation in Stuttgart in the field of factory planning and production management. There, he focused on data-driven process optimization for complex manufacturing lines using machine learning technologies.

Sandra Huster graduated from KIT in 2019 with a Master's degree in Industrial Engineering and Management. Since then, she has been working as a research associate at KIT, Institute for Industrial Production (IIP). She is part of the DeMoBat project team examining the industrial dismantling of traction batteries.

Ahmad Issa is a senior scientist at the Fraunhofer Institute for Manufacturing Engineering and Automation IPA in Stuttgart, Germany. He obtained his master's and Ph.D. degrees in Engineering from the University of Warwick, United Kingdom, and then joined the research team of the competence center "Digital Tools in Manufacturing" at the Fraunhofer IPA in 2016. His research focuses on digital transformation, Industry 4.0, innovation management and SME support ecosystems. Besides his involvement in national and international projects as a project manager, he is leading Fraunhofer IPA's team within an advanced manufacturing European Digital Innovation Hub in the Baden-Württemberg region.

Duygu Karabelli obtained her Ph.D. in Electrochemistry in 2011 from the University of Grenoble (France), after which she started her Postdoctoral research in the field of next-generation batteries at Fraunhofer ICT (Germany). In 2019, she joined the Center of Battery Cell Manufacturing at Fraunhofer IPA (Germany) as a project manager. Here, Dr. Karabelli supported and coordinated many national and international projects involving the manufacturing of both li-ion battery and all-solid-state li-ion battery technologies. Currently, she works as Head of Process Engineering for CUSTOMCELLS® at the company series production site in Tübingen (Germany).

Steffen Kiemel is project leader of the working group "Management of Sustainable Value Creation Systems" at the Fraunhofer Institute for Manufacturing Engineering and Automation IPA. His research priorities are holistic assessments of processes, products and manufacturing systems as well as the evaluation of strategies for circular economy. Furthermore, he is engaged in the research fields of resource management and efficiency as well as climate neutrality of industrial production. Prior to this, he studied industrial engineering with a focus on resource management at the University of Augsburg.

Jan Koller is a senior project manager at the Fraunhofer Project Group Sustainable Manufacturing in Bayreuth, a Fraunhofer Institute for Manufacturing Engineering and Automation IPA branch. Furthermore, he is a group leader at the Chair of Manufacturing and Remanufacturing Technology at the University of Bayreuth, Germany. In his work, he focuses on the topic of circular economy with special emphasis on remanufacturing. In national and international research and industrial projects, he addresses the challenges of remanufacturing and develops efficient, innovative and sustainable solutions.

Inga Landwehr is specialist topic manager for energy storage cells at the Fraunhofer IPA in Stuttgart, Germany. Her research focuses on electrode production processes, in particular the development of solvent-free dry coating of electrodes used in supercapacitors and lithium-ion batteries, as well as the accompanying characterization and post-mortem analysis of battery cells. Since 2014, she has worked first as a research associate in cooperation with and then as a project manager at the Fraunhofer IPA in the department of coating system and painting technology. Prior to this, she graduated from Friedrich-Alexander University Erlangen-Nuremberg in 2013 with a Master's degree in Chemical and Biological Engineering.

Jonas Lips is a research assistant at Fraunhofer IPA. He obtained his Master's degree in Mechanical Engineering from the University of Stuttgart in 2018 while focusing on the optimization of the dry coating process of energy storage electrodes. Since then, he has continued research on data-based process optimization of battery production processes and is additionally consulting companies in the field of quality management and risk assessments.

Anselm Lorenzoni is Sales Manager Battery Extrusion and Chemical Applications at Coperion GmbH in Stuttgart, Germany. In addition to his sales responsibilities, he manages and coordinates the topic of continuous extrusion of battery slurries across Coperion GmbH's globally operating sites. Previously, Dr.-Ing. Anselm Lorenzoni worked at the Fraunhofer Institute for Manufacturing Engineering and Automation IPA in Stuttgart, Germany in the field of quality and reliability management. He also completed a doctorate in the field of Industry 4.0 and tolerance management.

Florian Maier has been a research associate at the Fraunhofer Institute for Manufacturing Engineering and Automation since 2009. In the working group "IT Applications and Services in Production", he is responsible for IT and IoT structures in production, especially with regard to the integration of Industry 4.0 concepts, work preparation and follow-up and escalation systems, as well as components for data visualization and interaction between man and machine. After completing an apprenticeship and successfully graduating from Furtwangen University with a degree in Business Informatics, he successfully completed his Master of Business Engineering degree at the University of Stuttgart. In current research projects, he and his team are developing a framework for resilience strategies for manufacturing companies.

Pablo Mayer is a Co-Founder of the Fraunhofer spin-off plus10, which is developing and offering self-learning optimization software for fully automated and complex manufacturing lines. Between 2012 and February 2021, he worked as a research fellow at the Fraunhofer Institute for Manufacturing Engineering and Automation in Stuttgart in the field of automation and manufacturing technologies. There, he developed handling and assembly processes for semiconductor and battery manufacturing. He also focused on data-driven process optimization for complex manufacturing lines using machine learning technologies.

Dominik Nemec has an engineering degree in Process Engineering and Energy Technology. At the Fraunhofer Institute since 2001, he is currently the head of the Dispersing Technology Center. His research focuses on the process chain of dispersion technology, from the development of scientific fundamentals to industrial application. Slurry production and electrode manufacturing processes for battery cell production are some examples of his work topics.

Oliver Oechsle is a group manager at the Fraunhofer Project Group Sustainable Manufacturing in Bayreuth, a Fraunhofer Institute for Manufacturing Engineering and Automation IPA branch. He was previously a research associate at the Chair of Manufacturing and Remanufacturing Technology at the University of Bayreuth, where he completed his doctorate in 2014 on the topic of design and optimization of logistics systems and processes. In his work, he focuses on remanufacturing as well as on the design, optimization and management of circular and linear value creation systems. In national and international research and industrial projects, he deals with the challenges of remanufacturing in the areas of manufacturing, operational excellence, supply chain management and business models.

Martin Reisinger is a research group leader at the Fraunhofer Institute for Manufacturing Engineering and Automation in the Industrial Energy Systems department. The research of the group focuses on hybrid systems and energy data analysis. Previously, he worked as a process and IT consultant in the energy industry after graduating with a degree in environmental engineering.

Günther Riexinger studied mechanical engineering at the University of Stuttgart. From 2009–2012, he conducted research at the Fraunhofer IPA in the department of "Digital Factory". Since 2012, he has been project manager at the department of "Factory Planning and Production Management". His main expertise is in the fields of visualization, simulation and optimization of production systems as well as the design, evaluation and implementation of digitalization and technology solutions within production environments.

Sonja Rosenberg graduated from WFI – Ingolstadt School of Management Catholic University of Eichstätt-Ingolstadt (KU) in 2016 with a master's degree in Management Science focusing on supply chain management, production and logistics. She joined the DeMoBat project after working on several national and international research projects at the Institute for Industrial Production (IIP).

Brandon K. Sai leads the research Group of the Autonomous Production Optimization Research Group at the Fraunhofer Institute of Manufacturing Engineering and Automation IPA in Stuttgart. He has been a doctoral student at the Institute of Industrial Manufacturing and Management at the University of Stuttgart since 2019. His research interest is in behavioral models of production systems.

Martina Schiffer is a research associate at the Fraunhofer Institute for Manufacturing Engineering and Automation (IPA). Her research focuses on supply chain management and production planning and control. Before joining the Fraunhofer IPA, she worked for 3.5 years as a Process Engineer for a plant manufacturer for metallurgical and rolling mill technology.

Andreas Schlereth works as a research fellow in the group "Data and Application Services for Digital Production" at Fraunhofer IPA, located in Stuttgart, Germany. He received his bachelor's and master's degrees in industrial engineering and management from Karlsruhe Institute of Technology in Karlsruhe. His expertise lies in the modeling of discrete event simulation of production environments and project management with a focus on the energy sector. His research fields include the conception and implementation of Industrial IoT Platforms in battery manufacturing and realizing an energy-flexible platform for companies and markets.

Christian Schneider is a scientist at the Fraunhofer Institute for Manufacturing Engineering and Automation IPA and a Ph.D. candidate at the Institute for Energy Efficiency in Production (EEP) at the University of Stuttgart. His general research focus is on the optimal selection and prioritization of energy efficiency and decarbonization measures. Before starting his work for Fraunhofer, he earned his Bachelor's and Master's degrees in Industrial Engineering and Management from the Karlsruhe Institute of Technology (KIT).

Klaus Schöbel studied Manufacturing Engineering at Hochschule Aalen. He was working as a LEAN manager at Varta Microbattery before being appointed as the Head of Efficiency and Data Management at Varta Microbattery.

Frank Schultmann holds a Professorship of Business Administration at the Karlsruhe Institute of Technology (KIT) and is director of the KIT's Institute for Industrial Production (IIP) and the KIT's French-German Institute for Environmental Research (DFIU). He is also Professor of Complex Project Management for the University of Adelaide, Australia. He completed his doctoral thesis and his habilitation at the Faculty of Economics and Management at the University of Karlsruhe (now KIT). Prior to his present positions, he was Professor of Industrial Management at the University of Koblenz-Landau and holder of the Chair of Business Administration, Construction Management and Economics at the University of Siegen. His research interests include sustainable production and logistics, decision support, supply chain management and optimization, systems engineering and project management.

Lennard Sielaff is a research assistant focusing on plant and maintenance management in the department of sustainable production and quality at the Fraunhofer Institute for Manufacturing Engineering and Automation IPA. Previously, he worked for an automation components manufacturer as product manager for services after completing his bachelor's and master's degrees in mechanical engineering at the Karlsruhe Institute of Technology (KIT) and a business administration bachelor's degree at the LMU Munich.

Soumya Singh is part of the Center for Battery Cell Manufacturing at the Fraunhofer Institute for Manufacturing Engineering and Automation IPA in Stuttgart, Germany. Here, she is working on the simulation and optimization challenges for batteries and battery cell manufacturing. Along with that, she is pursuing the prospects of the concept of Digital Twins in the battery industry. Previously, Soumya Singh studied Electrical Engineering and Information Technology at the Technical University of Kaiserslautern, Germany.

Friedrich-Wilhelm Speckmann is Project Manager at the center for battery cell manufacturing at the Fraunhofer-Institute for Production Engineering and Automation. There, he is engaged in the active development of hydrogen technologies, especially in process simulation and modeling of the manufacturing process. He has been involved in the field of electric energy storage systems during his Ph.D. work at the Institute for Photovoltaics (ipv) at the University of Stuttgart. Prior to this, Dr.-Ing. Friedrich-Wilhelm Speckmann studied sustainable energy distribution with a focus on energy storage electric grid integration.

Michael Trierweiler has a Master's degree in Systems Engineering and works as a research associate at the Fraunhofer IPA at the department of "Factory Planning and Production Management". He is part of the research group "Assembly Planning and Data Driven Optimization". There, he is engaged in modular manufacturing systems and how to overcome their challenges via digital tools. In his Ph.D. project, he considers the automated planning of reconfigurations of such modular systems.

Johannes Wanner is part of the Center for Battery Cell Manufacturing at the Fraunhofer Institute for Manufacturing Engineering and Automation IPA in Stuttgart, Germany. Here, he is engaged in the battery cell testing, formation and the related cell quality issues in battery cell manufacturing. Previously, he studied Electrical Engineering and Information Technology as well as Electromobility at the University of Stuttgart, Germany.

Roman Weiner is a research assistant at the Fraunhofer Institute for Manufacturing Engineering and Automation (Fraunhofer IPA) in Stuttgart, Germany. Currently, he works there in the Factory Planning and Production Management department and supports various research topics on production planning and control. At the same time, he is studying for a master's degree in mechanical engineering with a focus on factory operations and logistics at the University of Stuttgart. Mr. Weiner is completing his master's thesis in cooperation with the Fraunhofer IPA on the topic of logistics business process design. Between 2017 and 2020, he worked for an international OEM in the automotive industry.

Ozan Yesilyurt is a research fellow of the group "Data and Application Services for Digital Production" at the Fraunhofer IPA in Stuttgart, Germany. He got his bachelor's and master's degrees in electrical engineering and information technology at the Technical University of Munich. His expertise lies in the databased production optimization and product smartification. His research field includes the conception and implementation of the IT system landscape for digital battery manufacturing and enabling energy flexibility with AGV batteries in manufacturing companies.

https://doi.org/10.1142/9789811245626_fmatter

Acknowledgment

The work presented in this book has been supported by the German research projects DigiBattPro 4.0, DeMoBat and S-TEC ZDB.

These projects have received funding from the State Ministry of Baden-Wuerttemberg for Economic Affairs, Labour and Housing Construction with a Grant No 017-100035|B7.R| (DigiBattPro 4.0), Grant No L7520101 (DeMoBat) and Grant No 017-102830 (S-TEC ZDB).

The Editors and Authors of this book are extremely grateful for the support from the Federal State of Baden-Wuerttemberg.

https://doi.org/10.1142/9789811245626_fmatter

Contents

https://doi.org/10.1142/9789811245626_0001

Introduction: Toward Sustainable, High-Quality Battery Cell Manufacturing

Max Weeber*, Julian Grimm†, Sabri Baazouzi‡, Michael Oberle§ and Kai Peter Birke¶

Fraunhofer Institute for Manufacturing Engineering and Automation IPA Center for Battery Cell Manufacturing, Nobelstraße 12, 70567 Stuttgart, Germany

*max.weeber@ipa.fraunhofer.de

†julian.grimm@ipa.fraunhofer.de

‡sabri.baazouzi@ipa.fraunhofer.de

§michael.oberle@ipa.fraunhofer.de

¶kai.peter.birke@ipa.fraunhofer.de

1. The Background

The severe effects of a carbon-fueled economy have been demonstrated to deteriorate the vital resources of our planet. In consequence, the use of renewable energy sources and cross-sectoral electrification has been promoted as a major strategy to reduce greenhouse gas emissions in the mobility, housing and industry sectors. On the path to mitigating the effects of anthropogenic climate change, the growing quantities of fluctuating renewable energy sources need to be handled by the energy system. However, the decarbonization of the power sector with fluctuating renewable energies requires an infrastructure that is able to connect utilities and end users in a flexible manner. In this scenario, diverse types of energy storage technologies, including hydro, thermal and electrochemical, play a vital role. In particular, lithium-ion-based electrochemical energy storage systems are expected to have a high penetration in the market. Their market deployment is expected to exceed 2 TWh worldwide, which corresponds to a tenfold increase over the next ten years.[1] These prospects are sustained by monthly announcements, revealing the installation of new lithium-ion battery (LIB) cell manufacturing capacities.

With more than 70% of total manufacturing capacities, Asian manufactures are expected to dominate LIB production for the years to come. However, the European annual manufacturing capacities are expected to grow to 230 GWh per year until 2025, an eightfold increase compared to 2020 levels.[2]

According to Benchmark Mineral Intelligence statistics, worldwide annual manufacturing capacities for battery cells are expected to surpass 2,582 GWh (Europe 452 GWh) by the year 2030.[3] Considering the growth potential of the European electric vehicle market only, including passenger cars, vans, trucks and buses, McKinsey expects the demand for European based battery cell manufacturing capacities to surpass 1,200 GWh in 2040.[4] Considering an average production capacity of 15 GWh per year and gigafactory, this would translate into a demand for 80 gigafactories in Europe only. However, given the strong competition with the Asian market, our estimates in this chapter assume that only 50% or 600 GWh of the demand for LIBs will effectively be manufactured in Europe by 2040.

The latest announcements of planned manufacturing capacities underline this accelerated development. Between late 2019 and mid-2020, the numbers announced in press releases almost doubled from 326 GWh/a to 613 GWh/a. Currently, the planned capacities exceed 1,000 GWh per year at their final extension levels (2030 and later) (see Fig. 1.).

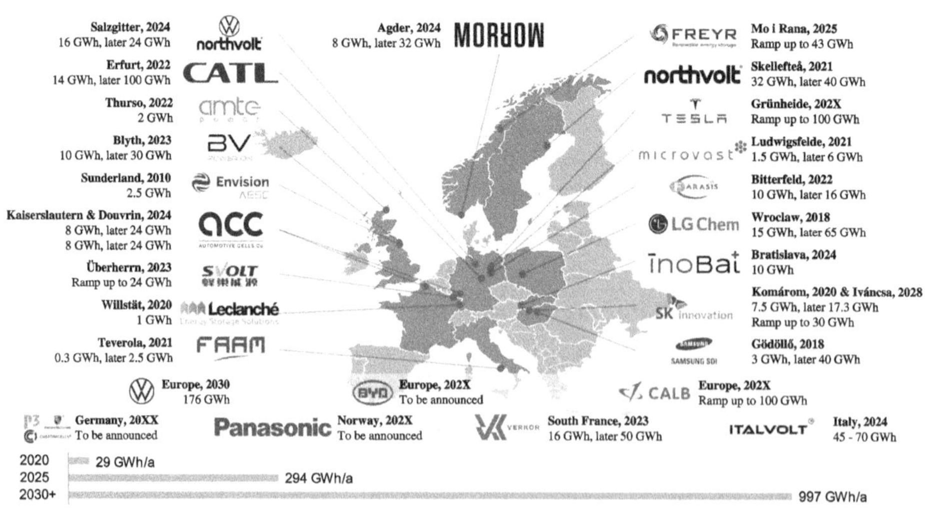

Fig. 1. Planned battery cell manufacturing capacities in Europe by April 2021 — Own illustration based on Ref. [5].

2. The Challenge

2.1. *Global increase in raw material demand for LIB*

Though demand prognoses are subject to future uncertainties, analysts expect the demand for raw materials used in LIB to increase by a factor of 18–20 for lithium, 17–19 for cobalt, 28–31 for nickel and 15–20 for most other materials from 2020–2050.[6] These growth scenarios bear the risk of creating potential supply shortages across the LIB raw material supply chain.[7] Alongside these availability concerns, supply dependencies are also expected to grow due to the concentration of certain mining and refinement activities in only a few countries.[8] What poses another risk is that many of the mining activities are located in politically unstable geographical regions. This also includes ethical concerns related to labor law violations, fair pay and occupational safety. There are many allegations of law infringements related to working conditions along the LIB raw material supply chain, which need to be termed as unacceptable under European standards.[9] Among others, the condition of workers in cobalt mines of the Democratic Republic of Congo is just one severe example.[10]

2.2. *Environmental impact of battery cell manufacturing*

In order to capture the full picture of the environmental impacts from battery cell manufacturing, a life cycle assessment (LCA) approach is required. Probably the most widely discussed example in this context is the comparison between material-intensive battery electric vehicles (BEVs) and fuel-intensive internal combustion engine vehicles (ICEVs).[11–13] Though LCA can be strongly biased by the underlying assumptions, the studies confirm the superiority of BEVs with respect to life cycle greenhouse gas emissions, mainly due to their lower emissions in the use phase. However, the results also underline the necessity to further advance the environmental friendliness of battery manufacturing and disposal in order to achieve improvements across all the impact categories of an LCA (climate change, resource depletion, human toxicity, etc.). For example, great uncertainties are still associated with the end-of-life phase of the battery.[14]

In this context, energy efficiency and material efficiency are two strategies that need to be adopted in order to push the advantage of BEVs over ICEVs from a life cycle perspective.

Today, the energy intensity of battery cell manufacturing varies between 50 and 65 $kWh_{el.}$ per kWh of produced battery capacity.[15] However, from a price per battery perspective, the total energy demand represents only 2–5% of the overall battery cell costs. This is about the same share as for labor costs in LIB

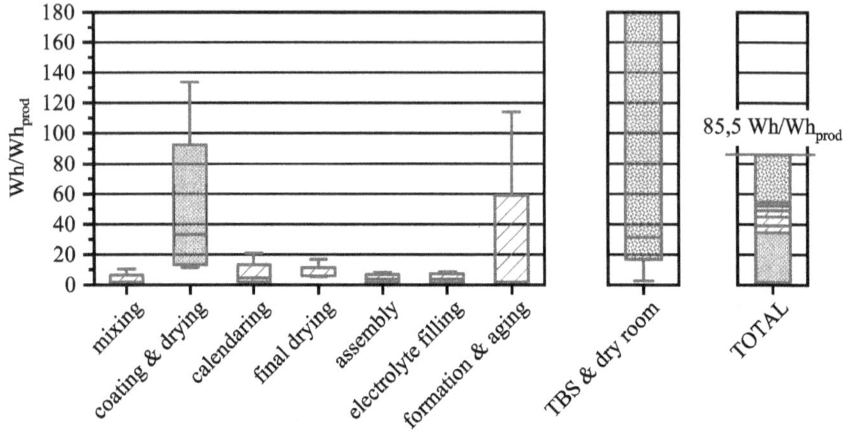

Fig. 2. Energy demand across the process chain for battery cell manufacturing — Own illustration with values from Refs. [16–18].

manufacturing.[2] Figure 2 illustrates the energy demand for individual process steps in battery cell manufacturing based on different references in the literature. According to recent findings, and also considering the energy demand for the factory's technical building systems and dry room operation. Together with the demand for additional energy sources apart from electricity, the specific energy demand in battery cell manufacturing is reported to be up to 85.5 kWh per kWh of produced battery capacity.

Combining the specific energy demand for battery cell manufacturing with the growth scenarios outlined in Section 1, it is possible to estimate the additional energy demand required for a growing battery cell manufacturing industry. Starting with 19,67 TWh in 2025 (EU battery cell manufacturing capacity: 230 GWh) and reaching 38,6 TWh in 2030 (EU battery cell manufacturing capacity: 452 GWh), the additional yearly energy demand will reach 51,3 TWh by 2040 (EU battery cell manufacturing capacity: 600 GWh), if only 50% of the projected European EV battery demand is manufactured inside the EU borders. The energy demand for battery cell manufacturing in 2040 will be equal to the energy generation capacity of seven nuclear power plants (see Fig. 3). In relation to the 2018 final energy demand of the European industrial sector, battery cell manufacturing will account for approximately 1.5% of the sector's final energy demand in 2040.[19] This estimation does not consider any improvements of energy efficiency in the operation of battery cell factories. It is also worth mentioning that the growing energy demand of a new battery industry will most probably be offset, at least partly, by the reduced energy demand of other industry sectors. On the bright side,

Fig. 3. Yearly energy demand for battery cell manufacturing in Europe by 2040 will be equal to the energy generated by 6,9 nuclear power plants (1-GW capacity per plant and a capacity factor of 85%).

more severe regulatory frameworks, including the EU carbon pricing system, will make selecting factory sites in a low-carbon energy infrastructure more appealing. Hence, the comparatively low CO_2 intensity of the European electricity mix presents a potential advantage for the competitiveness of a European cell manufacturing industry.[20] Process innovation such as low energy coating technologies and the use of mini-environments to reduce the need for dry room conditioning can further increase the productivity in future battery cell manufacturing factories.[21–23]

A lithium-ion battery cell in the 21700 format weighs between 60 and 70 g. It is composed of a cylindrical casing (aluminum or steel), an aluminum current collector with the cathode active material referred to as the positive electrode, a polymer separator, a current collector made of copper and graphite, referred to as the anode or negative electrode, and a liquid electrolyte. The maximum energy stored within a 21700 cylindrical cell varies between 13.15 Wh and 16.77 Wh.[24] Thus, a battery cell factory requires 4,800 t of raw materials to manufacture one GWh of battery cell capacity. However, the actual material demand is significantly higher due to wasted material from ramp-up and continuous production, which does not fulfill the quality criteria. Progress in material efficiency can be achieved through product design and process improvements, but it also requires an increase in production yield.[25] Material flow and process input-and-output analysis are methods that can assist during the assessment of material efficiency in battery cell manufacturing. In this context, improving the material yield of the coating/drying process has been identified as a significant lever to reduce wasted material and energy.[25]

2.3. *Quality in battery cell manufacturing*

The numbers on quality-related scrap rates and overall production yields vary considerably for different battery cell manufacturers. The total quality-related scrap rates are reported to be between 6 and 16%.[26–28] There is a general consensus among industry experts that further quality improvements are required in order to improve overall productivity, competitiveness and sustainability of the battery cell

Fig. 4. Yearly manufacturing waste equals the weight of almost six Eiffel towers.

Fig. 5. Yearly manufacturing waste requires the loading of approximately 1,050 cargo train wagons with a total length of 25 km.

manufacturing operations. Low yields and high scrap rates are found mainly in electrode production and formation.[29] Besides, the low yields in individual process steps can easily potentiate along the process chain and significantly lower the overall yield of a battery manufacturing factory.[30] For example, scrap rates of 5% in electrode manufacturing and cell assembly, 10% in formation and 2% in final assembly result in an overall yield of only 95% × 90% × 98% = 84%.[27] Moreover, quality improvements are crucial not only to raise the productivity of battery cell factories but also primarily to increase the longevity of batteries in the field. Consequently, improving and stabilizing quality in battery cell manufacturing can significantly improve the sustainability of a product life cycle that implies LIB technology.

Even if one considers further advances in process stability to be realized in the upcoming years, scrap rates as low as 2% will still lead to 57.500 t of manufacturing scrap per year in the EU in 2040 (assuming 600 GWh of battery cell manufacturing capacities — see Section 1).

In order to illustrate these huge numbers, we provide two relations. First, the annual weight of battery scrap would be the equivalent of almost six Eiffel towers (see Fig. 4).

Second, to transport this annual amount of scrap from A–B, one would require a cargo train with 1,050 wagons with a total length of 25 km (see Fig. 5).

Since manufacturing waste is not the only source of scrap material our society will have to deal with in the years to come, the next section will further stress the importance of sound battery end-of-life and recycling strategies.

2.4. *Battery waste and recycling*

There are strong political initiatives to establish a European battery industry that is able to supply the continent with sustainable "made in Europe" LIBs implying circular economy principles.[31,32] However, designing battery products considering their circular economy performance is still far from being industry practice. The difficulty associated with the dismantling of current EV battery systems underlines the fact that customers and industries are still stuck in traditional cradle-to-grave thinking based on a linear economic model that makes, uses and disposes. Generally, the sources of waste along a battery life cycle can be manifold. Starting from manufacturing waste generated during the ramp-up of electrode production to rejects after formation and aging, other sources of manufacturing waste include rejects in modules and packs or even final product assembly. Figure 6 outlines the participants of the battery supply value chain and their role in establishing a battery circular economy.

Recent studies expect that by 2030, 145 GWh (799,000 tons) of returning batteries will be directed to a reuse application, while 170 GWh (820,000 tons) will be recycled. However, out of this, only 16 % is predicted to be available for the European recycling industry.[33,34] The recycling demand for EV batteries in the EU only is forecasted to be 20,000 units or the equivalent of 3,600 tons in 2030.[35] Figure 7 contrasts the predictions with numbers on the available or planned recycling capacities in Europe. Though strong market growth is just expected for the years to come, even today, there is already a pressuring need to provide proper end-of-life treatment for different consumer LIBs, battery manufacturing "waste" and batteries that come from test vehicles or vehicles involved in accidents.

But, establishing a true circular economy demands more than just end-of-life material recycling. Extending the useful life of batteries in their individual applications or across different usages is crucial for achieving the goal of a battery circular economy that uses a waste hierarchy. In this context, all 3Rs or even 10R

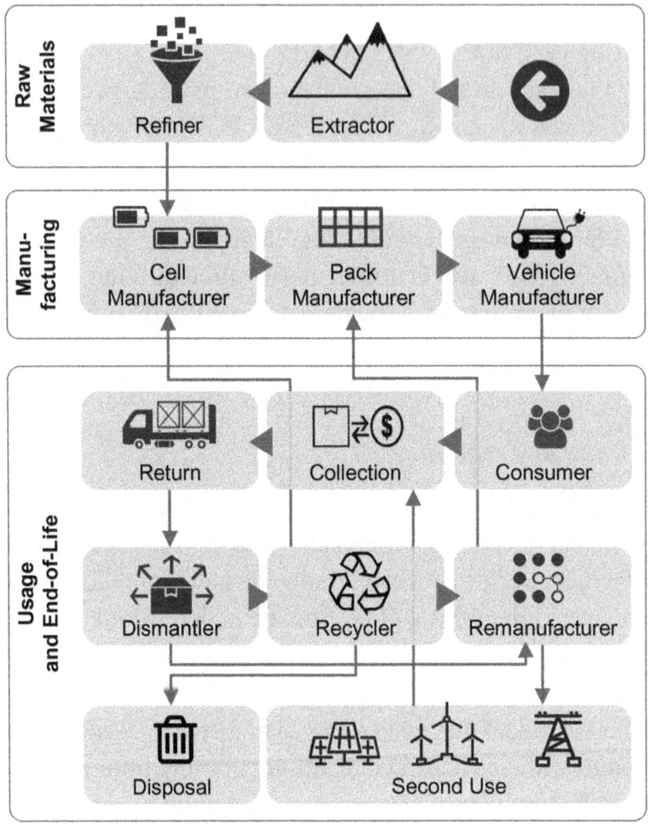

Fig. 6. Participants of the battery supply value chain and their role in establishing a battery circular economy.

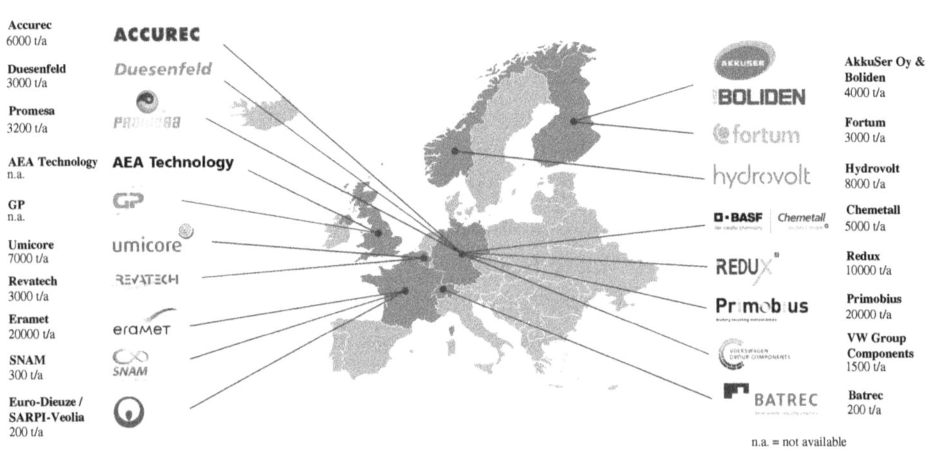

Fig. 7. LIB recycling capacities in Europe (installed and announced) (see also Part V, Chapter 19).

strategies (refuse, rethink, reduce, reuse, repair, refurbish, remanufacture, repurpose, recycle and recover) need to be considered and assessed for their environmental and economic benefits.[36,37] Recent investigations demonstrated that even the changeout of cells becomes an economically viable strategy at the end of a battery's first life.[38] Yet, further design improvements are required at the level of a battery system, including the design of connectors, busbars and frames to facilitate their repair, reuse and remanufacturability.[39] Unfortunately, the uncertainties related to the relevance of existing battery technologies in the future and concerns regarding intellectual property and warranty issues still prevent broad adoption of **repair** and **remanufacturing** practices by both OEMs and third parties.[40]

Making use of diversified end-of-life strategies, including the **repurposing** of batteries in second-life applications (e.g. in stationary application for energy storage and peak shaving), can mitigate their life cycle-related CO_2 emissions and other negative effects across the different impact categories of an LCA.[41,42] In fact, studies found that batteries that have served in EV for 15 years still maintain up to 70% of their original capacity and, if properly used, can last another 10 years in grid energy storage applications.[43] However, other studies also question the economic benefits of repurposing batteries in second-life applications before recycling due to high transportation costs and future advances in the performance of new batteries.[44,45] Besides, it is also possible that the number of available end-of-life batteries will most probably surpass the demand for economically beneficial second-life applications, especially in the years after 2030 (e.g. utility-scale energy storage).[46] Moreover, second-life application can even negatively affect the closed-loop recycling potential of battery materials in the upcoming decades since it postpones the time when primary battery material demand can be met with secondary material from battery recycling.[6]

Even if the strategies mentioned above can significantly increase a battery's useful life, one day, a battery will reach its end of life and will require material **recycling**. Unfortunately, today's commonly used battery recycling processes are still considered very resource-intensive, and some of them even show unsatisfactory recycling efficiencies.[47–49] For example, the recovery of lithium using a state-of-the-art recycling process and its reuse in battery cell manufacturing lead to significantly higher energy consumption and air emissions compared to the use of refined primary Lithium.[50]

Furthermore, many of the recovered materials are actually downcycled since neither quality nor price is superior to processing virgin materials originating from mining activities.[51] Many authors acknowledge the general possibility of reusing recycled battery materials in new batteries without any negative effects on the electrochemical performance.[52,53] However, there are still great challenges ahead considering the economics and quality of purification of active materials at a large scale.[54] Moreover, the decreasing content of valuable raw materials, especially cobalt, in future battery generations further challenges the business models of state-of-the-art battery recycling companies.[55]

3. The Approach

In our research, we analyze the prospects of transforming the battery cell manufacturing industry by applying the tools of digitalization. We identify current challenges, develop methods and evaluate their usability across a battery life cycle. The objective is to shape the development of an economically beneficial and environmentally friendly battery manufacturing sector in Europe. We envision battery manufacturers that operate gigafactories with a minimal environmental footprint using renewable energy sources and generating minimal waste. We expect the new battery manufacturing companies to always focus on reducing their overall resource demand without compromising on their competitive edge. Their success will be sustained by the superior quality of their product and the maximum flexibility of their manufacturing operations. Flexibility toward changing customer demands, disruptive technologies and unforeseen economic events will increase robustness and can present a competitive edge to a European battery cell manufacturing sector.

4. The Solution

Digitalization in its most diverse manifestations has already demonstrated its disrupting character in many industries and across different business segments. The promises associated with the digitalization of industries are manifold, and the expectations are high. Today, the question is not whether the battery industry has to adopt a digital agenda but rather how individual companies can effectively translate the promises into quantifiable results. Nevertheless, a digital transformation process is still associated with major uncertainty and decision-makers find it difficult to weigh benefits against associated challenges. The tricky question is to answer the following: How can battery companies actively shape this process for their individual benefit instead of being passively dragged alongside an overwhelming number of new technological developments?

We expect that digitalization will enable the flexible factory of the future, which is able to manufacture according to customer preferences and makes production changes including the integration of new machines at any time without substantial effort. Machine utilization levels will be higher, the consumption of resources will drop and there will be fewer rejects.

By using the tools of digitalization, the battery cell manufacturing process will undergo continuous optimization and improve its self-organizing capabilities. In the future, individual processes, complete process chains and even entire factories will autonomously change their parameter settings to offset various disturbances that influence stable production conditions.

Digital Twin technology will connect real-world products and processes with their virtual representation and assist to accelerate the development of new battery cell technologies and their respective manufacturing processes but also to build and ramp-up flexible gigafactories.

Traceability and information transparency across the entire battery value chain will increase and lay the foundation for cross-cutting business models and highly efficient recycling technologies necessary to establish a true battery circular economy.

The impact of digitalization on the battery industry can be manifold and we hope that our book will be able to contribute to a better understanding of its possibilities and limitations within the context of battery cell design and manufacturing.

References

1. BloombergNEF. Electric Vehicle Outlook 2020. New York: Bloomberg New Energy Finance; 2020 [cited 2021 Feb 10]. Available from: URL: https://about.bnef.com/electric-vehicle-outlook.
2. Bernhart W, Gabaldon D, Zheng R, Schmitt P, Hotz T, Braun S *et al.* Rising opportunities for battery equipment manufacturers; 2020.
3. Kumar V. Lithium-Ion Battery Megafactory Assessment. Benchmark Mineral Intelligence; 2020. Available from: URL: https://www.benchmarkminerals.com/wp-content/uploads/06252020-Vivas-Kumar-HyperChange-V4.pdf.
4. Eddy J, Pfeiffer A, van de Staaij J. Recharging economies: The EV-battery manufacturing outlook for Europe. McKinsey & Company 2019. Available from: URL: https://www.mckinsey.com/industries/oil-and-gas/our-insights/recharging-economies-the-ev-battery-manufacturing-outlook-for-europe.
5. Zenn R. Projection of installed capacities for battery cell manufacturing in Europe 2021 [cited 2021 Apr 26]. Available from: URL: https://twitter.com/ZennRoland/status/1353821694411026434/photo/1.
6. Xu C, Dai Q, Gaines L, Hu M, Tukker A, Steubing B. Future material demand for automotive lithium-based batteries. Commun Mater 2020; 1(1):1–10.
7. Olivetti EA, Ceder G, Gaustad GG, Fu X. Lithium-Ion Battery Supply Chain Considerations: Analysis of Potential Bottlenecks in Critical Metals. Joule 2017; 1(2):229–43.
8. Ahmad Mayyas, Darlene Steward, Margaret Mann. The case for recycling: Overview and challenges in the material supply chain for automotive li-ion batteries. Sustainable Materials and Technologies 2019; 19:e00087. Available from: URL: https://www.sciencedirect.com/science/article/pii/S2214993718302926.

9. Amnesty International. Amnesty challenges industry leaders to clean up their batteries; 2019 [cited 2021 Feb 10]. Available from: URL: https://www.amnesty.org/en/latest/news/2019/03/amnesty-challenges-industry-leaders-to-clean-up-their-batteries/.
10. Amnesty International. This is what we die for: Human rights abuses in the Democratic Republic of the Congo power the global trade in cobalt; 2016 [cited 2021 Feb 10]. Available from: URL: https://www.amnesty.org/en/documents/afr62/3183/2016/en/.
11. Bauer C, Hofer J, Althaus H-J, Del Duce A, Simons A. The environmental performance of current and future passenger vehicles: Life cycle assessment based on a novel scenario analysis framework. Applied Energy 2015; 157:871–83.
12. Tagliaferri C, Evangelisti S, Acconcia F, Domenech T, Ekins P, Barletta D *et al.* Life cycle assessment of future electric and hybrid vehicles: A cradle-to-grave systems engineering approach. Chemical Engineering Research and Design 2016; 112:298–309.
13. Del Pero F, Delogu M, Pierini M. Life Cycle Assessment in the automotive sector: a comparative case study of Internal Combustion Engine (ICE) and electric car. Procedia Structural Integrity 2018; 12:521–37.
14. Bothe D, Steinfort T. Cradle-to-Grave-Lebenszyklusanalyse im Mobilitätssektor: Metastudie zur CO2-Bilanz alternativer Fahrzeugantriebe. Forschungsvereinigung Verbrennungskraftmaschinen e.V. (FVV); 2020.
15. Davidsson Kurland S. Energy use for GWh-scale lithium-ion battery production. Environmental Research Communications 2019; 2(1):12001.
16. Schünemann J-H. Modell zur Bewertung der Herstellkosten von Lithiumionenbatteriezellen. Sierke; 2015.
17. Pettinger K-H, Dong W. When Does the Operation of a Battery Become Environmentally Positive? J. Electrochem. Soc. 2016; 164(1):A6274-A6277.
18. Yuan C, Deng Y, Li T, Yang F. Manufacturing energy analysis of lithium ion battery pack for electric vehicles. CIRP Annals 2017; 66(1):53–6.
19. Odyssee-Mure. Final energy consumption by sector in EU; 2021 [cited 2021 Apr 8]. Available from: URL: https://www.odyssee-mure.eu/publications/efficiency-by-sector/overview/final-energy-consumption-by-sector.html.
20. Helms H, Kemper C, Biemann K, Lambrecht U, Jöhrens J, Meyer K. Klimabilanz von Elektroautos: Einflussfaktoren und Verbesserungspotential; 2019.
21. Mauler L, Duffner F, Leker J. Economies of scale in battery cell manufacturing: The impact of material and process innovations. Applied Energy 2021; 286:116499.
22. Schünemann J-H, Dreger H, Bockholt H, Kwade A. Smart Electrode Processing for Battery Cost Reduction. ECS Transactions 2016; 73(1):153–9.
23. Thiede S, Turetskyy A, Loellhoeffel T, Kwade A, Kara S, Herrmann C. Machine learning approach for systematic analysis of energy efficiency potentials in manufacturing processes: A case of battery production. CIRP Annals 2020; 69(1):21–4.
24. Quinn JB, Waldmann T, Richter K, Kasper M, Wohlfahrt-Mehrens M. Energy Density of Cylindrical Li-Ion Cells: A Comparison of Commercial 18650 to the 21700 Cells. J. Electrochem. Soc. 2018; 165(14):A3284-A3291. Available from: URL: https://iopscience.iop.org/article/10.1149/2.0281814jes.
25. Wessel J, Turetskyy A, Cerdas F, Herrmann C. Integrated Material-Energy-Quality Assessment for Lithium-ion Battery Cell Manufacturing. Procedia CIRP 2021; 98:388–93.
26. Kirchhof M, Haas K, Kornas T, Thiede S, Hirz M, Herrmann C. Failure Analysis in Lithium-Ion Battery Production with FMEA-Based Large-Scale Bayesian Network: MDPI AG; 2020.
27. Hakimian A, Kamarthi S, Erbis S, Abraham KM, Cullinane TP, Isaacs JA. Economic analysis of CNT lithium-ion battery manufacturing. Environmental Science: Nano 2015; 2(5):463–76.
28. Brodd RJ, Helou C. Cost comparison of producing high-performance Li-ion batteries in the U.S. and in China. Journal of Power Sources 2013; 231:293–300. Available from: URL: http://www.sciencedirect.com/science/article/pii/S0378775312018940.

29. Sakti A, Michalek JJ, Fuchs ER, Whitacre JF. A techno-economic analysis and optimization of Li-ion batteries for light-duty passenger vehicle electrification. Journal of Power Sources 2015; 273:966–80.
30. Michaelis S, Rahimzei E, Kampker A, Heimes H, Offermanns C, Locke M *et al.* Roadmap Batterie - Produktionsmittel 2030. Frankfurt am Main: VDMA Batterieproduktion; 2020.
31. Capagemini invent. Fit For Net-Zero:: 55 Tech Quests to accelerate Europe's recovery and pave the way to climate neutrality.
32. EU. Green Deal: Sustainable batteries for a circular and climate neutral economy; 2020. Available from: URL: https://ec.europa.eu/commission/presscorner/detail/en/ip_20_2312.
33. Circular Energy Storage. The lithium-ion battery life cycle report. Circular Energy Storage; 2020.
34. Willuhn M. Europe's battery recycling quotas are blunt and a decade too late; 2020. Available from: URL: https://www.pv-magazine.com/2020/12/16/europes-battery-recycling-quotas-are-blunt-and-a-decade-too-late/.
35. Elementenergy. Batteries on wheels: the role of battery electric cars in the EU power system and beyond. enel; Iberdrola; TE Transport & Environment; Groupe Renault; 2019.
36. Kirchherr J, Reike D, Hekkert M. Conceptualizing the circular economy: An analysis of 114 definitions. Resources, Conservation and Recycling 2017; 127:221–32.
37. Potting J, Hekkert MP, Worrell E, Hanemaaijer A. Circular economy: measuring innovation in the product chain. The Hague: PBL Publishers; 2017. (vol 2544).
38. Mathew M, Kong QH, McGrory J, Fowler M. Simulation of lithium ion battery replacement in a battery pack for application in electric vehicles. Journal of Power Sources 2017; 349:94–104.
39. Kampker A, Wessel S, Fiedler F, Maltoni F. Battery pack remanufacturing process up to cell level with sorting and repurposing of battery cells. Jnl Remanufactur 2020; (11):1–23. Available from: URL: https://link.springer.com/article/10.1007/s13243-020-00088-6.
40. Hartwell I, Marco J. Management of intellectual property uncertainty in a remanufacturing strategy for automotive energy storage systems. Jnl Remanufactur 2016; 6(1):169–85.
41. Ahmadi L, Yip A, Fowler M, Young SB, Fraser RA. Environmental feasibility of re-use of electric vehicle batteries. Sustainable Energy Technologies and Assessments 2014; 6:64–74.
42. Richa K, Babbitt CW, Nenadic NG, Gaustad G. Environmental trade-offs across cascading lithium-ion battery life cycles. The International Journal of Life Cycle Assessment 2017; 22(1):66–81.
43. Neubauer J, Smith K, Wood E, Pesaran A. Identifying and Overcoming Critical Barriers to Widespread Second Use of PEV Batteries. Golden, CO (United States): Office of Scientific and Technical Information (OSTI); 2015.
44. McLoughlin F, Conlon M. Secondary re-use of batteries from electric vehicles for building integrated photo-voltaic (BIPV) applications 2015.
45. Zu C-X, Li H. Thermodynamic analysis on energy densities of batteries. Energy & Environmental Science 2011; 4(8):2614.
46. Engel H, Hertzke P, Siccardo G. Second-life EV batteries: The newest value pool in energy storage. Frankfurt, Detroit, San Francisco: McKinsey & Company; 2019. Available from: URL: https://www.mckinsey.com/industries/automotive-and-assembly/our-insights/second-life-ev-batteries-the-newest-value-pool-in-energy-storage.
47. Harper G, Sommerville R, Kendrick E, Driscoll L, Slater P, Stolkin R *et al.* Recycling lithium-ion batteries from electric vehicles. Nature 2019; 575(7781):75–86.
48. Ciez RE, Whitacre JF. Examining different recycling processes for lithium-ion batteries. Nat Sustain 2019; 2(2):148–56.
49. Velázquez-Martínez O, Valio J, Santasalo-Aarnio A, Reuter M, Serna-Guerrero R. A Critical Review of Lithium-Ion Battery Recycling Processes from a Circular Economy Perspective. Batteries 2019; 5(4). Available from: URL: https://www.mdpi.com/2313-0105/5/4/68.

50. Golroudbary SR, Calisaya-Azpilcueta D, Kraslawski A. The Life Cycle of Energy Consumption and Greenhouse Gas Emissions from Critical Minerals Recycling: Case of Lithium-ion Batteries. Procedia CIRP 2019; 80:316–21.
51. Beaudet A, Larouche F, Amouzegar K, Bouchard P, Zaghib K. Key Challenges and Opportunities for Recycling Electric Vehicle Battery Materials. Sustainability 2020; 12(14):5837.
52. Thompson DL, Hartley JM, Lambert SM, Shiref M, Harper GDJ, Kendrick E *et al.* The importance of design in lithium ion battery recycling – a critical review. Green Chem. 2020; 22(22):7585–603.
53. Zheng Z, Chen M, Wang Q, Zhang Y, Ma X, Shen C *et al.* High Performance Cathode Recovery from Different Electric Vehicle Recycling Streams. ACS Sustainable Chem. Eng. 2018; 6(11):13977–82.
54. Doose S, Mayer JK, Michalowski P, Kwade A. Challenges in Ecofriendly Battery Recycling and Closed Material Cycles: A Perspective on Future Lithium Battery Generations. Metals 2021; 11(2):291.
55. Niese N, Pieper C, Arora A, Xie A. The Case for a Circular Economy in Electric Vehicle Batteries. Boston Consulting Group (BCG); 2020 Sep 14 [cited 2021 Apr 9]. Available from: URL: https://www.bcg.com/de-de/publications/2020/case-for-circular-economy-in-electric-vehicle-batteries.

Part I

Smart Manufacturing in Battery Cell Manufacturing

https://doi.org/10.1142/9789811245626_0002

Chapter 1

Overview of the Battery Manufacturing Process

Duygu Karabelli* and Michael Oberle†
Fraunhofer IPA, Nobelstraße 12, Stuttgart, Germany
**duygukarabelli@gmail.com*
†michael.oberle@ipa.fraunhofer.de

Environmental sustainability begins with the manufacture of the individual material components, and continues with the design of energy- and material-efficient production systems and factories as well as reusability and recyclability of the product. Strengthening the quality and the flexibility of the battery production line is an important step for the establishment of competitive cell manufacturing. In order to derive decisive competitive advantages, all production chains of high lithium-ion cells and batteries should be optimized via industry 4.0 tools. This chapter will give an overview of current Li-ion battery manufacturing processes including the cost factors.

1. The Rise of Li-Ion Batteries

Although the battery history started with Benjamin Franklin in 1749, battery science and engineering significantly increased after Sony introduced the first commercial rechargeable lithium-ion batteries (LIBs) in 1991.[1,2] Since then, the LIBs have been one of the most important components of portable electronic devices, electric vehicles (EVs) and even for stationary storage devices. Among these listed segments, EVs are strong contributors to the expanding market of Li-ion cells because an EV battery system is much larger in ampere hour (Ah) as compared to consumer applications. Automotive traction batteries can contain up to 7,000–10,000 single cylindrical cells of the formats 18650-type (18 mm diameter, 65 mm height) or 21700-type (21 mm diameter, 70 mm height).[3] In consequence, the researchers have been working on novel materials, a combination of different cell

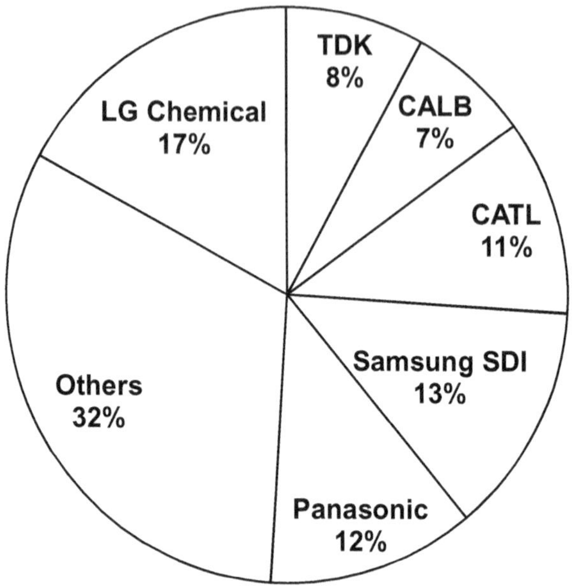

Fig. 1. Global Li-ion battery suppliers and their market share. The figure has been redrawn according to the study of Frost & Sullivan.[4]

chemistries and new cell designs in order to obtain better performances depending on their application areas.

In the global battery market, Li-ion cells have become the preferred choice for rechargeable lithium systems. Nearly all leading manufacturers of Li-ion batteries are located in Asia and are increasingly establishing manufacturing facilities in the USA and Europe to be able to serve the global market. Figure 1 shows the worldwide Li-ion suppliers and their market share in 2018.

However, the battery production lines and new business models are also a key to success. The whole picture of battery production includes building of a cell, module, pack and system, which makes the production chain more complex than one would assume (Fig. 2).

Typical xEV packs consist of hundreds to thousands of cells, connected in series and/or parallel (depending on the voltage and capacity requirements of the application) configuration within the modules, and modules are connected to create the battery pack.

It is known that the quality and the performance of the battery pack and system require a sophisticated design, starting from cell chemistry to cell housing, packing and BMS.[5] So far, there are three different cell formats, namely,

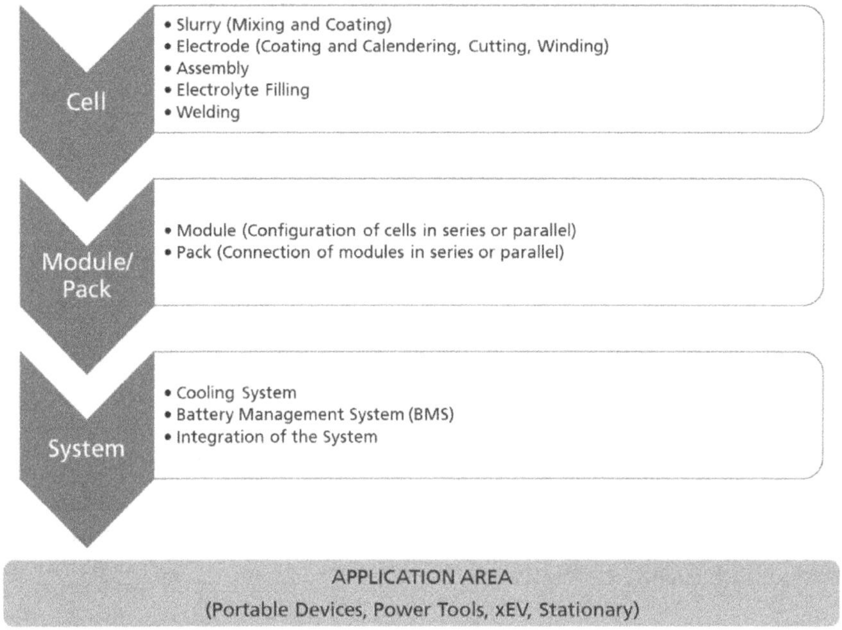

Fig. 2. Battery production chain from cell to system.[7]

cylindrical, prismatic and pouch cells[6] and each cell format brings along its own series of challenges.

Nevertheless, electrode production and battery cell formation are independent of the cell and module architecture.

2. Current Status and Challenges for Cell Manufacturing

The European battery cell manufacturing industry is on its rise. However, there are several challenges associated with the development of competitive, high-quality battery cell manufacturing capacities in prospective GigaFactories.

Asia is currently leading the lithium-ion battery cell manufacturing. Panasonic, LG Chem, Samsung SDI and CATL are very well established companies. From the European side, well-known cell manufacturers such Northvolt, Saft, Varta and Leclanché are aiming to increase their production capacities as well.

Although cylindrical and prismatic cells are two of the most dominant battery formats for EV applications, some EV manufacturers such as Nissan and Renault are using pouch cells. As a result, battery manufacturing companies provide battery cell formats according to the request of their customers. Selected companies and the cell formats that they are producing are given in the Table 1.

Table 1. List of selected companies with their cell types.

Company	Country	Cell Type
Panasonic	Japan	Cylindrical
LG Chem	South Korea	Pouch
Samsung SDI	South Korea	Prismatic
CATL	China	Prismatic
Leclanché	Switzerland	Pouch
Saft	France	Cylindrical
Varta	Germany	Cylindrical, Pouch

In total, there is an estimation of a global demand of 1,200 GWh battery capacity in the year 2030, which increases to 3,500 GWh by 2050.[8] In this chapter, the production process and the possible improvement within the cell manufacturing chain will basically be discussed.

2.1. *Electrode manufacturing*

LIBs have four most important components: Negative and positive electrodes (anode and cathode, respectively), an electrolyte solution and a separator. Each component has contributions on internal resistance, capacity, and energy and power density of the cells. The choice of electrode materials is the first step in order to reach certain capacities and the choice of cathode active material is the most challenging one.[9] Among many materials, the most popular 4V cathode materials are Nickel-Manganese-Cobalt (NMC) and Nickel-Cobalt-Aluminum (NCA), due to their higher capacity than other electrode materials (such as lithium iron phospahate and lithium manganese oxide), long life, acceptable safety, as well as voltage slope as function of the state of charge (SoC).[10] NMC and NCA electrodes are favorable not only for xEV applications but also for many other applications.[11–13] In order to get the highest performance from these materials, manufacturing parameters such as thickness, porosity and particle distribution have to be optimized which is directly related to their manufacturing processes.[14] Electrode preparation starts with slurry mixing, and continues with the coating of the slurry on a current collector and calendaring and drying. All these steps together are responsible for the electrode structure and therefore the cell performance as well. An overview of current electrode manufacturing steps are given in this section.

2.1.1. *Slurry mixing, coating and calendaring*

The main challenges for the Li-ion technology are related to its chemistry. Hence, the formulation of the anode and cathode plays an important role for the Li-ion battery characteristics.

Apart from the main active materials (Ni, Mn, Co, Al, Fe), the slurries consist mainly of carbon black or graphite (KS-6),[15] solvent (N-methyl-2-pyrrolidone, NMP, water), binder (poly-vinylidene fluoride, PVdF) and some additives.[16–18] The type of solvent defines the system as organic or aqueous.[19] The composition of each material, so called an electrode formulation, is very important to create a battery with improved electrochemical performances. Additionally, homogenous mixing and dispersing play important roles in electrode stability and quality (Fig. 3). Liu et al. reported that a good slurry should not be agglomerated, should have uniform particle distribution, should have good rheological characteristics such as fluidity to coat easily and should have high viscosity.[19] In order to reach these rheological properties, the mixer has to be chosen carefully and should ensure that all particles are distributed uniformly. Before the coating of the slurry on an electrode foil, particle size, viscosity and homogeneity of the slurry have to be checked.

It must be kept in mind that the mixer for the anode and cathode preparation has to be different due to different slurry formulations. The procedure of mixing can be discontinuous (with dissolver and/or kneader) or continuous (with an extruder) and each procedure causes different rheological and morphological properties. So far, the most common procedure of mixing is a discontinuous process,[17] and the ball mill mixer is an industrial reference mixer even though the

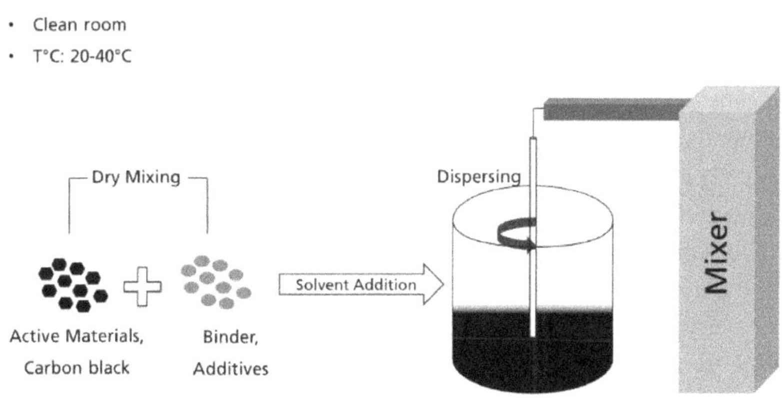

Fig. 3. Schematic illustration of a slurry preparation.

Fig. 4. Coating and calendaring of the electrode foils: (a) continuous coating (b) intermittent coating (for pouch cells).

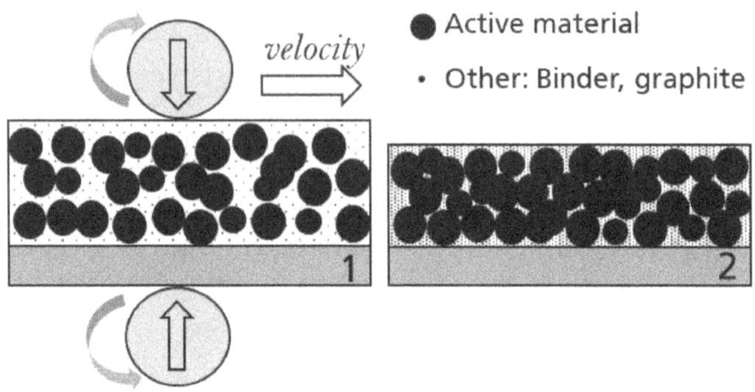

Fig. 5. Schematic illustration of the calendaring process and the morphology of the electrode (1) before and (2) after calendaring.

energy consumption of this mixer is very high.[19] The continuous production of the slurry and the interlinking with the coating process are possible solutions for reduction of the mixing time.[7]

Once the slurries are ready to be applied, the aluminum and copper foils are coated with cathode and anode slurries, respectively. The old-fashioned double-layer electrode manufacturing is based on a non-simultaneous technique in order to overcome problems related to adhesion forces.[20] Most commonly, the coating band is directed to a dryer before the other side of the foil is coated (Fig. 4).[15]

In recent years, simultaneous slot-die coating has gained attention in order to manufacture double-layered cathodes simultaneously.[15] The slurry is delivered onto a substrate via a narrow slot with upper and bottom feed. This method is capable of achieving extremely uniform films with precise thickness over large areas.[20]

Another coating process is based on the "no-solvent" technique. In order to reduce the amount of solvents, toxicity and costs, researchers have been paying attention to dry coating.[21,22] Nevertheless, the lack of solvent can make it difficult to obtain a thin layer as well as good adhesion and stability. Some works have stated that after coating, the electrode surface can peel off from the foil during winding.[23] The optimization of the dry coating is thereby indispensable.

After the drying process, the electrodes are compressed by rollers (calendaring) in order to make them uniform and modify the surface structure (Fig. 5).

Depending on the force of the rollers and calendaring velocity, the thickness and the porosity of the electrode coating can be altered. Meyer *et al.* described that calendaring the cathode and the anode with the same parameters causes completely different results.[24] When calendaring force increases, the electrode is more compact and hence has less porosity. Zheng *et al.*[25] observed calendaring effects on NMC 111 cathode. It has been reported that, when the calendaring force is not high and the electrode has high porosity, the coating can very easily break due to less contact between active and inactive (adhesion) materials. Good contact between all particles creates a continuous network which helps to increase electronic conduction (through percolation theory).[26] On the contrary, when the compression is too much, internal stress increases and electrochemical deterioration may occur due to the structural destruction. Additionally, the wettability of the electrode decreases if the electrode does not have sufficient porosity. The anode should have higher porosity compared to the cathode because Li^+ ion intercalation takes place in the anode. Li^+ ions should merge easily into the anode during charging.[24]

In order to increase electrode quality, in-line measurements (force, thickness, band width) can be integrated to the calendaring process[7] which helps to identify product and process parameters and optimize the quality.

2.1.2. *Slitting and final drying*

The last steps of electrode manufacturing are cutting the electrodes according to cell type, welding tabs and final drying.

Various cutting solutions, namely, laser, knife/die cutting and punching, are available in battery manufacturing industries (Fig. 6). As the electrodes are coated on a long band, later on they must be cut and divided into several smaller foils depending on the application.[27–30] In a conventional way, knives (dies) are mostly used for cutting Li-ion battery electrodes, but the quality and stability of the process are not of a high level.[27,30]

Recently, laser technology has been developing for this process as an alternative technology and is industrially available.[31] Laser cutting offers key benefits of a "contact-free" process and flexibility. Depending on the thickness of the electrode, the cutting speed can be adjusted by increasing laser power.[32] However, it should be considered that the active material can be damaged in case of high cutting speeds and high laser power. Also, contamination in the cell may occur due to metal spatters and active material particles induced by the laser–material interaction.[27]

The final step of the electrode manufacturing is "drying". The speed, temperature and the atmosphere of the drying process have a significant impact on the

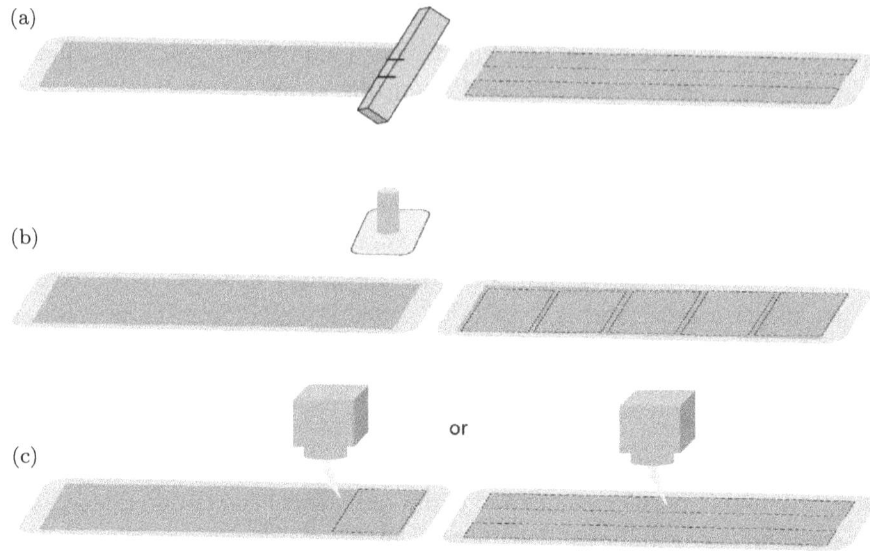

Fig. 6. Different electrode slitting methods (a) die cutting, (b) punching and (c) laser cutting.

electrode quality.[33–36] Depending on the drying parameters, particle distribution and thus the porous structure may vary.

As the NMP is the most common solvent used in cathode slurries, the heat needed to dry the electrode can be generated around 130 kW (average temperature around 120°C, under vacuum).[37,38] As it can be predicted, the drying process does have a major impact on battery cell manufacturing costs as well, due to high energy requirements and long process time.[39]

The long process time can be reduced by increasing the drying speed. This speed is generally optimized by applying higher temperature, but one should consider that high temperatures can cause adhesion problems which result in cracks on the electrode surface.[39]

For more cost reduction, during the drying process, NMP can be removed and collected for recycling which directly affects the battery cost with 96% of the recycling rate.[38]

2.2. *Cell assembly*

The standard format for cylindrical cells is 18650, which is known as the "old type" battery in Tesla EVs. Tesla is replacing its battery technology with the new 21700.[40]

EVs such as Nissan Leaf and BMW i3 equip battery packs with prismatic rather than cylindrical cells.[41] The electrode-separator assembly shows slight differences depending on the cell type. The electrode and separator are wound in a round or flat manner, depending on the cylindrical or prismatic cell format. In pouch cells, single sheets of electrodes are stacked and separators are stacked or z-folded.[42]

After packaging, the cell is activated with *electrolyte filling*. This step can be considered as a critical process because some parameters such dosing method, temperature and pressure have a direct effect on the cell quality. The electrodes must be completely wet, not only at the surface but also through pores. That is why this process is performed under pressure (either pressure by inert gas or under vacuum) in order to create a capillary effect to push electrolyte into the pores.[43,44] Machinery and equipment of an electrolyte-filling process vary by the cell format and are not flexible.

The filled cells are stored for hours, which is known as the soaking time (*wetting*). This time allows a homogenous electrolyte distribution between electrodes and separator layers.

2.3. *Cell finishing*

2.3.1. *Formation*

After wetting, the cell is ready for the formation process. This process includes several charge-discharge cycles (at different C-rates, depending on the cell chemistry) in order to create a solid electrolyte interphase (SEI). The SEI plays a central role in ensuring the reversibility of the Li^+ intercalation/ de-intercalation process in the following battery life cycles. The SEI layer has electrical insulating properties, but allows the transport of Li^+ and prevents the electrolyte from further degradation (Fig. 7).

The sequence of initial charge/discharge cycles and rest periods is required to produce a stable and uniform SEI layer. This time-consuming process represents an industrial bottleneck and requires specific instrumentation and a large area of the battery manufacturing plant, which has a significant impact on the manufacturing costs. In addition, cell formation can take up to 24–32 hours,[45] resulting in high energy consumption and capital costs. Therefore, formation represents a conflict of objectives between process time and product quality, and an acceleration of the cell formation step, with uniform cell quality, is highly desirable.[46]

2.3.2. *Aging*

The formation process is followed by an aging procedure, which is the final and quality control step. The cells are stored between 14 and 21 days in shelves.[34]

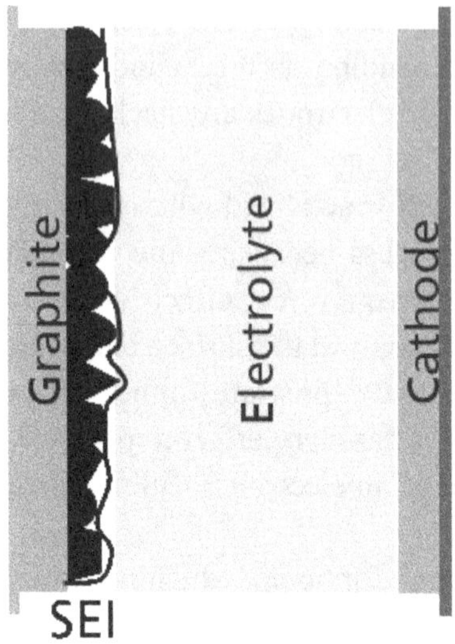

Fig. 7. SEI layer in LIB cell.

After this period, if the cells show no change in terms of their electrochemical behaviors (such as capacity over the entire state of charge-SoC range and impedance), they are considered fully functional and can be used in modules. The cells which do not meet the full requirements are sorted according to their electrochemical characteristics and can be used for alternative applications or rejected.

Since this process is also a time-consuming process, it has a direct effect on the production costs. According to the study of Kwade *et al.*, the formation and aging process equals 30% of investment costs.[34]

2.4. *Module and pack assembly*

Battery modules are the cell-to-cell connections including electronic management components and sensors. The battery pack assembly is the module-to-module construction which additionally includes a battery management system (BMS), a cooling system and power electronics.

In automotive applications, the integrated cell numbers in each module and whole pack are different. For example, Tesla model S has a 85-kWh battery pack consisting of 16 modules and in total 7.104 li-ion battery cells,[41] whereas BMW i3 has a 42-kWh battery pack with eight modules and as a whole 96 Li-ion cells. Each module in a pack contains the same number of cells.

Modules are protecting battery cells from damage. A primary task in order to build a module is the cell connection. The tabs of each cell (positive and negative tabs) can be connected together in series or in parallel and then to the terminals. Two terminals on the module are used for module connections in a pack.

Depending on a cell type, sensors in a module can be either only temperature sensor or temperature and pressure sensors which allow the BMS to control safety limits.

Another important component of a module is cooling channels. Depending on the cooling type (air or liquid), a cooling element flows between cells in order to avoid thermal runaway. Among the cooling systems, liquid cooling systems can be considered as the most efficient;[47–49] however, their complexity, cost and potential leakage risk limit their use. When the battery pack uses direct liquid cooling, mechanical protection with good insulation and sealing must be performed. It must also be kept in mind that the liquid will add extra weight and volume to the battery pack. Tesla Model S and Chevrolet Volt use an active liquid cooling system,[49] whereas Nissan Leaf uses air cooling.[50]

After the attachment of the modules, the necessary thermal and electrical management systems are integrated and a whole system is placed in a battery pack housing. A sealed pack is tested if a gas and/or liquid leakage occurs.

The first charge of a pack can be much more complex than that of a single cell or a module, and an LIB Pack requires a delicate management system due to the overcharged and discharged battery states. As a battery pack consists of many cells, BMS has a very vital task: a charge balance among the battery cells. An overcharged cell has a high risk of explosion and a deeply discharged cell can reduce the cycle life of the battery cell.[51] As soon as the battery pack is adjusted to the required SoC, it can be shipped to an EV manufacturer for its integration.

The following subsections will introduce battery construction solutions from a research project (LIBELLE*) and a start-up (EMECTRIC). Both approaches are based on sector-specific methodologies.

2.4.1. *LIBELLE*

Battery development for electromobility (e-mobility) is driven by continuous energy density enhancement (Wh/kg, Wh/l). This applies for weight and volume as well as for the costs of batteries in €/kWh. If more energy can be stored by using the same raw materials, batteries become more cost effective.

Lately, the focus has been more and more on volumetric energy density. Here, Li-ion cells show a unique property since volumetric energy density is double that of gravimetric energy density. This is why Li-sulfur (Li-S) cannot be a competitive technology anymore for applications in electric vehicles. Even Li-air (on a realistic cell level, not by theoretical calculation) may not be able to compete anymore in this field including all future progress of Li-ion.

Generally, the focus of energy density improvement is still extremely targeted on the cell level. The latest potential advancement can be seen in the renaissance of Li-metal solid-state cells. The level of standardization is also developing and there are standards such as 18650 and 21700 and the latest 4680 for cylindrical cells, and for prismatic ones that standardization committees are at last accepting.

However, the situation on battery levels (assembled cells) is surprisingly lacking in focus. As a rule of thumb, only half of the energy density remains on the battery level for most of the batteries which are actually installed in electric vehicles. This means that, from 200 Wh/kg on the cell level, only 100 Wh/kg is available on the battery level. For volumetric energy density, it is often less than one third. Additionally, those batteries are not easily recyclable.

The reasons may be seen in battery constructions which satisfy first of all the safety and lifetime. First-generation Li-ion batteries for hybrid vehicle applications show a very rigid housing which withstands harsh deformation from outside, and thus satisfies safety demands. For the lifetime of the battery, usually welded connections of the cells are employed. All this makes recycling very difficult.

However, looking at the tremendous potential of energy density enhancement by avoiding senseless losses from cell to battery level, new construction methods are necessary. It is important to consider that energy density enhancement of Li-based cells will be more and more exhausted within the next decade.

The Tesla, BMW i and Opel Ampera all have a fully or partially electric drive. Tesla uses cylindrical cells, around 10,000 of them. The batteries in the BMW i models contain prismatic or cuboidal cells. Since these types of cells can be built relatively large, only around 100 cells are required to achieve high nominal capacity. The Ampera's battery, on the contrary, contains pouch cells, which externally resemble a vacuum pack for coffee. Battery production for electric vehicles strongly depends on the vehicle model and is based on manual work. Each manufacturer does it differently for each model. So, it is hardly possible to actually achieve value creation with a battery that is generally assumed to have the current concepts.

It is important to consider what happens to the energy density during the transition from the individual cell to the overall system. As the energy density directly affects the range of an electric vehicle, designing a battery is an important key to achieving higher ranges. However, a suitable battery design must favor production, repair and recycling.

Here, the project "LIBELLE", founded by the "VECTOR", shows new directions. The project has been undertaken by the Electrochemical Energy Storage Systems team at the University of Stuttgart in Germany, under the supervision of

Fig. 8. Illustration of self-supporting cell arrays.

Professor Kai Peter Birke. The basic idea is to use the cell as a self-supporting part. One cylindrical cell may fail under static load tests. However, a multiple array easily sustains the automotive standard deformation test conditions from different directions (x, y, z) for traction batteries. This is an important result, which has already been achieved in this project. Such a cell array is depicted in Fig. 8.

On the module level, 180 Wh/kg can still be preserved and this shows the power of the concept of self-supporting cell arrays. The final battery housing thus has no tasks to prevent deformation and can be constructed with very lightweight materials. The cooling board will be made out of modern plastics. The final target of the project is to reduce the loss of energy density from a factor 2 (if gravimetric energy density from a cell is divided through energy density of the battery) to at least 1.5. The level from cell to module (as shown in Fig. 8) is actually in the regime of 1.2.

Additionally, a recycling concept on cell level is successfully employed. The modules can be completely disassembled since the cells are screwed on and clamped. Here, experimental results show that this way of cell connection is fully competitive with the conventional cell welding methods. However, these results still have to be verified over lifetime.

Generally, battery production concepts still show a huge lack of industrialization which, surprisingly, lacks the focus applied to standardization efforts on cell level. A Giga-Factory usually means a cell factory. Looking at the cost structure, the ratio between cell and battery costs should be about 80:20 or even more

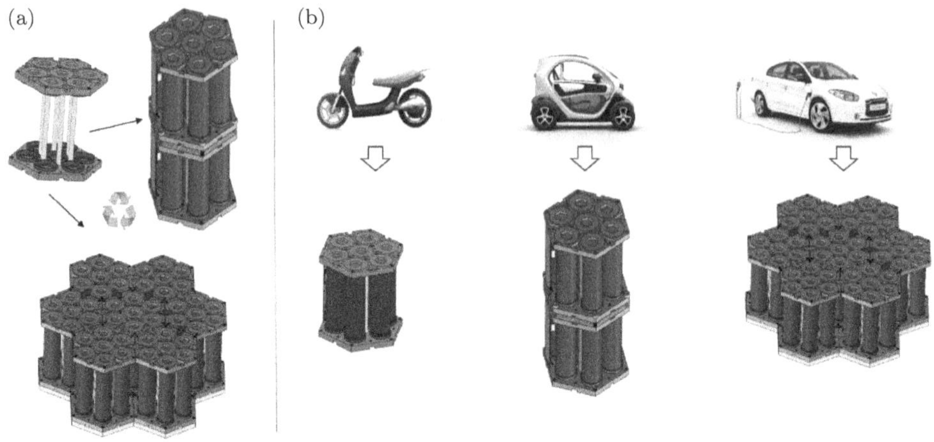

Fig. 9. (a) Modular block building concept and (b) the design for different applications.

for the cell. In reality, we find 40:60 and even less than 40, which is consequently one of the cost drivers of actual batteries. Again, the reason is not the cell, but the battery. If such issues are consequently addressed, cost friendly mass production on a battery level and system costs of less than 100 €/kWh should to be feasible. This becomes competitive if efforts for combustion engines to pass emission rules increase continuously. Also, simple assembly concepts can be applied and the cell becomes a central construction element — as presented in the LIBELLE project. This may help to achieve the cost targets. Therefore, the assembly concepts for the cell are hugely important. In addition, a modular block-building concept (Fig. 9(a)) to address many applications, from light mobility over HEV, PHEV and EV to heavy-duty applications (Fig. 9(b)), is equally essential. This is just tuned by employing higher-power or high-energy tuned cells (which have the same size). The heat conduction is carried over the common metal board where the cell poles are mounted (see Fig. 8). These common boards are then connected to a slim cooling board (not depicted). Its dimension and the cooling media throughput finally determine the temperature of the battery and can thus be adjusted to the different above-mentioned applications.

2.4.2. *Modular battery system EMECTRIC*

EMECTRIC addresses the challenges especially in aviation applications with a safe, flexible and economic battery system (Fig. 10). The safety concept with non-flammable interstitial material between cells and single cell fuses, the modularity

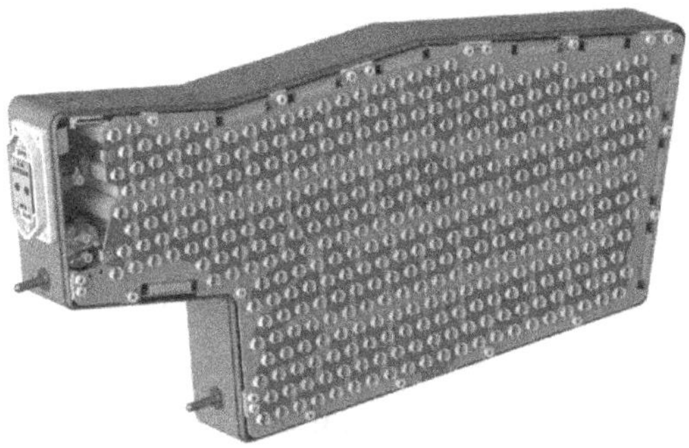

Fig. 10. EMECTRIC 96S4P battery module.

on the cell level and the assembly concept, which are addressing the needs of the aviation market, are key technologies of the system. The idea of the system came from the electrification of a self-starting glider, which should be equipped with an electric propulsion system. As there was no battery system available that met the needs for the challenging assembly space in the glider body or wing, as well as appropriate safety provisions, the EMECTRIC battery system was developed to enable electrification with a tailored battery system at low initial investment especially for prototypes and small series.

2.5. *Electrification in aviation — Status quo*

With electrically powered aircraft, a new market is currently emerging that could develop enormous potential in the coming years. A current study by the management consultancy Roland Berger concludes, "Is aircraft electrification the next chapter in aviation? It is not a question of if, but when" and counts 215 Projects of electric aviation worldwide with +~30% in 2019.[52]

Compared to conventional drive systems, electric drives offer the potential for greater safety and lower emissions while at the same time lowering operating and maintenance costs. Only the energy density of the batteries and concerns about safety of lithium-ion technology prevented it from becoming more widespread in the past. Current lithium-ion cells have already achieved sufficient energy densities for gliders with auxiliary drives, motor gliders, sports and training aircraft. Regional and short-haul flights can also be carried out fully electrically using a battery system with a gravimetric energy density of >200 Wh/kg on a system level. Overflights of the Alps with sufficient safety reserves had already been achieved in 2015 by DLR, for example, with their motor glider known as the "e-Genius", with batteries having a significantly lower energy density than it is possible today.

No matter if the battery is electric or hybrid electric, there will be an increasing need for batteries in aviation.

2.5.1. *Challenges in battery development in aviation applications*

The battery market becomes more and more complicated — many different cell types and formats, high safety requirements and regulations. The effort to measure and gather relevant data is comparably high, for example, cycle tests of various cells with various parameters take up a lot of time, space and capital.

2.5.1.1. Utilization of space

The efficiency of electric airplanes mostly depends on their aerodynamics, and in order to increase the efficiency for future electric flight, fuselages will become narrower and wings are designed to be thinner and more stretched, which reduces the available space for battery installation (Fig. 11). Additionally, not all the available space in the airplane can be utilized as battery compartments, as the center of gravity must remain within tight boundaries and airframe loads need to be distributed to avoid excess structural weight in certain areas.

2.5.1.2. Safety/Certification

The aviation market is highly safety sensitive, which translates to high safety demand and tight regulations. For batteries, there are certain minimum operational standards that require designs to survive very low probability events with catastrophic consequences, e.g. thermal runaway of two adjacent cells at the same time.[53]

In order to pass these tests, a battery engineer can follow different design strategies (Fig. 12): One is to stop the thermal runaway propagation at an early stage, ideally only allowing the two triggered cells to run away. This can be achieved by adding a thermal barrier between cells and/or quick dissipation of the resulting heat buildup and by minimizing electrical discharge of parallel-connected cells through the affected cells.

A second design strategy can be to try to contain the full or partial thermal runaway propagation, e.g. by having a container around the battery that can sustain the thermal and mechanical loads of a full propagation.

Fig. 11. Scaled composites of "Proteus" high-efficiency aircraft with a narrow fuselage and a wing with high aspect ratio — Wikipedia.org.

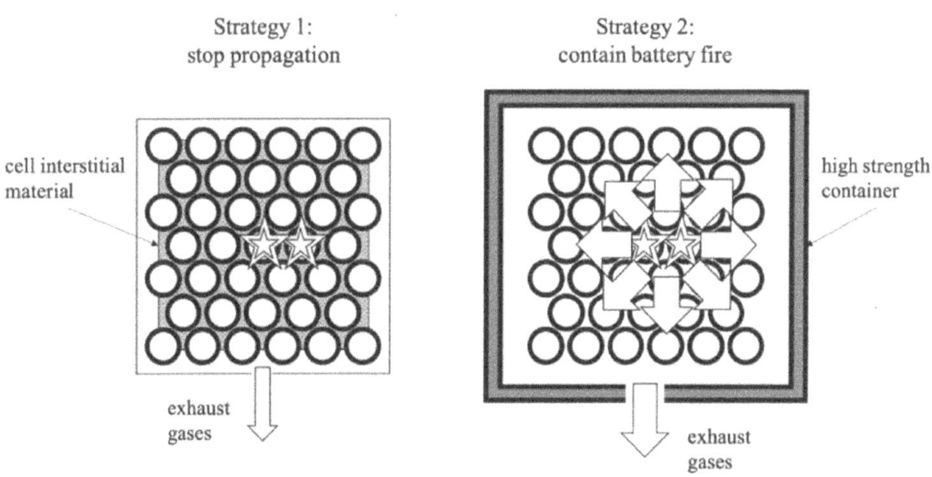

Fig. 12. Battery safety strategies for certification.

The main challenges during these tests are the high temperatures of several hundred °C and the high pressures inside the battery compartment caused by the expelled gases, resulting in a volume increase in the order of 550 times the runaway cell volume.[54]

For the EMECTRIC battery system, strategy 1 was followed by wrapping each cell into a non-flammable material that contained chemically bound water, which is released as water vapor at temperatures above 200°C (Fig. 13). This stops the propagation by consuming thermal energy during the endothermic phase change.

2.5.1.3. Weight/Energy density

The gravimetric energy density of available battery cells tops out at around 260 Wh/kg nominal, of which ca. 90–95% is usable after losses at discharge rates anticipated in contemporary electric or hybrid aircraft concepts.

By building these cells into battery systems, a mass overhead is added to the cell weight in the form of excess mechanical, electrical and thermal structure (see Fig. 14).

In order to optimize this system, these individual overheads can be minimized either individually or by utilizing synergy effects between them, e.g. by using the mechanical structure as a heat sink or by directing structural load through the cell casing. A way to further reduce the overhead can be to increase the observed scope of the battery by integrating it into the airplane structure, as an Airbus research

Fig. 13. EMECTRIC cell unit composed of an 18650 cell and the cell holder parts.

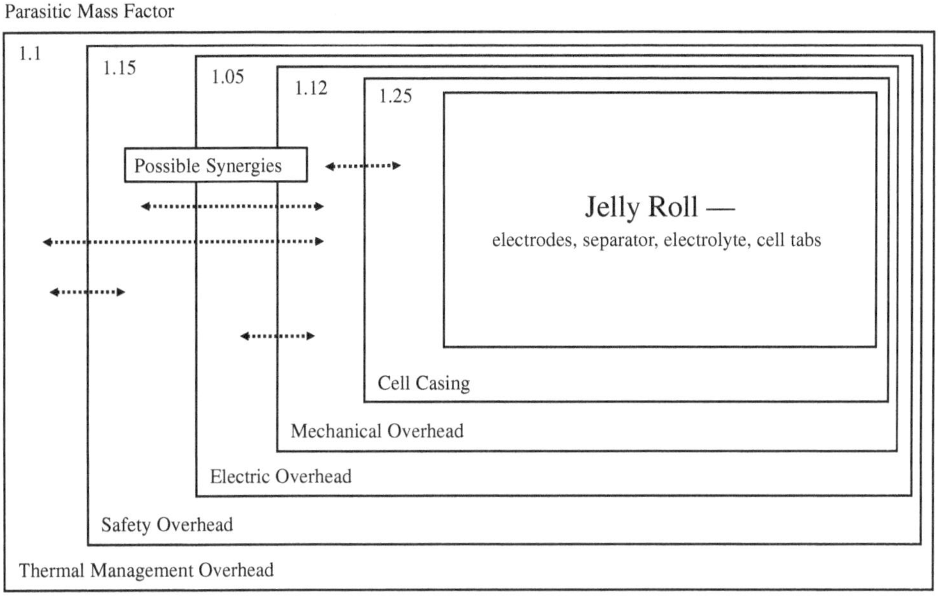

Fig. 14. Battery system mass composition.

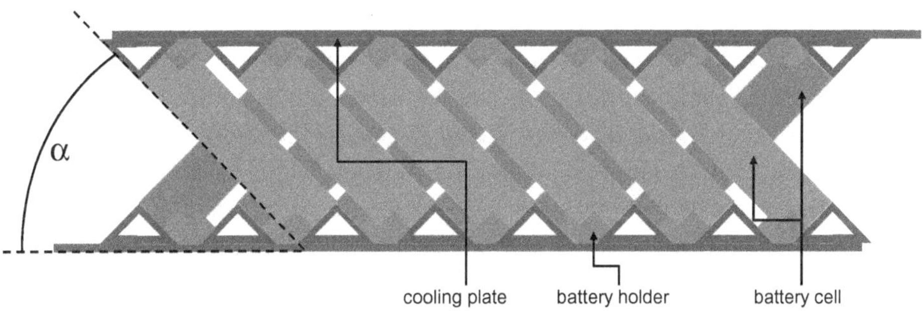

Fig. 15. Battery for structural integration (Own representation adopted from Airbus[a]).

concept suggests.[a] In this design, the housings of the cylindrical cells are used to transfer structural loads, potentially resulting in a negative structural overhead of the battery design (see Fig. 15).

The power density is an important factor for airborne batteries, since the mean discharge rates are significantly higher than in cars. The power density depends largely on the cell type, but is also influenced by the battery layout, conductor

[a]Europäisches Patentamt. EP 3 544 083 A1 — Batterieanordnung zur Lasttragenden Strukturellen Integration von Batterien in ein Fahrzeug (September 2019).

design, conductor material, etc. and can be improved by applying battery cooling as well as heating at cold temperatures. The thermal management in turn needs additional structures and increases the mass overhead, leaving less weight budget for cells.

2.5.1.4. Production and investment cost

Most electrification projects need an individually tailored battery system which cannot be replaced by a standard module. Therefore, it is important to not require expensive tooling to keep the initial investment low. This is especially important since the initial costs are split over low unit counts — as the production volumes in aviation are several magnitudes lower than in the Automotive industry.

2.5.1.5. Cell type

Choosing the right cell type for a given application is an important factor and influences the size, energy density, power density and safety of the system. There are mainly three different cell formats available, the cylindrical cell, the prismatic cell and the pouch cell, all having advantages and disadvantages regarding these performance metrics. For the EMECTRIC system, the cylindrical cell type was chosen for cost and availability reasons, but mainly because it was anticipated to lead a maximum gravimetric energy density in system level gravimetric energy density due to minimal total mass overhead required for battery safety and certification. The factors that decide the amount of required mass overhead are the individual cell mass, the cell failure modes during thermal runaway and the mechanical structure required to ensure a safe operation of the battery. Cylindrical cell housings provide mechanical rigidity against external and internal loads, e.g. from active material swelling, at a minimal required mass due to their optimal geometric form with regard to pressure differentials. In order to provide a pouch cell with the same mechanical protection, a higher amount of overhead mass is required. Another advantage of cylindrical cells is the available small cell size, which means that a smaller percentage of the battery is going into a thermal runaway condition when one cell has a malfunction/is triggered for testing, resulting in less needed safety overhead overall. Cell designs are always a compromise and the drawbacks of cylindrical cells are inherently lower power densities and shorter potential lifetimes when compared to pouch or prismatic cells.

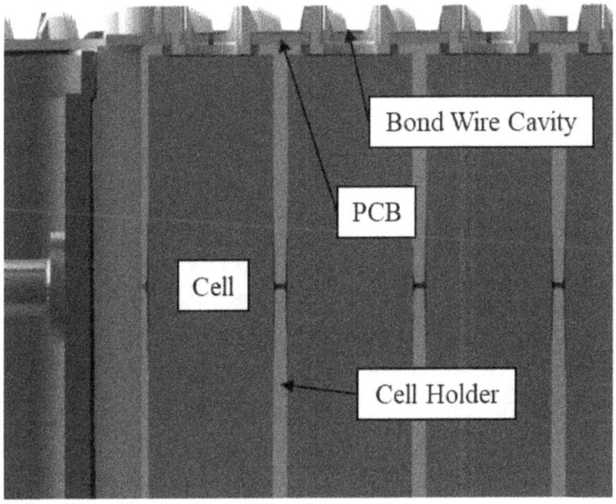

Fig. 16. Cell assembly cross section.

2.5.2. *Flexibility*

The modularity of the EMECTRIC battery system lies initially in the cell fixation. Thanks to a modular cell holder, which is made of a single injection molded part and an easily customizable PCB base plate, various battery geometries can be produced in all room dimensions with low development costs and little manufacturing effort (See Fig. 16). In this way, the installation space can be optimally utilized, and economical production of prototypes and small series of tailor-made battery packs can be realized without investment in tools.

This leads to geometric freedom and a high volumetric/gravimetric energy density of the system. Figure 17 show an example of an assembly space in a glider body — with the EMECTRIC system and the flexibility on the cell level, 72% more cells can be added to the system compared to a semi-modular system.

The following examples show the possibilities of battery integration in the airplane. The battery system in Fig. 18 shows a self-starting 18/15 m-glider with a 10,5-kWh redundant battery system and provides a Battery capacity for a self-start and an additional 125-km range.

The battery installation in the wings allows for an appropriately sized battery to be installed in a glider without the need to recertify the plane due to changed loads on the wing root. The pilot is even able to carry more luggage than with the previous fuselage installed combustion engine. The challenge of this project was that the battery has to bend together with the wing, not disturbing the dynamics of the airplane. To allow for this flexibility, the connections between the cell packs

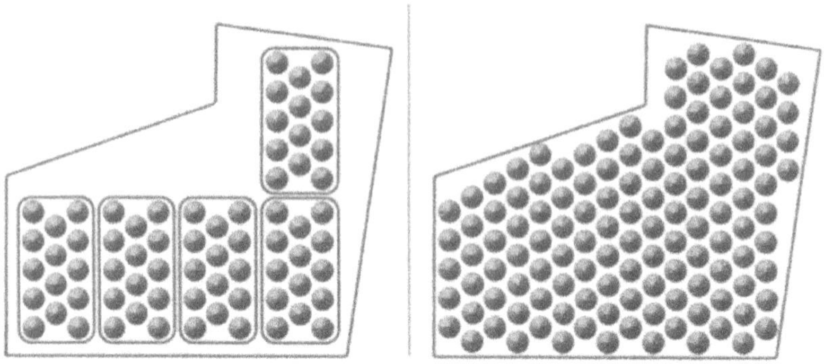

Fig. 17. Comparison of a semi-modular system integration and the modularity on the cell level (EMECTRIC).

Fig. 18. AS 34 Me with EMECTRIC wing battery (10.5 kWh).

allow for +–1° of bending, resulting in a maximum angle of 47° between the battery ends.

For another project, fuselage batteries were designed that fit inside an extremely narrow section of the Jonker Sailplanes JS3 high-performance glider. The battery system is a redundant configuration of two fully independent 350V (96S4P) batteries. This allows for a flexible use as self-starter with both batteries

Fig. 19. Bond connections on EMECTRIC battery system.

installed and as a "sustainer" with only one battery installed. Each battery weighs ca. 23 kg and contains a nominal 4,2 kWh of energy.

2.5.3. *Cell contacting*

The cells are contacted by wire bonding. In addition to the contacting of the cells, these bonds act as thermal fuses and break in the event of an internal or external short circuit (See Fig. 19). This means that the main contacts are free of high voltage potential in the event of danger, and the affected cell is electrically and thermally isolated. The technology of cell contacting by means of wire bonds has been applied successfully at the electric car manufacturer Tesla for years. The process is associated with a high level of process reliability, the possibility of redundancy through several bonds per cell, a high degree of automation and low costs. In this process, the conductor elements of the battery can also be made of aluminum, which has significant weight and cost advantages over copper or nickel.

2.5.4. *Battery management system*

The modular battery management system was specifically developed for the use in aircraft, and monitors the battery with regard to individual voltages, temperature and humidity and is able to save these data for maintenance as well as transmitting in real time to other systems via an adaptable CAN interface. The BMS also controls the electromagnetic switches in each battery module in order to be able to disconnect the other drive components in case of a malfunction and protects the battery from deep discharge or overcharging in the event of failure of other components. Further functionalities involve several analog safety circuits that do not require a functional IC to work and the SOC and SOH estimation algorithms.

2.5.5. *Certification*

A thermal runaway test according to the RTCA DO-311A test specifications shows that a thermal runaway of individual cells can be reliably stopped with the presented actions. The external structure of the wing shown in Fig. 20 experienced effectively no thermal loads during the test. For the test, a 480-cell wing battery was preheated up to 65°C and then three cells on different positions were triggered by heating them with a heating wire. The spread of the thermal runaway of one cell was successfully prevented. The flame-retardant housing and the degassing openings in the bottom of the housing, through which the gasses are guided outside of the fuselage, released all expelled gases and were not seriously damaged.

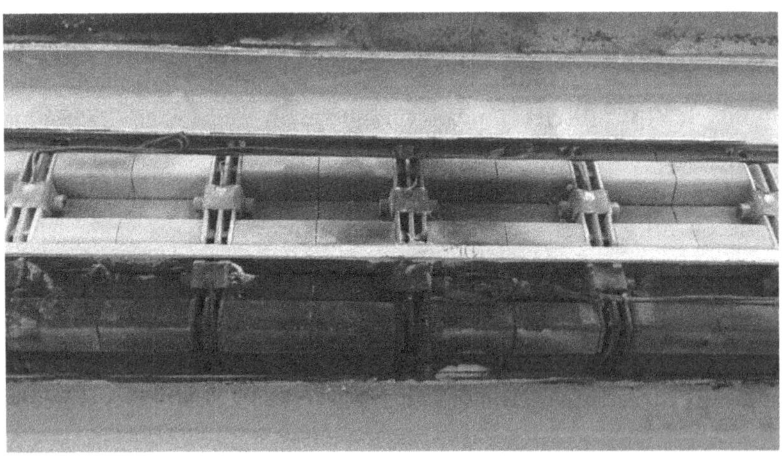

Fig. 20. Inside of the battery after the thermal runaway test — Highly damaged but no propagation.

2.5.6. *Economic efficiency and recycling*

The modular construction reduces costs effectively, as no tooling is needed for individual geometries and prototypes and small series can be produced quickly, which reduces the time to market. The system also uses raw materials which are available and economically scalable. Furthermore, almost all connections in the EMECTRIC system are reversible, and individual cells or even the entire cell stack can be replaced. This allows for economical repair and makes the system easy to recycle. The 18650 cells can be reused in other applications such as home storage in their second life.

3. Cost Factors of Battery Manufacturing

As it was described briefly in the Section 2, battery manufacturing is a highly complex process and the production line is strongly dependent on each process. The total cost of battery cell production depends on several factors including raw material and energy costs, consumed energy and production region.

Associated with today's technology, battery production is still very expensive. For example, a battery pack cost can represent up to 50% of the entire vehicle cost.[55] As there is huge competition for reduction of battery production costs, the manufacturers are seeking new cell chemistries and novel production technology approaches without compromising quality.

Berckmans *et al.*[56] calculated a cost and sales price of a battery and predicted up to the year 2030 by using process-based cost modeling of several battery chemistries.

For an NMC 6:2:2 cell, they have reported that the cost of 432 $/kWh is dominated by material cost representing 65% of the overall cost. The total sales price goes up to 670 $/kWh (including the profit margins of the producer and middleman's sales price). However, Tesla claims to be below 190 $/kWh and the future battery market has a cost target for below 150 $/kWh (according to McKinsey).

In order to derive price competitive advantages, all steps within the value chain of LIB cell manufacturing need to be optimized, standardized and automated. A lack of standardization and automation in battery assembly and disassembly processes is one of the important reasons for low-yield and high-cost production. Taking into account that rejects amounting to 5% may occur due to inadequate product quality, it would result in 18-GWh/year battery cells, which is equivalent to 70,000 tons of cells.

With the tools of digitalization like digital twins (DTs) and traceability, battery manufacturers and other industrial companies have the opportunity to optimize their product quality and improve the resource efficiency and consequently the economic viability of battery cell manufacturing.

References

1. Scrosati B. History of lithium batteries. J Solid State Electrochem. 2011; 15(7–8):1623–30.
2. Nishi Y. Lithium ion secondary batteries; past 10 years and the future. J Power Sources. 2001;100:101–6.
3. Quinn JB, Waldmann T, Richter K, Kasper M, Wohlfahrt-Mehrens M. Energy density of cylindrical li-ion cells: a comparison of commercial 18650 to the 21700 cells. J Electrochem Soc. 2018;165(14):A3284. Available from: https://iopscience.iop.org/article/10.1149/2.0281814jes/pdf
4. Frost & Sullivan. Global lithium-ion (Li-ion) batteries market, forecast to 2025; Market Report 2019 Oct.
5. Saw LH, Ye Y, Tay AA. Integration issues of lithium-ion battery into electric vehicles battery pack | Elsevier Enhanced Reader. J Cleaner Prod. 2016;113:1032–45.
6. Miao Y, Hynan P, Jouanne A von, Yokochi A. Current Li-ion battery technologies in electric vehicles and opportunities for advancements. Energies. 2019;12(6):1074.
7. Roadmap Batterie Produktionsmittel 2030. Frankfurt am Main; 2018.
8. Tsiropoulos I, Tarvydas D, Lebedeva N. Li-ion batteries for mobility and stationary storage applications. JRC Science for Policy Report. 2018.
9. Dash R, Pannala S. RETRACTED ARTICLE: Theoretical limits of energy density in silicon-carbon composite anode based lithium ion batteries. Sci Rep. 6(1):1–7. Available from: https://www.nature.com/articles/srep27449.pdf
10. Brand M, Gläser S, Geder J, Menacher S, Obpacher S, Jossen A, *et al.* Electrical safety of commercial Li-ion cells based on NMC and NCA technology compared to LFP technology. World Electr Veh J. 2013;6(3):572–80. Available from: https://www.mdpi.com/2032-6653/6/3/572/pdf
11. Thielmann A, Sauer A, Wietschel M. Gesamt-roadmap lithium-ionen-batterien 2030. Karlsruhe: Fraunhofer-Institut für System- und Innovationsforschung ISI; 2015.
12. Dominko R, Fichtner M, Perraud S, Tarascon J-M, Vegge T, Winter M. Inventing the batteries of the future; 2019 Nov.
13. Barré A, Deguilhem B, Grolleau S, Gérard M, Suard F, Riu D. A review on lithium-ion battery ageing mechanisms and estimations for automotive applications. J Power Sources. 2013;241:680–9.
14. Schmidt O, Thomitzek M, Röder F, Thiede S, Herrmann C, Krewer U. Modeling the impact of manufacturing uncertainties on lithium-ion batteries. J Electrochem Soc. 2020;167(6):60501.
15. Chen L-C, Liu D, Liu T-J, Tiu C, Yang C-R, Chu W-B, Wan C-C. Improvement of lithium-ion battery performance using a two-layered cathode by simultaneous slot-die coating. J Energy Storage. 2016;5:156–62.
16. Dreger H, Bockholt H, Haselrieder W, Kwade A. Discontinuous and continuous processing of low-solvent battery slurries for lithium nickel cobalt manganese oxide electrodes. J Elec Mater. 2015;44(11):4434–43.

17. Wenzel V, Nirschl H, Nötzel D. Challenges in lithium-ion-battery slurry preparation and potential of modifying electrode structures by different mixing processes. Energy Technol. 2015;3(7):692–8.
18. García A, Culebras M, Collins MN, Leahy JJ. Stability and rheological study of sodium carboxymethyl cellulose and alginate suspensions as binders for lithium ion batteries. J Appl Polym Sci. 2018;135(17):46217.
19. Liu T-J, Tiu C, Chen L-C, Liu D. The influence of slurry rheology on lithium-ion electrode processing. Printed Batteries Mater Technol Appl. 2018;63.
20. Diehm R, Kumberg J, Dörrer C, Müller M, Bauer W, Scharfer P, *et al. In situ* investigations of simultaneous two-layer slot die coating of component-graded anodes for improved high-energy li-ion batteries. Energy Technol. 2020;8(5):1901251.
21. Al-Shroofy M, Zhang Q, Xu J, Chen T, Kaur AP, Cheng Y-T. Solvent-free dry powder coating process for low-cost manufacturing of LiNi1/3Mn1/3Co1/3O2 cathodes in lithium-ion batteries. J Power Sources. 2017;352:187–93. Available from: www.sciencedirect.com/science/article/abs/pii/S0378775317304457
22. Kirsch DJ, Lacey SD, Kuang Y, Pastel G, Xie H, Connell JW, *et al.* Scalable dry processing of binder-free lithium-ion battery electrodes enabled by holey graphene. ACS Appl Energy Mater. 2019;2(5):2990–7.
23. Haselrieder W, Westphal B, Bockholt H, Diener A, Höft S, Kwade A. Measuring the coating adhesion strength of electrodes for lithium-ion batteries. Int J Adhes Adhes. 2015;60:1–8.
24. Meyer C, Bockholt H, Haselrieder W, Kwade A. Characterization of the calendering process for compaction of electrodes for lithium-ion batteries. J Mater Process Technol. 2017;249:172–8. Available from: https://www.sciencedirect.com/science/article/abs/pii/S0924013617302054
25. Zheng H, Tan L, Liu G, Song X, Battaglia VS. Calendering effects on the physical and electrochemical properties of Li[Ni1/3Mn1/3Co1/3]O2 cathode. J Power Sources. 2012;208:52–7. Available from: https://www.sciencedirect.com/science/article/abs/pii/S0378775312003023
26. Guzmán G, Vazquez-Arenas J, Ramos-Sánchez G, Bautista-Ramírez M, González I. Improved performance of LiFePO4 cathode for Li-ion batteries through percolation studies. Electrochimica Acta. 2017;247:451–9. Available from: https://www.sciencedirect.com/science/article/pii/S001346861731410X
27. Jansen T, Blass D, Hartwig S, Dilger K. Processing of advanced battery materials — Laser cutting of pure lithium metal foils. Batteries. 2018;4(3):37.
28. Demir AG, Previtali B. Remote cutting of Li-ion battery electrodes with infrared and green ns-pulsed fibre lasers. Int J Adv Manuf Technol. 2014;75(9):1557–68. Available from: https://link.springer.com/article/10.1007/s00170-014-6231-7
29. Lee D, Patwa R, Herfurth H, Mazumder J. Parameter optimization for high speed remote laser cutting of electrodes for lithium-ion batteries. J Laser Appl. 2016;28(2):22006.
30. Lee D. Investigation of physical phenomena and cutting efficiency for laser cutting on anode for Li-ion batteries. Appl Sci. 2018;8(2):266.
31. Pfleging W. A review of laser electrode processing for development and manufacturing of lithium-ion batteries. Nanophotonics. 2018;7(3):549–73.
32. Herfurth HJ, Patwa R, Pantsar H. Laser cutting of electrodes for advanced batteries. Proceedings of LPM2010 — The 11th International Symposium on Laser Precision Microfabrication; 2010.
33. Westphal B, Bockholt H, Gunther T, Haselrieder W, Kwade A. Influence of convective drying parameters on electrode performance and physical electrode properties. ECS Trans. 2015;64(22):57–68.
34. Kwade A, Haselrieder W, Leithoff R, Modlinger A, Dietrich F, Droeder K. Current status and challenges for automotive battery production technologies. Nat Energy. 2018;3(4):290–300.

35. Jaiser S, Kumberg J, Klaver J, Urai JL, Schabel W, Schmatz J, *et al.* Microstructure formation of lithium-ion battery electrodes during drying — An *ex-situ* study using cryogenic broad ion beam slope-cutting and scanning electron microscopy (Cryo-BIB-SEM). J Power Sources. 2017;345:97–107. Available from: https://www.sciencedirect.com/science/article/pii/S0378775317301374
36. Kumberg J, Müller M, Diehm R, Spiegel S, Wachsmann C, Bauer W, *et al.* Drying of lithium-ion battery anodes for use in high-energy cells: influence of electrode thickness on drying time, adhesion, and crack formation. Energy Technol. 2019;7(11):1900722.
37. Ahmed S, Nelson PA, Gallagher KG, Dees DW. Energy impact of cathode drying and solvent recovery during lithium-ion battery manufacturing. J Power Sources. 2016;322:169–78.
38. Wood DL, Quass JD, Li J, Ahmed S, Ventola D, Daniel C. Technical and economic analysis of solvent-based lithium-ion electrode drying with water and NMP. Drying Technol. 2018;36(2):234–44.
39. Susarla N, Ahmed S, Dees DW. Modeling and analysis of solvent removal during Li-ion battery electrode drying. J Power Sources. 2018;(378):660–70.
40. Quinn JB, Waldmannz T, Richter K, Kasper M, Wohlfahrt-Mehrens M. Energy density of cylindrical Li-ion cells a comparison of commercial 18650 to the 21700 cells. J Electrochem Soc. 2018;165:A3284–A3291.
41. Sharma A, Zanotti P, Musunur LP. Enabling the electric future of mobility: robotic automation for electric vehicle battery assembly. IEEE Access. 2019;7:170961–91.
42. Schröder R, Glodde A, Aydemir M, Seliger G. Increasing productivity in grasping electrodes in lithium-ion battery manufacturing. Procedia CIRP. 2016;57:775–80.
43. Knoche T, Zinth V, Schulz M, Schnell J, Gilles R, Reinhart G. In situ visualization of the electrolyte solvent filling process by neutron radiography. J Power Sources. 2016;331:267–76.
44. RWTH Aachen V. Produktionsprozess Einer Li-Ionen-Batteriezelle. 2019.
45. Moretti A, Sharova V, Carvalho DV, Boulineau A, Porcher W, de Meatza I, *et al.* A comparison of formation methods for graphite//LiFePO 4 cells. Batteries & Supercaps. 2019;2(3):240–7.
46. An SJ, Li J, Daniel C, Mohanty D, Nagpure S, Wood DL. The state of understanding of the lithium-ion-battery graphite solid electrolyte interphase (SEI) and its relationship to formation cycling. Carbon. 2016;105:52–76. Available from: https://www.sciencedirect.com/science/article/pii/S0008622316302676
47. Jiaqiang E, Han D, Qiu A, Zhu H, Deng Y, Chen J, *et al.* Orthogonal experimental design of liquid-cooling structure on the cooling effect of a liquid-cooled battery thermal management system. Appl Therm Eng. 2018;132:508–20. Available from: https://www.sciencedirect.com/science/article/pii/S1359431117353115
48. Chen D, Jiang J, Kim G-H, Yang C, Pesaran A. Comparison of different cooling methods for lithium ion battery cells. Appl Therm Eng. 2016;94:846–54. Available from: https://www.sciencedirect.com/science/article/pii/S1359431115010613
49. Malik M, Dincer I, Rosen MA, Mathew M, Fowler M. Thermal and electrical performance evaluations of series connected Li-ion batteries in a pack with liquid cooling. Appl Therm Eng. 2018;129:472–81. Available from: https://www.sciencedirect.com/science/article/abs/pii/S1359431117344393
50. Yang S. A review of lithium-ion battery thermal management system strategies and the evaluate criteria. Int J Electrochem Sci. 2019;6077–107.
51. Kim J, Shin J, Chun C, Cho BH. Stable configuration of a Li-ion series battery pack based on a screening process for improved voltage/SOC balancing. IEEE Trans Power Electron. 2012;27(1):411–24.
52. Berger R. Aircraft electrical propulsion: The next chapter of aviation?

53. RTCA Inc. RTCA DO-311: Minimum operational performance standards for rechargeable lithium batteries and battery systems battery systems; 2017 Dec 19.
54. Maloney T. Lithium battery thermal runaway vent gas analysis. 2016.
55. Zhou C, Qian K, Allan M, Zhou W. Modeling of the cost of EV battery wear due to V2G application in power systems. IEEE Trans Energy Convers. 2011;26(4):1041–50.
56. Berckmans G, Messagie M, Smekens J, Omar N, Vanhaverbeke L. Cost projection of state of the art lithium-ion batteries for electric vehicles up to 2030. Energies. 2017;10(9):1314.

https://doi.org/10.1142/9789811245626_0003

Chapter 2

Technologies Enabling Smart Manufacturing

Ozan Yesilyurt*, Andreas Schlereth†, Sascha Gärtner‡, Ahmad Issa§ and Michael Oberle¶
Fraunhofer IPA, Nobelstraße 12, Stuttgart, Germany
**ozan.yesilyurt@ipa.fraunhofer.de*
†andreas.schlereth@ipa.fraunhofer.de
‡sascha.gaertner@ipa.fraunhofer.de
§ahmad.issa@ipa.fraunhofer.de
¶michael.oberle@ipa.fraunhofer.de

The shift from traditional manufacturing to smart manufacturing has been initiated by Digital Transformation. The transformation is being carried out through the gradual implementation of new technologies such as cyber-physical systems, artificial intelligence, 5G and digital twins. Using these novel technologies results in far-reaching opportunities for production in the future. These include increased productivity in factories and greater resource efficiency with regard to the batch size one manufacturing strategy.

1. Introduction

The increasing demand for customized products, better product quality and lower production costs has led to a global competition between manufacturing companies, and along with the demand for customized products, it has led to an emerging transformation of manufacturing.[1,2] Today's centralized production systems are static and thus unable to react agilely, flexibly and quickly enough to changing requirements and markets.[3] To overcome these challenges, a paradigm shift toward decentralized and networked production processes must take place. In order to maintain the competitiveness of companies, it is necessary not only to increase flexibility and reduce costs but in particular to also have smart manufacturing processes that can act independently and autonomously.[4]

This chapter gives a brief overview about the technologies and enablers that are crucial for the paradigm shift from a static manufacturing environment to a decentralized autonomous manufacturing environment. First, the definition of digital transformation in manufacturing is introduced and the specific challenges for digital transformation in battery manufacturing are addressed. Subsequently, technologies and enablers of digitalization in production such as Cyber Physical Systems, Cloud Computing, 5G, Artificial Intelligence, Blockchain, Digital Twins, Augmented Reality, Virtual Reality, Robotics and AGVs are illustrated. Finally, the Gartner Hype Cycle is used to provide an outlook on the expected added value of some of the technologies. At the end of the chapter, possible challenges that may occur during the implementation of the technologies in a manufacturing environment are described.

2. Definition of Digital Transformation in Manufacturing

In order to define what digital transformation means, it is first essential to explain how the terms production and manufacturing differ from each other. Further, it is necessary to define the terms digitalization and automation. Afterward, these terms will be used to describe the concept and challenges of digital transformation battery cell manufacturing.

2.1. *Definition of production and manufacturing*

Manufacturing is defined by the processing or conversion of raw materials in several process steps by various machines and systems. The end product of manufacturing is a product that is ready for further processing, e.g. assembly.[5] Production is a broader term that considers the transformation of raw materials and products into a finished product. While only tangible elements are considered in manufacturing, production also considers non-tangible elements such as financial activities.[6] Thus, by definition, manufacturing is included in production. Since the overlap of the terms is very strong and in this book battery cell production exclusively considers the value-adding processes, the terms manufacturing and production are used synonymously from now on.[7]

2.2. *Digitalization and automation*

The overarching goal of Industry 4.0 is fully networked and highly automated production. For a better understanding of the difference between digitalization and automation, the meaning of these two terms will now be explained in more detail.

Automation in production is realized by the information transfer between the manufacturing process and the automation equipment. A concatenation of numerically controlled manufacturing equipment can be realized in automation with the help of information and communication technology. In this case, the use of information and communication technologies is referred to as digitalization.[8]

In general, however, the distinction between digitization and digitalization must be introduced. The transfer of analog information into digital information is considered as digitization in the literature.[9] In the context of production, for example, a production drawing in paper form can be cited as analog information. The digital counterpart represents the production drawing as a graphic on a screen.

The term digitalization is broader and encompasses the processing of data across companies by means of advanced digital technologies. These types of technologies change the processes as a whole and even enable the introduction of novel business models.[9]

Digitalization in production is perceived as a paradigm shift enabling a revolutionary approach to increase productivity and product quality while minimizing costs. The term "Industry 4.0" has been selected to highlight the revolutionary impact of the fourth industrial revolution. Previous industrial revolutions were marked by the transition from hand production methods to machines in the 18th century, the "Technological Revolution" and development of the production line in the 19th century, and the computer age in the 20th century.[10]

2.3. *Digital transformation in general*

The development of new technologies is not the only factor in digitalization of a factory according to the principles of Industry 4.0. The strategies and processes for the step-by-step implementation of digital technologies in the real production environment are referred to as digital transformation. The German Federal Ministry for Economic Affairs and Energy (BMWi) gives an even more comprehensive definition and describes digital transformation as intelligent networking between people, machines and industrial processes.[11]

The digital transformation can be described in four different stages. This stage model is shown in Fig. 1.[12] The first stage is digitization, in which digital images of analog processes are created. NC-technology or adoption of an MRP/ERP system is an example of digitization in production. In the second stage, processes are digitally modeled. For example, digitally created designs are used as the basis for calculating the tool path of a CNC drill, which is then converted to control commands for a milling machine. In the third stage, all value-adding processes are networked within and beyond the organizational boundaries with high-bandwidth communication

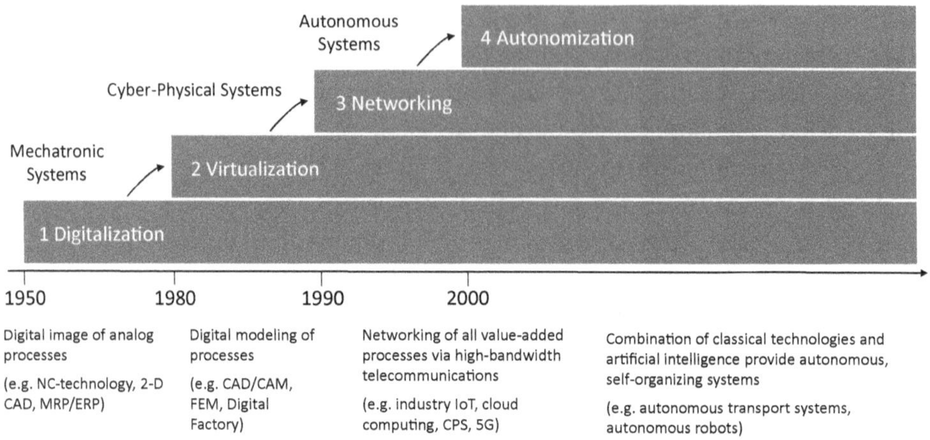

Fig. 1. The development stages of digital transformation according to (Bauernhansl 2018).

leveraging, industrial IoT, CPSs and 5G network. In the fourth and highest stage, autonomous systems are realized by combining classical technologies and artificial intelligence.

Digital transformation in production is therefore an ongoing process of change in which conventional processes and technologies are constantly enhanced by new digital technologies.

2.4. *Brief overview of digital transformation in battery production*

There are three main processes of battery cell production, the electrode production, cell assembly, and cell formation and quality control.[15] In the first process step, the electrodes are manufactured. This step consists of the following sub-processes: mixing of the active material, coating and drying of the carrier material as well as calendaring and slitting of the coated electrode foil.[15] In the second process step, the cells are assembled. This process step includes the sub-processes of assembly, rolling of the electrodes, welding of the arresters and insertion of the electrolyte and electrodes into the package.[15] In the last process step, the cells are formatted and their quality is checked. The sub-processes are formation, degassing and sealing, aging and the final quality control of the cells.[15]

Currently, there are different types of technology and heterogeneous distribution of competencies in battery production.[13] As a result, the complete production line consists of equipment from different manufacturers as well as legacy systems from different providers. In a study to create a roadmap for digitalized battery cell production, workshops were conducted, which showed that the process steps for the actual battery cell production (electrode manufacturing) pose the greatest challenge. They also showed that the process steps for module and pack assembly are less critical.[14]

There are a number of challenges facing the digital transformation of the battery production processes mentioned above. Some of these challenges can be solved with the help of digitalization and automation as follows[15,16]:

- Reducing the throughput times of production processes;
- Increasing the process stability and speed;
- Increasing the battery quality, while reducing battery manufacturing costs;
- Standardizing the interfaces of IT systems used within battery production.

In addition, the optimization potential of factory digitalization is well known among battery manufacturers, as they are ready to carry out the necessary changes in their factories to adopt Industry 4.0 technologies and become a "smart factory".[15] As the degree of digitalization increases, the amount of data in the battery manufacturing industry will rise.[17] In combination with the technologies and enablers of digitalization described in the following section, the data can be used profitably to optimize battery production systems in the long term toward the realization of smart factories for battery production.

3. Technologies and Enablers of Digital Transformation in Production

In this section, the most relevant technologies and enablers of digitalization in production are addressed, starting with cyber-physical systems (CPSs) and cloud computing, while more specific topics, such as 5G or robotics and AGVs are also described with a focus on their application within the production environment.

3.1. *Cyber physical systems*

Cyber-physical systems (CPSs) are not only objects, devices, buildings and transportation but also production plants or logistics components, which contain embedded systems that are able to communicate via the Internet and use Internet services. CPSs can sense their environment directly with their corresponding sensor technology, store data and evaluate it with the aid of globally available data and services. Additionally, they can interact with the physical world with the help of actuators. CPSs can ultimately build autonomous and decentralized networks and optimize themselves independently. In interaction with humans, they can solve problems independently.[18]

In the bigger picture, a smart factory uses CPSs to organize itself in a decentralized manner in near real time. It is a central feature to gather the data in real time from the factory. Through this real-time data, it is possible to merge the real world with the virtual world, to create a virtual image of reality permanently.[18]

CPS platforms then form the basis for connecting the Internet (network of people) with the Internet of Things and Services networks.[18] In order to achieve this, however, the CPSs must go through various stages of development as follows:

The first stage is a passive one where technology is used to provide basic functionality, for example, applying RFID chips on objects to provide unique identification. The system itself is not yet intelligent and does not have storage or evaluation capabilities. To provide intelligence, central services have to be applied. The second stage is then realized through the adoption of active sensors and actuators, which still would have a precisely defined and relatively small range of functionality. The third stage will be the intelligent network-capable systems, which consist of several actuators and sensors that already have a degree of internal intelligence with a corresponding interface, while being able themselves to establish communication with other systems. This leads to the fourth stage, the so-called System of Systems. Here, CPSs can independently combine their individual capabilities. This enables them to develop entirely new capabilities and provide services themselves.[18]

3.2. *Cloud computing*

Cloud computing is a decisive factor in the networking of components. Due to their ability to be accessed via the Internet and their service-oriented character, cloud services can be accessed easily via their interfaces in order to use other services. This feature is also reflected in the development of Industry 4.0: Production plants and their components have standardized interfaces that allow easy and homogeneous access to machine functions and information.[18]

The development of Industry 4.0 and Cloud Computing is closely related. Due to the high degree of networking of production facilities, new tools are required to monitor and manage the production plants. Cloud infrastructure is particularly suitable for these tasks because it has high computing and storage capacities. Due to their central position as central data storage, in which all information converges, and with their interfaces to production facilities through which production processes can be controlled, cloud structures are among the critical infrastructures for Industry 4.0. A decisive factor for the acceptance and widespread use of cloud technologies to improve business processes in Industry 4.0 is that the technology is trustworthy and secure, but at the same time easy and efficient to use.[18]

In cloud systems, data from production plants are analyzed and converted into reports and dashboards that can provide a precise overview of all connected systems. At the same time, they provide an extended view into the machine and component level. Production processes are monitored and stored in the cloud and displayed graphically. Cloud systems can be used to plan and intervene in active processes or even make automated decisions and send corresponding control commands to the production plants. This allows a level of synchronization between the production facilities and thus minimizes phases of insufficient capacity utilization when cloud systems work together in a process.[18]

3.3. *Mobile communication standard — 5G*

5G is the successor of the currently used 4G mobile networks. In comparison to 4G, the new standard builds on higher-frequency bands. The lower-frequency band, which is important for indoor and nation-wide coverage, starts at "only" around 600 MHz, but the middle-frequency band is located around 3.6 GHz and the high-frequency band, required in particular for very high data rates, at around 24 to 28 GHz (mmWave). The specific used frequencies differ from country to country as they were assigned by the International Telecommunication Union (ITU).[19]

5G provides new possibilities for many sectors due to advanced capabilities including the manufacturing sector. Generally, three application dimensions are defined for 5G, illustrating these new capabilities.[19] Table 1 provides an overview and lists the most important properties for each individual application.

Table 1. Overview of the different applications of 5G technology and their most important features.

Enhanced Mobile Broadband (eMBB)	Massive Machine Type Communications (mMTC)	Ultra Reliable Low Latency Communications (URLLIC)
• Data volume maximum 10 Tbit/s/km^2 • Peak data rate of 10 Gbit/s	• Up to one million IoT devices per km^2 • Optimized energy consumption of 1% related to LTE systems	• One-way latency of maximum 1 ms • Generally reliability of 99.999%

All of these applications play an important role in manufacturing. URLLIC will be important for reliable real-time communications, for example, on the field level between PLC and sensors and actors. In addition, many devices on the shop floor need to communicate, which mMTC will address. For some applications, e.g. video data streaming in quality control or time series data collection, data transfer rate per user is also a key requirement. eMBB will address this requirement with around 10Gbit/s for a user and 10Tbit/s capacity for the cell.

The following advantages for manufacturing come into place with the use of 5G[20]:

- Total data coverage in factories.
- Low costs for devices and energy consumption.
- Higher flexibility.
- Boosting digitization phase by connecting previously isolated system and equipment to gather data and information.
- Enabling factories to be smart and ready for other technologies like augmented reality.
- Denial of wired solutions.

Although these advantages offer great possibilities, the realization is currently in phase 2 of 3 (started with eMBB in phase 1) and will take about a year to be finished.[21]

3.4. *Artificial intelligence*

The definition of artificial intelligence (AI) varies historically. While intelligence is often defined in terms of fidelity to human performance, other sources choose a more abstract and formal definition of intelligence in terms of rationality.[22]

AI is defined as a branch of intelligence science and essentially includes the fields of natural and artificial intelligence. The discovery of the intelligent behaviors of living systems is defined in science as natural intelligence. Artificial intelligence, on the other hand, encompasses both the science and technology of creating intelligent software systems and machines.[23,24] Developments in natural intelligence led to artificial neural networks, genetic algorithms or colony optimization.

Machine learning is a subfield of artificial intelligence and describes the ability to improve performance based on past experience.[22] Over the years, rising computer power and the creation of the World Wide Web led to a vast amount of datasets, which is often called big data. These datasets consist of words of text, images, social network data and more. Learning algorithms, especially machine learning, are specifically based on these large amounts of data to make reasonable decisions.[22]

In the context of production, 29% of AI implementations today are used to predict when machinery and equipment will need maintenance. Other specific applications include video and image recognition through deep learning. This makes it possible, for example, to detect the failure of machine components or to evaluate the quality of workpieces and products. Predictive IoT analytics are used to improve worker safety or cost reductions.[25]

3.5. *Blockchain*

The Blockchain technology is of great relevance for the digitalization of services and processes and many different application areas outside the financial sector and above all independent of crypto currencies. While the automation potential associated with smart contracts is particularly important for the Internet of Things, it is essential for applications in the supply chain or for proof of origin of the irreversibility of managed transactions.[26]

Since the publication of Satoshi Nakamoto's paper in 2008[27] and the creation of the first Bitcoins in early 2009, both crypto currencies and Blockchain technology have received increasing attention. The reasons for this are multiple characteristics. Documents and assets can be uniformly coded to prevent forgery. The transfer between senders and recipients can be stored as a transaction in the Blockchain. Further, the storage of transactions is irreversible and traceable based on distributed consensus and encryption. Another advantage is that transactions are verified in a peer-to-peer network (P2P) rather than by a central authority. This leads to Smart Contracts which offer the possibility to describe and execute complex transactions and their boundary conditions. They enable both the automation of simple processes in the Internet of Things and new governance models through the establishment of alternative forms of organization.[26]

Thus, the application potential of blockchains goes far beyond crypto currencies. After the Internet of Things, the technology can initiate a new generation of the Internet of Values or Trust.[26]

3.6. *Digital twins*

The digital twin was defined first by the National Aeronautics and Space Administration (NASA). One definition of a digital twin is an integrated multiphysics, multi-scale, probabilistic simulation of a vehicle or system. This uses the best available physical models as well as sensor values and historical data to reflect the life of its twin.[28] Another definition of digital twin is a system which consists of data and information about each real asset as well as different models at different levels of granularity.[29]

The digital twin is applied in different areas. The focusing areas of the digital twin are manufacturing, simulation (virtual commissioning and robotic simulation) and the product itself.[30] To realize digital twins in manufacturing, there are several technologies on different levels of granularity available. In addition to individual production goods, machines and plants can also be modeled as digital twins in manufacturing. Standardized concepts such as the Asset Administration Shell are suitable for this use case.[31] The AAS provides standardized reference models to simplify data exchange and the self-description of assets (machines, plants, processes, etc.).

The usage of the digital twins provides benefits in manufacturing as follows: increased productivity and efficiency, reduced complexity, time saving, lower costs, identification of new business fields, increasing quality and reduced risks.[32]

3.7. *Augmented reality and virtual reality*

A human–CPS interface can be created with the help of virtual and augmented reality. Virtual Reality (VR) enables the user to simulate the behavior of a cyber-physical production system and to explore it in an interactive way by reproducing as realistic a picture of the production process as possible.[18]

Augment Reality represents the computer-aided extension of human perception by using virtual objects. Relevant information can be displayed directly in the field of vision of the worker. This is accomplished through the use of mobile platforms such as smartphones, tablets and smart glasses, which will be the most important tools for handling CPSs and the information they provide in the future.[18]

VR and AR technologies support actors in production in various application scenarios, for example, not only in maintenance (i.e. servicing, inspection, repair and improvement) of production facilities by providing interactive, virtual instructions but also during the monitoring of production processes and quality control by context-sensitive retrieval and provision of information, e.g. regarding the status of a CPS. Another scenario is the planning and (co-)simulation of production processes, e.g. by predicting the behavior of CPSs (e.g. traverse movements of linear axes or material flow of products).[18]

3.8. *Robotics and AGVs*

Due to the universal applicability and the possibility to configure the robot and its peripheral components for a specific production task, robotics is an ideal application for Industry 4.0 technologies.[18] Robotics is often used to automate the transformation of materials in manufacturing. This is called the material handling system (MHS) which operates continuously or under demand. Various solutions such as belt, roller and vertical conveyors, elevators, conveyor robots and automated guided vehicles (AGVs) are available for material transport.[33]

The primary task of automated guided vehicles is the internal transport of goods, or more precisely material, to the designated location. The transport of persons is therefore not one of the actual tasks of AGVs.[34]

When the manufacturing space is limited and flexibility is important, AGVs are particularly effective for manufacturing applications.[33] They also provide benefits such as saving labor costs,[35] increasing productivity,[36] reducing energy consumption[37] and increasing safety.[38] Therefore AGVs are often used by manufacturers to realize MHS.

In the context of Industry 4.0, systems that integrate sensors, actuators and cognition are called cyber-physical systems. AGVs are an example of such complex elements of production. However, the complete flexibility of these automation systems can only be exploited to the maximum if they are integrated into the Industry 4.0 production plant and their autonomy is increased.[18]

4. Summary and Outlook

This chapter gives an overview of digitalization and digital transformation in production. First, important terms are defined to explain the digital transformation in manufacturing. Then, a specific overview of the digital transformation in battery manufacturing is given. After that, the important technologies and enablers of digitalization are described.

The chapter introduces the major factors to enable smart manufacturing for manufacturing companies and establishes an application-oriented reference to battery manufacturing. At this point in time, the industry is still in different stages of digital transformation. This has been triggered by multiple factors. The digital transformation places high demands on the processes in the company and on the existing IT infrastructure. Frequently, this requires the use or development of new processes, interfaces or APIs, as well as investment in more powerful infrastructure.[39]

Furthermore, manufacturing companies are dependent on the output of production. Intervention in production during the digital transformation process must not lead to a reduction in productivity. Along with this criterion, manufacturers must also comply with budget and resource constraints.[39] Last but not least, the human factor must be taken into account. The introduction of new digital methods, processes or technologies must ultimately meet with a high level of acceptance so that they can be implemented sustainably.[40]

Which technologies are implemented in the course of digital transformation depends heavily on the industry and the intended use. The Gartner Hype Cycle puts the expectation of new technologies in relation to the time horizon.[41] In the context of production and manufacturing, technologies or topics such as Differential Privacy, Responsible AI, Embedded AI and Explainable AI can be of great benefit. Differential Privacy favors transparent training of neural networks and ensures privacy of individual or corporate data.[42] This will lead to increased adoption of these algorithms that will ultimately automate production. Responsible AI will make it possible to develop artificial intelligence with responsible behavior. Especially when combined with the increased use of collaborating robots, this can lead to an increase in productivity. The current problem that many people see is that the processes in a neural network as a black box are solved by explainable AI. This makes the decision-making of neural networks and thus, for example, the actions of robots transparently comprehensible. This also results in increased user acceptance.[43]

It remains to be seen how technology trends will develop in the future and which new topics will emerge. However, it is a fact that traditional production in its current form is no longer competitive and the digital transformation must therefore take place as quickly as possible.[44]

References

1. Erol S, Jäger A, Hold P, Ott K, Sihn W. Tangible Industry 4.0: A scenario-based approach to learning for the future of production. Procedia CIRP 2016; 54:13–8.
2. Fatorachian H, Kazemi H. A critical investigation of Industry 4.0 in manufacturing: Theoretical operationalisation framework. Prod Plan Control 2018; 29(8):633–44.
3. Moghaddam M, Cadavid MN, Kenley CR, Deshmukh AV. Reference architectures for smart manufacturing: A critical review. J Manuf Syst 2018; 49:215–225. Available from: https://www.sciencedirect.com/science/article/pii/S0278612518301043.
4. Nodehi T, Jardim-Goncalves R, Zutshi A, Grilo A. ICIF: An inter-cloud interoperability framework for computing resource cloud providers in factories of the future. Int J Comp Integr Manuf 2015; 1–11.
5. Westkämper E, Löffler C. Strategien der Produktion. Berlin, Heidelberg: Springer Berlin Heidelberg; 2016.
6. Warnecke H-J. Der Produktionsbetrieb 2. Berlin, Heidelberg: Springer Berlin Heidelberg; 1995.
7. Schenk M, Wirth S, Müller E. Fabrikplanung und Fabrikbetrieb: Methoden für die wandlungsfähige, vernetzte und ressourceneffiziente Fabrik. 2., vollständig überarbeitete und erweiterte Auflage 2014. Berlin: Springer Vieweg; 2014.
8. Heinrich B, Linke P, Glöckler M. Grundlagen Automatisierung: Sensorik, Regelung, Steuerung. 2., überarbeitete und erweiterte Auflage. Wiesbaden: Springer Vieweg; 2017. (Lehrbuch).
9. Neugebauer R, editor. Digitalisierung. Berlin, Heidelberg: Springer Berlin Heidelberg; 2018.

10. Bauer W, Hämmerle M, Bauernhansl T, Zimmermann T. Future Work Lab. In: Neugebauer R, (ed.), Digitalisierung. Berlin, Heidelberg: Springer Berlin Heidelberg; 2018. p. 179–195.
11. Bundesministerium für Wirtschaft und Energie. Digitale Transformation in der Industrie; 2020 [cited 2020 Nov 16]. Available from: https://www.bmwi.de/Redaktion/DE/Dossier/industrie-40.html.
12. Bauernhansl T. Die digitale Transformation. Wie werden wir morgen leben und arbeiten? Stuttgart-Hohenheim; 2018 [cited 2020 Sep 30]. Available from: http://publica.fraunhofer.de/dokumente/N-490477.html.
13. Heimes HH. Methodik zur Auswahl von Fertigungsressourcen in der Batterieproduktion [Zugl.: Aachen, Techn. Hochsch., Diss., 2014]. 1. Aufl. Aachen: Apprimus-Verl.; 2014. (Edition Wissenschaft Apprimus; Vol. 2014, p. 41).
14. Kampker A, Burggraf P, Deutskens C, Heimes H, Schmidt M. Process alternatives in the battery production. In: 2012 Electrical Systems for Aircraft, Railway and Ship Propulsion. IEEE; 2012. pp. 1–6.
15. VDMA Batterieproduktion. Roadmap Batterie-Produktionsmittel 2030: Fraunhofer ISI; PEM der RWTH Aachen; VDMA Batterieproduktion; 2016 [cited 2020 Oct 1]. Available from: https://www.researchgate.net/publication/328583673_Roadmap_Batterie-Produktionsmittel_2030_-_Update_2018.
16. Kubler K, Verl A, Riedel O, Oberle M. Simulation-assisted run-to-run control for battery manufacturing in a cloud environment. In: 2017 24th International Conference on Mechatronics and Machine Vision in Practice (M2VIP). IEEE; 2017. pp. 1–6.
17. Turetskyy A, Thiede S, Thomitzek M, Drachenfels N von, Pape T, Herrmann C. Toward data-driven applications in lithium-ion battery cell manufacturing. Energy Technol. 2020; 8(2):1900136.
18. Vogel-Heuser B, Bauernhansl T, Hompel M ten. Handbuch Industrie 4.0 Bd.4. Berlin, Heidelberg: Springer Berlin Heidelberg; 2017.
19. Zielinski E, Stanczak S, Schulz-Zander J. Leitfaden 5G-Campusnetze – Orientierungshilfe für kleine und mittelständische Unternehmen: Konzepte, Begriffe, Betreibermodelle und Auswahlkriterien für Produktion und Logistik mit Übertragbarkeit auf weitere Domänen wie Medizin-Campus/Krankenhäuser, Häfen, Bergbau, Baustellen und Landwirtschaft; 2020 [cited 2020 Dec 14]. Available from: https://www.bundesregierung.de/breg-de/service/publikationen/leitfaden-5g-campusnetze-orientierungshilfe-fuer-kleine-und-mittelstaendische-unternehmen-1752148.
20. Temesvári ZM, Maros D, Kádár P. Review of mobile communication and the 5G in manufacturing. Procedia Manufacturing 2019 [cited 2020 Dec 14]; 32:600–12. Available from: https://www.sciencedirect.com/science/article/pii/S235197891930294X.
21. O'Donnell B. Look out, here comes 5G, Phase 2; 2020 [cited 2020 Dec 14]. Available from: https://www.forbes.com/sites/bobodonnell/2020/07/08/look-out-here-comes-5g-phase-2/.
22. Russell S, Norvig P. Artificial Intelligence: A Modern Approach. 4th ed. Pearson; 2020.
23. Negnevitsky M. Artificial Intelligence: A Guide to Intelligent Systems. 2nd ed. Harlow: Addison-Wesley; 2005.
24. Wang L. From intelligence science to intelligent manufacturing. Engineering 2019 [cited 2020 Dec 14]; 5:615–8. Available from: https://www.engineering.org.cn/en/10.1016/j.eng.2019.04.011.
25. Columbus L. 10 Ways AI is improving manufacturing in 2020: Forbes; 2020 [cited 2020 Dec 15]. Available from: https://www.forbes.com/sites/louiscolumbus/2020/05/18/10-ways-ai-is-improving-manufacturing-in-2020/?sh=1aa7d7601e85.
26. Prinz W, Rose T, Osterland T, Putschli C. Blockchain: Verlässliche Transaktionen. In: Neugebauer R, editor. Digitalisierung. Berlin, Heidelberg: Springer Berlin Heidelberg; 2018. pp. 311–319.

27. Nakamoto S. Bitcoin: A peer-to-peer electronic cash system; 2008 [cited 2020 Nov 25]. Available from: https://bitcoin.org/bitcoin.pdf.
28. Shafto M, Conroy M, Doyle R. TA 11: Modeling, Simulation, Information Technology, and Processing Roadmap: National Aeronautics and Space Administration; 2010.
29. Talkhestani BA, Jazdi N, Schloegl W, Weyrich M. Consistency check to synchronize the Digital Twin of manufacturing automation based on anchor points. Procedia CIRP 2018; 72:159–64. Available from: http://www.sciencedirect.com/science/article/pii/S221282711830324X.
30. Negri E, Fumagalli L, Macchi M. A review of the roles of Digital Twin in CPS-based production systems. Procedia Manufacturing 2017; 11:939–948.
31. Federal Ministry for Economic Affairs and Energy. Details of the Asset Administration Shell: Part 1 The Exchange of Information between Partners in the Value Chain of Industrie 4.0. Berlin: Plattform Industrie 4.0; 2020 November 2020.
32. Biesinger F, Weyrich M. The facets of Digital Twins in production and the automotive industry. In: 2019 23rd International Conference on Mechatronics Technology (ICMT): Salerno, Italy, 23–26 October 2019. Piscataway, NJ: IEEE; 2019. pp. 1–6.
33. Martinez-Barbera H, Herrero-Perez D. Development of a flexible AGV for flexible manufacturing systems. Industrial Robot 2010; 37(5):459–468.
34. VDI-Gesellschaft Produktion und Logistik. Automated Guided Vehicle Systems (AGVS). Berlin: Beuth-Verlag; 2005. [cited 2020 Feb 2]. Available from: https://www.vdi.de/richtlinien/details/vdi-2510-fahrerlose-transportsysteme-fts.
35. Abhijit G, Grasman SE. Simulation-based optimization for determining AGV capacity in a manufacturing system; 2009.
36. Negahban A, Smith JS. Simulation for manufacturing system design and operation: Literature review and analysis. J Manuf Syst 2014; 33(2):241–261.
37. Acciaro M, Ghiara H, Cusano MI. Energy management in seaports: A new role for port authorities. Energy Policy 2014; 71:4–12.
38. Duffy VG, Wu FF, Ng PP. Development of an Internet virtual layout system for improving workplace safety. Comput Ind 2003; 50(2):207–230.
39. Zangiacomi A, Pessot E, Fornasiero R, Bertetti M, Sacco M. Moving towards digitalization: A multiple case study in manufacturing. Prod Plan Control 2020; 31(2–3):143–157.
40. Albukhitan S. Developing digital transformation strategy for manufacturing. Procedia Comput Sci 2020; 170:664–671.
41. 5 Trends Drive the Gartner Hype Cycle for Emerging Technologies, 2020; 2021 [cited 2021 Apr 15]. Available from: https://www.gartner.com/smarterwithgartner/5-trends-drive-the-gartner-hype-cycle-for-emerging-technologies-2020/.
42. Agrawal M, Du D, Duan Z, Li A. Theory and applications of models of computation: 5th international conference, TAMC 2008, Xi'an, China, April 25–29, 2008; Proceedings. Berlin: Springer; 2008. (Lecture Notes in Computer Science; vol 4978). Available from: http://www.springerlink.com/openurl.asp?genre=issue&issn=0302-9743&volume=4978.
43. Barredo Arrieta A, Díaz-Rodríguez N, Del Ser J, Bennetot A, Tabik S, Barbado A *et al.* Explainable Artificial Intelligence (XAI): Concepts, taxonomies, opportunities and challenges toward responsible AI. Information Fusion 2020; 58:82–115.
44. Vogelsang K, Liere-Netheler K, Packmohr S, Hoppe U. Barriers to digital transformation in manufacturing: Development of a research agenda; 2019. Available from: http://128.171.57.22/handle/10125/59931.

https://doi.org/10.1142/9789811245626_0004

Chapter 3

Current State of Digitalization in Battery Cell Manufacturing

Sascha Gärtner* and Michael Oberle†
Fraunhofer IPA, Nobelstraße 12, Stuttgart, Germany
**sascha.gaertner@ipa.fraunhofer.de*
†michael.oberle@ipa.fraunhofer.de

This chapter describes two methods to determine the current state of digitalization in battery cell manufacturing. The objective is to identify a possible discrepancy between research and industrial companies. For this purpose, job advertisements of various battery cell manufacturers are searched for relevant keywords in the context of digitalization. This keyword search will also be applied to scientific literature databases to reflect the state of research. Subsequently, the data are systematically evaluated and analyzed. The result shows a clear offset between industry and research. While industrial companies often address conventional problems like process control and process planning, research is largely focused on the topics of Big Data, IoT and Artificial Intelligence.

1. Introduction and Motivation

The demand for batteries for energy storage is increasing rapidly due to the further development of mobile devices and electric vehicles.[1] The increasing global demand for battery cells is impressively demonstrated by the registration numbers of battery-powered electric cars.[2]

Regardless of the industry, high demands are placed on battery cells by consumers. In addition to a high degree of individualization, end customers particularly demand high quality and thus battery cell runtime and durability. To meet these requirements, the production of battery cells must be as flexible, efficient and effective as possible.[3]

In the course of Industry 4.0, the digitalization of processes in companies is an elementary component for meeting these goals. A high degree of digitalization enables battery cell manufacturers to exploit hidden potential in production and thus build a flexible, adaptive and efficient production system.[4] This allows time to market and delivery times to be shortened, ensuring competitiveness and enabling the company to grow at the same time.[5]

The digitalization of the manufacturing process presents companies with unexpected challenges. Often, the necessary area-wide infrastructure is missing, so that the elementary basis for the introduction of digitalization is not given. In addition, implementation requires a high level of domain knowledge and an investment in costs. All in all, these are the factors that determine a company's current level of digitalization.[5]

The points mentioned also imply that battery cell manufacturers are at different levels of digitalization. In order to provide improved recommendations for action in the future and to better understand the needs of the customer, this article analyzes the level of digitalization in the production of well-known battery manufacturers. The current state of research in relation to digitalization in battery cell manufacturing is also researched as a reference.

2. Approach

Two different approaches are taken to research the current state of digitalization. For the current state of research, relevant publications in the context of battery manufacturing are searched with a list of keywords. To determine the current state of battery manufacturers, their job offers are matched with relevant keywords. The following sections describe the scope and details of how the respective research was conducted.

2.1. *State of research — Analysis of scientific articles*

To determine the current state of research, a systematic search for relevant publications on the portals ACM, IEEE, Google Scholar, Science Direct and JSTOR will be carried. The search period ranges from 2010 to 2020. A list of search strings containing relevant keywords in the context of digitalization is prepared in advance. This list contains about 60 keywords which are combined via AND conditions. A relevant publication must therefore contain the search strings. Table 1 shows an excerpt from the entire list of search strings.

Each of the databases mentioned is searched using the defined search strings. The most relevant 50 search hits per strings and database are taken as a maximum. A total of 649 search hits are obtained for the queries with the defined list of search strings. The relative ratio of the number of hits per database is shown in Fig. 1.

Table 1. Excerpt of different search strings when searching for relevant publications in the context of battery cell manufacturing.

Query	Search String		
1	"battery production"	AND	"digitalization"
2	"battery manufacturing"	AND	"digitalization"
3	"battery production"	AND	"machine learning"
4	"battery manufacturing"	AND	"machine learning"
…	…	AND	…
60	"accumulator"	AND	"PLC"

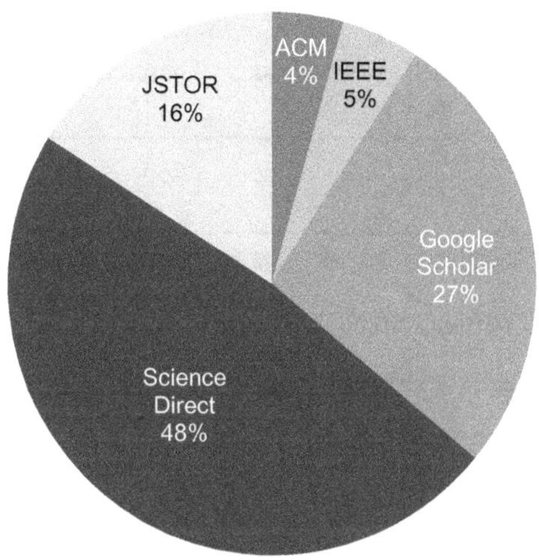

Fig. 1. Relative number of relevant search hits on the databases considered (taking into account a maximum of 50 search hits per search string and database).

In the following step, the publications are reviewed in order to classify only relevant articles from the total number of search hits. For this purpose, it is checked whether the publications are specifically related to the battery production. If this is the case, the paper is classified as relevant. After the analysis, the total number of 649 search hits is reduced to a total of 105 relevant publications that are directly related to battery cell production.

The terms of the search strings are now assigned to the appropriate subject areas. As a topic area, general terms are selected that describe the content of several search strings. Table 2 shows the subject areas considered.

Based on Table 2, the relevant publications are assigned to the corresponding topic area — duplicate assignment is excluded. The distribution across the subject areas is shown in Fig. 2.

Table 2. Subject areas and their specific explanation for the classification of relevant publications.

Subject area	Definition/Criteria
Big Data	Large amounts of data information that are generated or evaluated
Machine Learning	Data preparation, analysis, training, prediction and application of artificial intelligence methods
IoT	Automation of plants through, e.g. machine-to-machine communication — Step toward self-organizing production
Simulation	Process, sequence or material flow simulation, also machine simulation (e.g. physical or chemical processes)
MES	Connection ERP system with processes in production — Automated operations production planning
PLC	Manufacturing process is controlled through a programmable logic controller
ERP	The ERP system supports all business processes carried out within a company

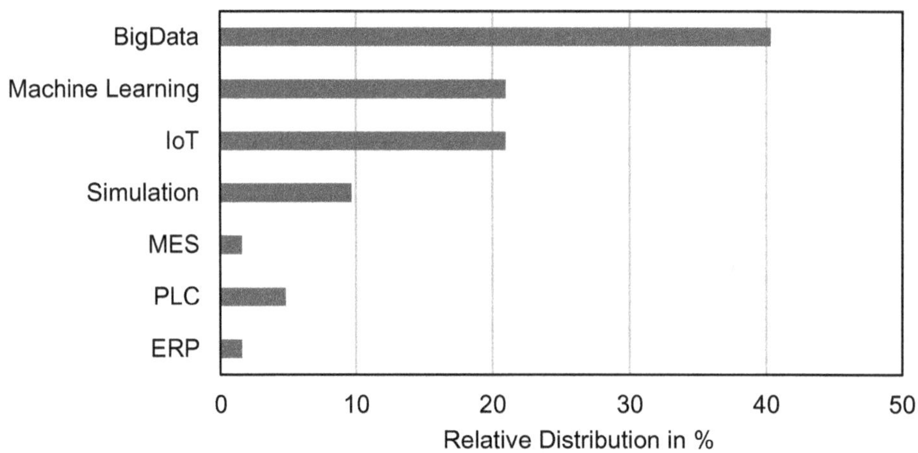

Fig. 2. Relative distribution of relevant papers per subject area, 105 publications in total.

2.2. *State of manufacturers — Analysis of job ads*

Now that the basis for analyzing the digitalization status of research has been established, the next step is to query the status of manufacturing companies.

The goal is to find out the current status of battery cell manufacturers by searching for keywords in their job ads. For this purpose, first a list of twenty manufacturers (cf. Table 3) is created whose ads are to be searched afterward. The list includes established global players as well as smaller companies and start-ups. The company data are taken from the respective company homepage, the

Table 3. Battery cell manufacturers examined as part of the job advertisement evaluation (in alphabetical order).

Manufacturer	Turnover in Euro (€)	Employees (Total)	Founded
Contemporary Amperex Technology GmbH	4.000.000	37	2014
CustomCells Itzehoe GmbH	n.a.	24	2012
EAS Batteries GmbH	n.a.	30	2016
Envision Energy CoE GmbH	n.a.	22	2015
Farasis Energy Europe GmbH	n.a.	60	2019
Leclanché GmbH	15.559.000	87	2000
Leclanché SA	14.998.000	250	1909
LG Chem Europe GmbH	633.871.000	127	2005
Microvast GmbH	12.670.000	20	2016
NorthVolt AB	11.622.000	108	2015
Northvolt Zwei GmbH & Co. KG	n.a.	n.a.	2019
Saft Groupe SA	9.492.000	6	1918
Samsung SDI Battery Systems GmbH	352.349.000	307	2009
Samsung SDI Europe GmbH	24.050.000	77	2011
Skeleton Technologies GmbH	n.a.	62	2013
VARTA Consumer Batteries GmbH & Co. KGaA	292.197.000	959	2002
VARTA Microbattery GmbH	98.495.000	765	2001
VARTA Storage GmbH	9.857.000	50	2012

commercial register or the financial database Amadeus. The data are primarily from the past fiscal year 2020. In some cases, data from earlier years were also considered due to missing annual reports, but not older than fiscal year 2018.

The distinction between large companies, medium-sized companies and start-ups is made in order to determine the status of the entire industry. Start-ups often have more innovative ways of thinking and strategies than large established companies with static structures. Larger companies, on the other hand, usually have greater financial resources and numbers of employees, which means that a greater workload can be achieved. The progress of digitalization can therefore be driven forward by different approaches.

In addition, a keyword list with search terms is also created here. These terms are found in different contexts with digitalization in production. In addition to the application of current information technologies, the search terms describe, for example, the infrastructure, information technology interfaces or digitalized/analog processes in production. This means that different levels of digitalization can be addressed.

A total of 280 keywords are used for the search. The job advertisements of the above companies (cf. Table 3) are searched using the keyword list in the period

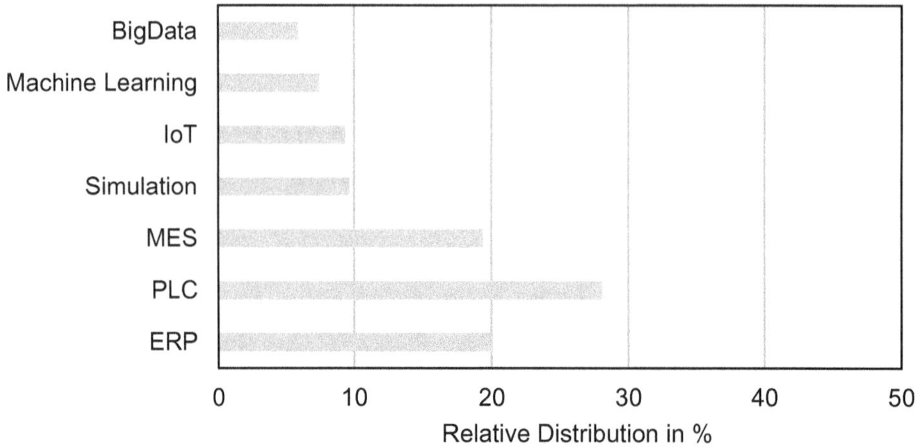

Fig. 3. Relative distribution of relevant papers per subject area, 160 job ads in total.

from July 2018 to July 2020. The search includes job ads from a total of nine different countries. These are Germany, Austria, Switzerland, Belgium, France, the Netherlands, Luxembourg, Sweden and Denmark. In addition, search hits from all job levels are included in the evaluation.

The search hits for the given conditions result in a total of 597, including 189 IT jobs. Due to partly generally formulated search terms, these 597 are filtered in more detail. In this context, 437 job ads are classified as not relevant. This leaves a total of 160 relevant job ads in the area of digitalization in manufacturing over all twenty battery cell manufacturers. Looking at the topics addressed in the job advertisements, the following distribution can be derived (cf. Fig. 3).

3. Evaluation and Interpretation

After conducting and processing the searches, the current digitalization status of research and battery manufacturers is now estimated. As mentioned at the beginning, the comparison of the two areas is the primary goal.

Figures 2 and 3 can be used to interpret the results. If the two bar charts are superimposed, the graph in Fig. 4 is obtained. In addition to the individual trends, the discrepancy between the state of research and industry can be noted in particular. While the publications in the period from 2010 to 2020 are mainly addressed to solutions relating to networked and highly automated production, industrial companies deal with problems at a lower level of digitalization.

One reason for this is that new technologies usually involve greater financial risk than the use of existing technologies. Medium-sized companies also lack

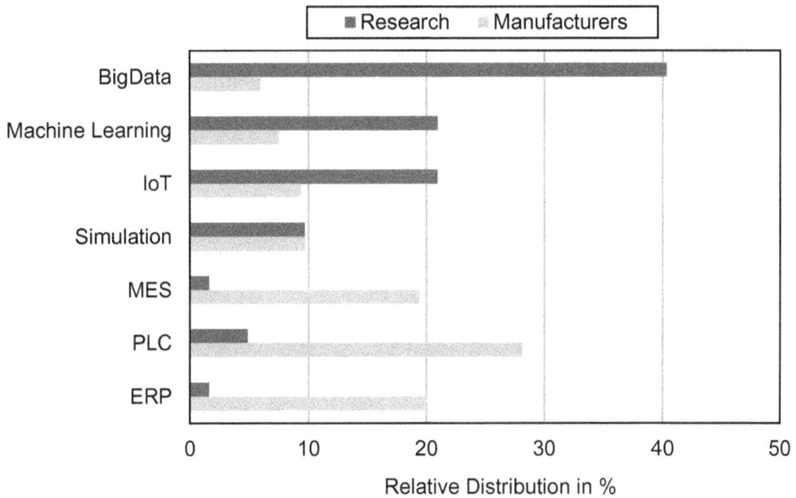

Fig. 4. Comparison between research and battery manufacturers regarding the addressed subject areas.

Table 4. Listing of the different digitalization levels and their description.

Level	Description
Level 1	*Digital Data Recording*: such as numerical controls (NC), material requirements planning (MRP), sensor data connection
Level 2	*Digital Modeling*: Digital factory, manufacturing execution system (MES), enterprise resource planning (ERP)
Level 3	*Networked Value Chain*: Internet of things (IoT), 5G, cloud computing
Level 4	*Autonomy*: Artificial intelligence, machine learning

suitable strategies and expertise in new technological areas.[6] In addition, there are different perceptions of the terms digitalization and Industry 4.0. While large industrial companies see Industry 4.0 as self-organizing networked production using Big Data and Machine Learning, smaller companies often only understand it as the use of production planning systems or the conversion of analog formats into digital formats.[7]

In order to explicitly distinguish the difference between research and the battery manufacturers, the first step is to divide them into digitalization levels. With the help of the digitalization levels, the chronological history and the exact status can be better represented. Table 4 describes the levels and their underlying conditions.

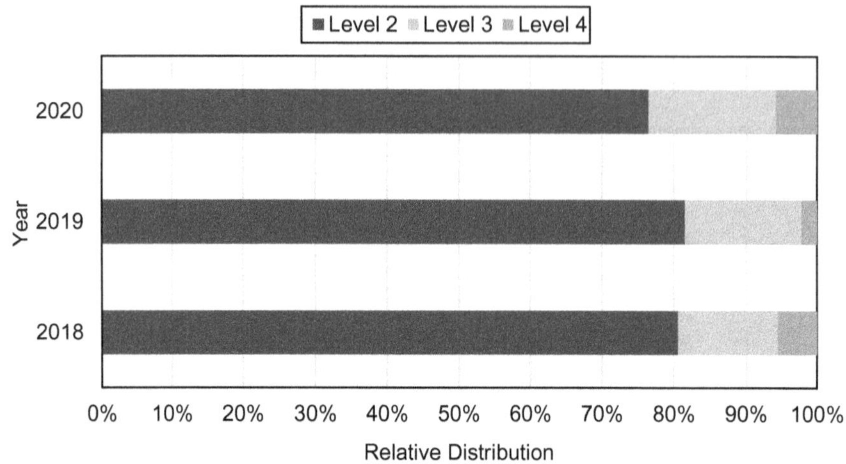

Fig. 5. Relative distribution of job ads regarding their digitalization level per year.

Based on their content, the relevant job ads are assigned a digitalization level according to Table 4. Depending on the publication year of the job ad, the different digitalization levels are shown in Fig. 5. The horizontal axis shows the relative distribution of the different digitalization levels per year.

This illustration shows that the distribution from 2018 to 2020 remains almost constant. The most frequently advertised jobs are those that address the second level of digitalization. The focus of production at this level is in particular the stable manufacturing process and supply chain through the use of MES and ERP systems.

While the digitalization level is addressed almost constantly each year at around fifteen percent, Level 4 fluctuates by a few percentage points and is considered by far the least.

To further analyze the existing discrepancy between research and battery manufacturers, the number of relevant publications per year can be presented (Fig. 6). It is obvious that there is a certain latency period between research results and their application in practice. This can be confirmed by the graph in Fig. 6 and various studies on the transfer of knowledge from research to industry.[6,8,9]

Figure 6 also shows a steady increase in publications since 2017, which means that the topics of digitalization in the context of battery cell production were only addressed to a small extent in research before 2017. Accordingly, little knowledge can have flowed into the manufacturing companies up to the present time. In addition to knowledge transfer, the successful implementation of findings from basic research also requires validation of the findings in the practical production

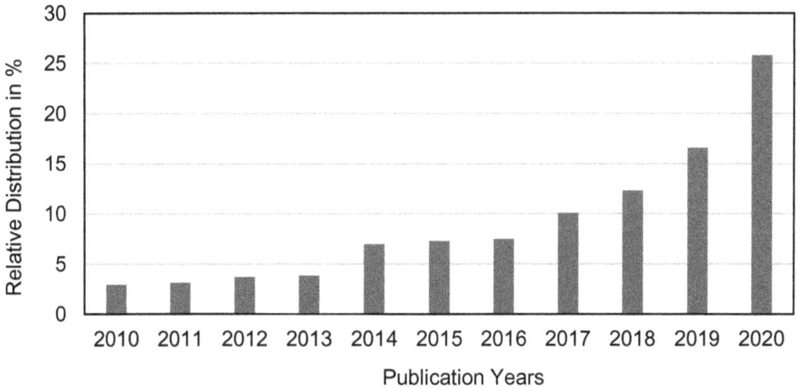

Fig. 6. Relative distribution of publications in context of digitalization over the past 10 years.

environment. This is because reliable processes are absolutely necessary to avoid failures at a later stage.

The application or implementation of new methods is therefore initially carried out on pilot plants rather than in series production. However, setting up these plants is a cost-intensive and high-risk factor.

In summary, the evaluation and interpretation of the results show that there is a discrepancy between the digitalization status of research and battery manufacturers. According to Fig. 5, a large proportion of battery manufacturers are still involved in the planning, management, control and monitoring of production processes (MES). Research, on the other hand, has been addressing more and more topics dealing with a higher degree of digitalization since 2017. For a holistic statement, companies of different sizes and years of existence were considered in the analysis.

4. Conclusion

A general problem is the different goals of industrial companies compared to research. While science wants to develop solutions that are as comprehensive as possible, industrial companies, on the other hand, need specific and individual solutions for their problems. In addition, high demands are placed on new methods in the production environment. The efficiency and effectiveness of a production plant are in the foreground.

At the beginning of this chapter, it was explained why the use of new methods and technologies entails financial and functional risks. The knowledge gap between industry and research must be closed through cooperation projects between practice-oriented research associations and battery manufacturers.

The analysis in this chapter clearly shows in which topics and at which level of digitalization battery manufacturers still need to act. Problems of the industry partners can be addressed more precisely and communication can be improved. The orientation of the research association can therefore be based on this analysis in order to better solve the goals of the battery manufacturers and to set priorities.

References

1. Asif A, Singh R. Further cost reduction of battery manufacturing. Batteries 2017; 3(4):17.
2. Statista. Worldwide number of electric cars | Statista; 2021 [cited 2021 Mar 23]. Available from: https://www.statista.com/statistics/270603/worldwide-number-of-hybrid-and-electric-vehicles-since-2009/.
3. Lödding H, Riedel R, Thoben K-D, Cieminski G von, Kiritsis D, (eds.). Advances in production management systems: The path to intelligent, collaborative and sustainable manufacturing; IFIP WG 5.7 International Conference, APMS 2017, Hamburg, Germany, September 3–7, 2017; proceedings. Cham: Springer; 2017. (IFIP advances in information and communication technology; vol. 514).
4. Zhong RY, Xu X, Klotz E, Newman ST. Intelligent manufacturing in the context of Industry 4.0: A review. Engineering 2017; 3(5):616–630.
5. Siedler C, Dupont S, Zavareh MT, Zeihsel F, Ehemann T, Sinnwell C *et al.* Maturity model for determining digitalization levels within different product lifecycle phases. Prod Eng Res Devel. 2021; 15:431–50.
6. Projektträger Jülich; Statustagung Schifffahrt und Meerestechnik. Statustagung Schifffahrt und Meerestechnik: Tagungsband der Statustagung 2010. Hannover, Jülich: Technische Informationsbibliothek u. Universitätsbibliothek; PTJ; 2010. (Schriftenreihe Projektträger Jülich). Available from: https://edocs.tib.eu/files/e01fn11/653804849.pdf.
7. Schröder C. Herausforderungen von Industrie 4.0 für den Mittelstand. Bonn; 2016. (Gute Gesellschaft – soziale Demokratie # 2017plus).
8. Warschat J, (ed.). Transfer von Forschungsergebnissen in die industrielle Praxis: Konzepte, Beispiele, Handlungsempfehlungen; Zusammenfassung der Ergebnisse des Projektes Entwicklung von Transfermechanismen für die effiziente und nachhaltige Verbreitung von Forschungsergebnissen in die industrielle Praxis am Beispiel Mechatronik. Stuttgart: Fraunhofer Verl.; 2013. Available from: http://www.bookshop.fraunhofer.de/buch/238468.
9. Maicher L, Radic M, Dijk S, Große C. Zukunftschance Digitalisierung — Ein Wegweiser; 2015.

Part II

Digital Planning in Battery Cell Manufacturing

https://doi.org/10.1142/9789811245626_0005

Chapter 4

Digital Planning of Battery Cell Factories

Julian Grimm* and Max Weeber†
Fraunhofer Institute for Manufacturing Engineering and Automation IPA
Nobelstraße 12, 70569 Stuttgart, Germany
**julian.grimm@ipa.fraunhofer.de*
†max.weeber@ipa.fraunhofer.de

In this chapter, digital planning of battery cell manufacturing is discussed. Therefore, factory planning frameworks and participants are described. Furthermore, general interaction challenges in factory planning and challenges specific to battery cell manufacturing are presented. Digital tools to counteract those challenges are introduced. The challenges arising when using those tools are outlined and solution approaches are given and evaluated.

1. A Toolbox for Digital Assisted Factory Planning

The economic, environmental and social performance of a factory is very much determined by the initial planning process. Planning is the first stage of a factories' life cycle, while the others are construction, ramping up, operation and reutilization/dismantling.[1]

Based on the product to be manufactured, the intended manufacturing capacity as well as related supply chain restrictions, factory planning involves the search for an adequate factory site, layout and building design. It also involves the selection of manufacturing and related auxiliary equipment, the planning of the technical building systems and the design of a demand-oriented energy supply concept. The early stages of a planning process already determine a major share of investments required to execute the project during construction as well as subsequent ramp-up and operation costs (Fig. 1).

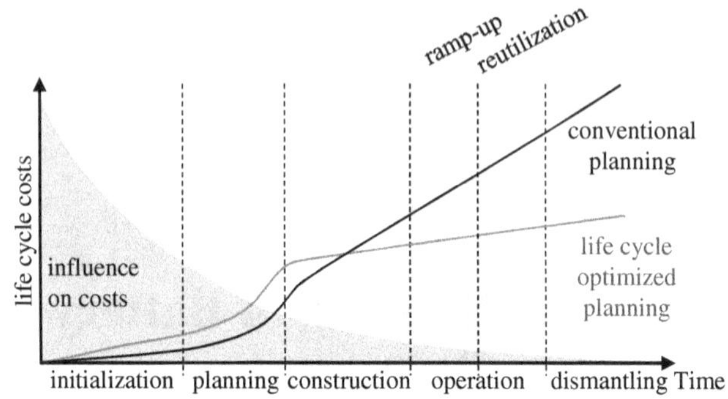

Fig. 1. Influence of factory planning process on costs along a factories life cycle according to Ref. [2].

Factory planning involves several planning activities and involves many internal planning participants and external stakeholders. The challenge is to coordinate this complex process in a timely and content-related manner. An efficient planning process is able to provide the right information, at the right time to the right person.

To reach this, planning in every single stage of the factory planning process needs to be carefully interconnected horizontally with other planning stages and vertically across the different planning objects. Good projections on how individual design decisions influence the different planning objectives need to be provided during early design stages, and close-to-reality insights also need to be provided on how these decisions impact factory performance after being operational. Conventional planning has only limited forecasting abilities, which does not meet the demand of today's complex factory planning. Digital planning technologies have the potential to analyze the system's behavior, make the planning results and decisions more accurate and thus improve the factory planning process.

Within factory planning, there are several planning objectives including the general project objectives, time, costs and quality, as well as factory-specific objectives such as economic and resource efficiency, adaptability, interconnectivity and attractiveness. Based on public interest, the ongoing climate debate and the aim to produce "green" batteries in the future, one central objective during the planning of battery cell factories is the sustainable, resource- and energy-efficient operation. A comparison of current values for the energy intensity of battery cell manufacturing and with the forecast on future manufacturing capacities gives a hint of the energy demand of the battery cell manufacturing industry (see preface). This underlines the need to focus on sustainability together with techno-economic objectives during the planning of new factories for battery cell manufacturing. Digital planning tools have the capability to explore and leverage the potentials on the path toward future green batteries.

In the following, several factory planning frameworks are described. Based on one example, the planning stages and parties involved in the process are characterized. Given the complexity of a greenfield planning project, general challenges related to the interaction of the planning participants are presented in detail and the characteristics of planning a battery cell factory are specified. To overcome those challenges, digital planning tools are presented the simulation of which is described in more detail. The potential problems of using digital tools are outlined and different solutions to counteract these problems are proposed.

1.1. *Factory planning frameworks and planning participants*

A standard for factory planning is given in VDI 5200. It distinguishes between different reasons for planning: design or new planning, which is often referred to as greenfield planning, replanning or brownfield planning, dismantling and revitalization.[3]

The industry of battery cell manufacturing has just started to grow in Europe, but is likely to increase tenfold by the end of the decade. Because of it being a relatively new industry, the current planning tasks can be characterized by a greenfield approach. This describes the design of a factory on a new location without any restrictions.[3] Furthermore, planning is not based on experience as knowledge is being built up and experiences are gathered along the course of the first planning projects. Considering everything, a standardized approach is useful, especially when planning battery cell manufacturing factories on a green field.

Standard approaches include the sequential, integral, holistic, cooperative and synergetic approaches, all of which are described in the following. The sequential approach of Ref.[3] is a normative standard for factory planning. A sequence of tasks has to be fulfilled for the action of factory planning: scoping of objectives, establishment of the project basis, concept planning, detailed planning, preparation and monitoring of realization, ramp-up support and project closeout. The integral approach is setting up a fully integrated planning team in the beginning in order to meet the needs for coordination and communication along the life cycle of a factory.[4] The holistic approach considers the components, product, technology, organization, equipment, staff and finances holistically.[5] The planning of the factory is derived top-down. Each level of detail is planned at a time and with system orientation considering interactions and interdependencies. In the cooperative approach, processes and planning participants are synchronized, avoiding a

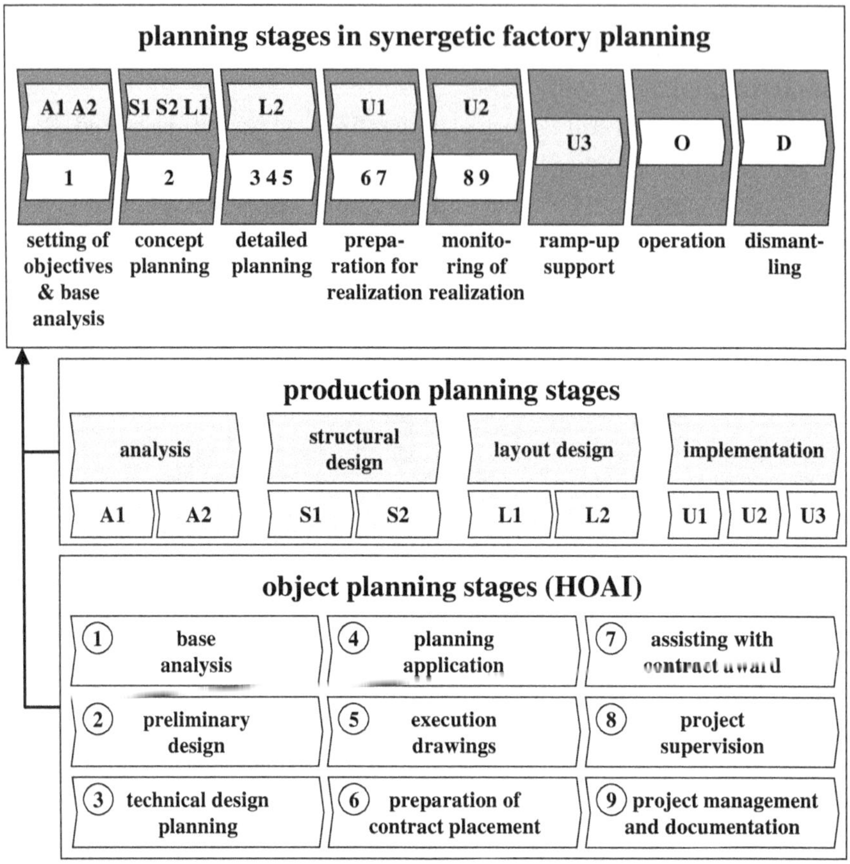

Fig. 2. Synergetic factory planning stages according to Ref. [8].

strictly sequential approach.[6,7] The synergetic approach is based on the combination of processes and decisions of all participants, aiming for synergies between the process and the spatial view.[8]

It can be seen that the decision on a specific factory planning approach influences the way single aspects of a planning process are prioritized, how the process of planning is implemented and how the participants involved in the planning process communicate with each other.

In the following, the synergetic factory planning approach is described in greater detail (Fig. 2). The stages of operation and reutilization/dismantling are added, in order to represent all stages of a factory's life cycle.

The design tasks in a synergetic approach are considered with respect to the room or building (spatial view) and the process being performed in the specific room (process view) for all of the following planning stages. Both the design of the manufacturing system and the building are integrated into one single approach. The first stage ("objectives & base analysis") analyzes the planning objectives and determines the planning baseline. In the design of the manufacturing system, the planning objectives are set and a process analysis is conducted. For the building design, a basis analysis is performed. In the subsequent "concept planning", the structure and dimensions of the building are established. In addition, a rough manufacturing layout as well as a preliminary building design is developed. In the following "detailed planning" stage, the detailed layout for manufacturing is planned. This planning stage also determines the technical specifics of the building and develops the drawings necessary for the approval procedures and the execution of the project. In the following planning stage, "preparation for realization", the tender for the manufacturing equipment and the execution of the building construction is prepared and executed. "Monitoring of realization" includes monitoring and project supervision of both manufacturing and building design. The "ramp-up support" assists in ramping up the manufacturing equipment and the technical building systems.[8]

All single planning stages need to contribute to the overall success of the project. Therefore, necessary interactions and interdependencies of each planning stage with different levels of detail need to be implemented and put into practice.

In the factory planning approach, many planning participants are involved in each planning stage (Fig. 3).

The building owner defines the requirements together with the users and makes a decision. The project manager leads, controls and steers the project. The design of the manufacturing system, involves different planners for the individual functional areas, such as manufacturing (process chain and assembly planning, production planning and scheduling), logistics (material flow, warehousing and

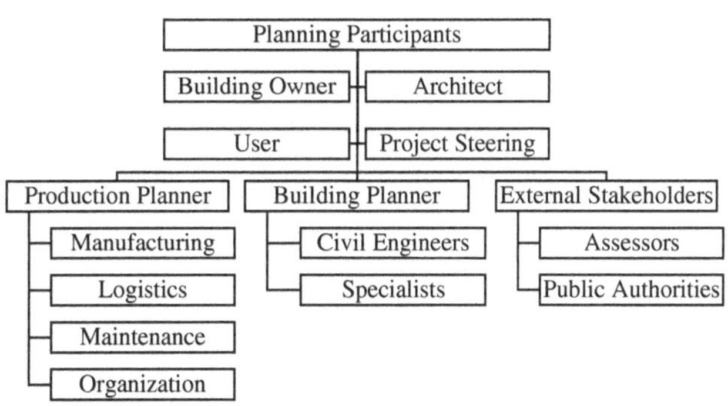

Fig. 3. Participants involved in factory planning.

commissioning), factory organization (staff planning and organizational development) and maintenance (accessibility, maintenance system). In building design, the participating parties are architects (floor planning, interior and exterior design and outdoor facilities), civil engineers (statics, structure and engineering structures) and specialist planners (supporting structure, building physics, technical building systems TBS, heating, ventilation and air-conditioning systems HVAC).

All parties involved are participating in the planning process at different points in time and with different intensities. For this reason, they need to be coordinated throughout the different planning stages in a way that the information requirement and supply meet the demand of the overall project.

1.2. *Characteristics and challenges of planning a battery cell factory*

The factory planning approaches, stages and participants outlined in Section 1.1. underline the complexity involved in factory planning projects. The general challenges in the interaction of all parties involved in a factory planning project are described in the following. In addition, the characteristics and peculiarities of planning a battery cell factory are given.

On a high level, there are three major interaction challenges in factory planning: exchange of information, coordination and cooperation of involved parties, and project monitoring.[9]

The challenge related to the exchange of information may have variable reasons, one being the unavailability of information.[10] Available information can still be unknown for the individual user due to different reasons. The reasons for this can be a missing information flow or an information flow with the wrong participants due to the complex planning process. Another reason is the timespan between when a piece of information is generated and when it is used. Often, most attention is put on the execution of single planning tasks, and interdependencies gain little attention or are disregarded. This may lead to conflicts at the intersections between different planning participants and/or planning stages.[11] Information management currently does not meet the holistic requirements.[12] Considering information exchange, a mutual, clearly defined and well-functioning information exchange within planning participants and planning stages is necessary.[13]

A second general challenge is the coordination and cooperation of involved participants. The risk of planning local optima, which are inefficient for the overall project, may arise when different professionals work separately on single, isolated planning tasks and during individual planning stages.[11] When integrating participants, miscommunication may arise due to different backgrounds, e.g. professional backgrounds, expertise or experiences with different planning processes.[13] Besides these interpersonal challenges, the coordination of data exchange is also challenging.[5] Changes may lead to obsolete data being shared with other planning participants or different maturity levels not being communicated properly (final version vs. work status). Consequently, coordination and cooperation are a major challenge and highly necessary in factory planning.[13,14]

With respect to project monitoring, the related high-level challenges can be summarized as follows. Shorter innovation and product life cycles require shorter response times to unexpected changes in the planning process resulting in additional planning activities, which have not yet been considered. This potentially delays the planning schedule.[5] The beginning of a project may already reveal possible challenges as long as planning objectives are defined inappropriately. The interdependencies between technical, economic, commercial and legal planning should also be taken into account.[13] Due to the different activities in each of the segments, the right information needs to be received at the right time. Because of this, the interdependencies have a high risk of leading to challenges, e.g. offerings with wrong economic key figures or contracts with missing technical specification. In order to reach the project objectives, the different planning results from all levels of detail in both manufacturing and building have to be merged. The planning process thus requires substantial effort from project management with respect to organization, coordination and documentation.[15] A clear project structure and steering is necessary and the project management needs to ensure that planning is seen and handled in the general context with respect to the different planning participants and stages.[13]

The general challenges described in the previous section can be illustrated by the specific characteristics of planning a battery cell factory.

Given the small number of battery cell factories in Europe, there is only some experience and operational data available to be used during the greenfield planning of battery cell factories. The forecasted rise in demand for planning, construction and operation of battery cell factories offers the opportunity to design state-of-the-art factories considering the latest advances in digitalization and sustainability.

The interaction between different planning participants complicates the planning of a battery cell factory in particular, as there are many participants involved in detail in each stage. In most cases, a well-defined demand for a specific product is the reason for a factory planning project. In contrast, battery cell manufacturing serves a younger and broader demand. Battery cells for future applications cannot yet be considered standardized products. Future applications include batteries for BEV and stationary batteries for households, industry or energy supply system. As of today, different OEMs install different battery cell and pack types in their vehicles. This indicates that the market is still in a process of consolidation, without many standards being established and adopted on a broad basis. Yet, there is great effort in research and development to improve today's battery cells and systems. As the product requirements are not frozen yet, product development is an additional and important participant in a project for planning a battery cell factory. It becomes evident that product development needs to become a substantial part in the factory planning process, thus increasing the amount of planning participants.

Building upon a well-defined product, production planning can derive manufacturing and logistics requirements. This results in manufacturing and logistics equipment (size, performance). Going into more detail, the examples of energy system planning and planning objectives illustrate further challenges specific to battery cell manufacturing.

In energy system planning, energy conversion equipment is designed using the yearly load profiles for the individual energy demand (power, heat on different temperature levels, compressed air, etc.). The load profiles define the required performance over time, which has to be considered when deducing the dimensions of energy conversion and storage equipment. The challenge in battery cell manufacturing is to forecast the load profile, as there is no measured data. Forecasting a load profile is essentially depending on the future operation and machine data, which has to be forecasted as well. To generate an hourly forecast of the power demand, each machine state has to be known, which is influenced by the production program which in turn is influenced by the expected sales and customer orders. Besides, product characteristics need to be known to roughly design manufacturing equipment. Knowing the machine type and state, information on the machine has to be gathered with detailed machine data for the different machine states (operation, standby, partial load, etc.) for all types of effective (e.g. electricity) and auxiliary (e.g. compressed air) energy sources. Integrating all those pieces of information, a load profile can be forecasted. It becomes obvious that, in this context, designing the energy system with the individual energy conversions systems has a high amount of uncertainty and further increases the complexity during factory planning.

Another example is the prioritization of planning objectives. Due to the high energy demand of battery cell manufacturing, current objectives in planning like the classical time, cost and quality as well as adaptability and innovation capability[8] need to be extended with sustainability objectives like climate neutrality or energy efficiency. With the ongoing improvements in cell chemistries and different cell types, battery cell factories need to be flexible in order to allow the adoption to new battery cell designs, chemistries and formats. As a result, the planning objectives need to be clearly stated, well communicated and followed precisely by all participants involved in the planning process.

Summarizing, the big issue in planning battery cell factories is the lack of information, which calls for additional efforts to improve communication, coordination and cooperation. This applies to all planning stages and includes all participants involved in the planning process. In addition, there is a need for a project setup which respects the general conditions and restrictions and sets feasible and consistent planning objectives.

As described above, the design of a battery cell factory incorporates and even intensifies the general challenges. In the following, digital technologies are presented to counteract those challenges and to support a greenfield planning process while designing state-of-the-art factories concerning digitalization and sustainability.

1.3. *Digital technologies to support factory planning*

Digital technologies can be implemented to overcome the challenges associated with the exchange of information, coordination and cooperation of involved parties as well as the monitoring of a project's time, costs and degree of completion. In the following, an overview of different digital planning tools is given. Challenges arising from using these tools in the planning process are presented together with recommendations for their successful implementation using the example of simulation.

Databases and the corresponding database management systems are fundamental to solve data management holistically. Within a data management system, data are organized and managed in order to use and store data efficiently and consistently for a long time.[16] A common data environment supports the collaborative generation, administration and exchange of data and information between different users and various digital tools.[17] Besides, it is beneficial for data consistency and can be used across all life cycle stages of a factory.[18] Databases are fundamental to other digital solutions, like Building Information Modeling (BIM) or simulation.

Modeling and simulation represent a system with dynamic processes in a modeling environment, while aiming to generate understanding and knowledge of the system.[19] After modeling a process or factory environment, different scenarios can be implemented in order to analyze the behavior of the modeled system and generate helpful information for decision-making. Examples of the application of simulation in the context of factory planning are material flow, process or energy system planning. The benefits of using simulations include the understanding of the systems' behavior, the detection of interdependencies, and the possibility to support communication of the simulation findings through visualization.

The technology of virtual reality enables the planner to see 2D and 3D models of the planned factory, building or production system before it is built. Using visualization, the cooperation between the different planners can be supported. For example, planners can use virtual reality during meetings in order to jointly develop new designs and promptly visualize and verify the changes applied to the virtual representation of the planning objective.[20] Additionally, project management can be kept up to date using this technology.

Augmented reality expands virtual reality through placing virtual objects into the real world. By augmenting the real world like this, the user gets a more realistic representation of the planned design. In factory planning, augmented reality can be used to project a factory design on a property or manufacturing equipment into an existing building. It can also support maintenance planning.[21]

Digital twin describes a technology which represents a physical system in a virtual dynamic system. Digital twins have the ability to acquire and process data in real time.[22] With the help of automated data exchange, the virtual representation communicates with the real object. In factory planning, the technology can be implemented already in the planning stage as a virtual shadow and be improved until the operation stage to represent the physical system of a machine, process chain, building or a whole factory. Specific use cases for a digital twin are the planning and evaluation of a manufacturing system with the help of automated data acquisition[23] or the support of production planning and control.[24]

Building information modeling (BIM) can be described as a "shared knowledge resource for information about a facility forming a reliable basis for decisions during its lifecycle".[10] The fundamental principle is to connect all participants during design and construction of an object and to gather, store and distribute data centrally.[14] There are different levels in BIM[14]:

- 3D BIM: 3D building model (geometry + specific parameters).
- 4D BIM: information on time for schedule planning.
- 5D BIM: information on costs for cost planning.
- 6D BIM: information for handling and maintenance during the life cycle.

There are several software solutions for the implementation of a BIM model, e.g. Autodesk REVIT, Graphisoft Archicad, TEKLA or SketchUp. BIM is widely used in the planning stage, but planning participants can transfer the central building information model to the subsequent factory life cycle stages.[10,8] The benefits of BIM are widely described in the literature and include the control of the rising amount and complexity of information,[8] the possibility to connect the modeled buildings with a simulation tool for analysis and optimization[25] or the option to have different planners working together in a way which decreases the amount of mistakes caused by an inconsistent database.[26]

Using digital technologies has the potential to support involved participants in different stages of a factory planning process. However, some challenges may arise when using these tools. This paragraph analyzes the occurring problems and gives solution approaches and recommendations to overcome those problems using the example of modeling and simulation, as these are an integral part of many digital technologies. The methodology combines a literature analysis on potential problems when using the tool of modeling and simulation with Ishikawa diagrams to identify the problem causes. Subsequently, the problem causes of all problems are clustered and solution approaches to overcome the (clustered) problem causes are derived. Finally, the solution approaches are evaluated based on expert interviews. The results are presented in the following.

Using Ishikawa diagrams, the analysis of problems related to model conception, development and usage identified different problem causes. Some root causes deal with the aspects of communication, cooperation, professional knowledge, objectives and project management mentioned above.

Yet, there are also some technical challenges when using simulations. Being based on data, it becomes clear that most of the problems are caused by data availability, consistency and exchange. Problems related to data availability can occur when there is simply no data available to be used for simulation. If there are data, a possible problem is data access. Data can be classified as internal, which is why it is not shared or the access can be complicated due to the lack of a central platform or server. This results in diverse points of access or isolated systems. If data are available and accessible, it still can be inconsistent, redundant or in a disadvantageous structure. Further, data exchange is a main problem caused by inconsistent or missing interfaces, communication protocols or simply the necessity of a manual import or export of data. Besides, usability problems are occurring due to difficulties with or within the simulation tool. Possible reasons might be that the wrong tool was chosen or that a complex user interface makes the usage more difficult. As a result, the model is not subdivided into subsystems, complicating the process of different people working on the same model at the same time.

In order to overcome the problems related to data availability, consistency, exchange and usability, six solution approaches have been identified.

The first one is implementing a standardized approach for a common understanding of benefits, procedures and fields of application for different simulation techniques. Knowledge in these areas is helping to involve planning participants and generate helpful simulation use cases. Standardization also helps during the development of connectable models (inter-time and inter-content). Besides standardizing the simulation approach itself, it is helpful to standardize its integration in project management, e.g. during the definition of planning objectives and the introduction of standardized data management.

The precondition for standardization in data management is a central database, which consequently is the second solution approach. Implementing a central database with a data management system allows access to every simulation practitioner and modeling software. This in turn helps to unify and use existing data properly, always representing the current status of the project. As a result, outdated or inconsistent data can be avoided and data availability can be improved.

The project work can be facilitated even more by implementing automated connections to widely used programs. These automated connections require defined interfaces, which secure connectivity and enable data exchange. Standardized interfaces between different software can be implemented by using, for example, the IFC-standard. Establishing each data source and sink with an appropriable interface facilitates data exchange, and import and export can be automated.

Automated connections further require data consistency between the connected systems, for example, different software or physical systems. Data consistency can be reached by using standardized communication protocols, e.g. OPC-UA. With defined interfaces and a consistent data transfer to other models or real equipment, additional data can be gathered.

Having implemented a standardized approach together with a central database and interfaces, the access to a superordinate application like a BIM system can be beneficial. If introduced as a central source of information, BIM can improve the exchange of consistent data across all stages of a factory planning project. One example is the exchange of 3D-models, which can be exchanged with all participants working on the project as well as across different planning stages.

The last solution approach is enabling simulation technologies to be cloud ready. In today's practice, engineers access modeling and simulation software from stand-alone workstations and save their projects data on a file server. However, in many cases, it is not possible to work together at the same time on the same model. Implementing a cloud-ready structure is a future requirement to further improve the cooperation between modeling practitioners and to speed up the planning process.

In order to validate the stated solution approaches, six expert interviews have been carried out with professionals dealing with different kinds of simulation. The results are shown in Fig. 4(a), where usage frequency (horizontal axis),

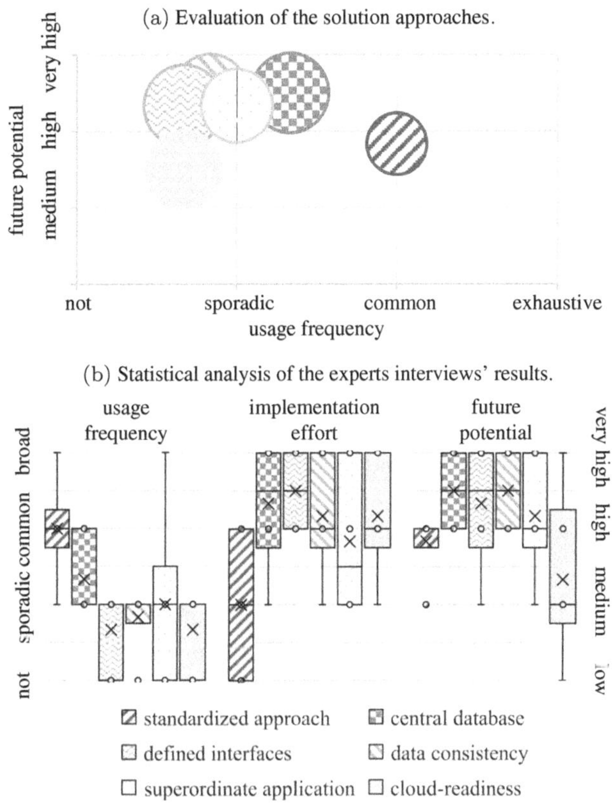

Fig. 4. (a) Evaluation of the solution approaches based on experts' interviews; (b) Statistical analysis of the experts' interview results.

implementation effort (bubble size) and future potential (vertical axis) are evaluated. Usage frequency describes the experts' evaluation of the current market penetration. The implementation effort shows the subjective assessment of the required efforts to establish a solution approach. Future potential characterizes the beneficial potential that a solution approach can have in the future after being implemented. A statistical analysis of the results is shown in Fig. 4(b).

Experts responded that a standardized approach is implemented already quite frequently, but still has a high potential with a moderate implementation effort. That is why the objective should be the comprehensive usage and exploitation of existing standardized approaches. Considering the central database and the defined interfaces, it can be seen that they are rarely implemented. Given the very high future potential, these solution approaches should be considered when carrying out a simulation. Yet, because of the high implementation effort, a cost–benefit analysis should be performed to identify individual advantages. Cloud-ready simulation and the access to superordinate applications tend to have a higher standard deviation combined with widely differing minimum-maximum values. This indicates that these solution approaches do not have a common definition yet. Research has to standardize and evaluate those approaches. Because of this, no recommendation is given for dealing with these two solution approaches.

Given the general character, the authors believe that the described problems of simulations can also be generalized for many other digital tools. Thus, the solution approaches should be helpful and give hints for improvements when using any digital tool.

Digital technologies have the potential to overcome challenges related to the exchange of information, the coordination and cooperation of involved parties and the monitoring of planning projects. In order to overcome challenges when using those technologies, it is recommended to implement a standardized approach with a digital backbone, which offers available, consistent and exchangeable data. Not only should each project member be qualified in order to be able to support this backbone, but there should also be a clear responsibility in the planning process, which keeps the digital backbone usable and promotes its features. In addition to the potential within the planning process, this concept of a digital backbone can help to overcome the information gaps between the design and the operation stage by interlinking respective sets of data. Thus, information is made connectable and usable.

Regarding the different scales of a greenfield factory planning, including the process chain, infrastructure and building, a holistic approach with consistent fundamentals should be implemented. One approach extending simulation to multiple scales is multi-scale simulation, which will be described in the following chapter. Implementing a solid multi-scale simulation approach with a digital backbone is assumed to be beneficial in terms of planning accuracy and results.

References

1. Schenk M, Wirth S, Müller E. Fabrikplanung und Fabrikbetrieb. Berlin, Heidelberg: Springer Berlin Heidelberg; 2014.
2. Bednar T, Bleicher F, Dür F, Eder K, Emrich S, Flatz T, *et al.* Roadmap für Energieeffiziente Fertigungsbetriebe. Wien; Konsortium Projekt INFO; 2013.
3. Verein deutscher Ingenieure. Fabrikplanung: Planungsvorgehen. Berlin: Beuth Verlag; 2011.
4. Heidemann A, Kistemann T, Stolbrink M, Kasperkowiak F, Heikrodt K. Integrale Planung der Gebäudetechnik: Erhalt der Trinkwassergüte — Vorbeugender Brandschutz — Energieeffizienz. Berlin: Springer Vieweg; 2014.
5. Pawellek G. Ganzheitliche Fabrikplanung: Grundlagen, Vorgehensweise, EDV-Unterstützung. 2. Aufl. Berlin: Springer Vieweg; 2014.

6. Grundig C-G. Fabrikplanung: Planungssystematik — Methoden — Anwendungen. 6., neu bearbeitete Auflage. München: Hanser; 2018.
7. Wiendahl HP, Reichhardt J, Hernandet R. Kooperative Fabrikplanung: Wandlungsfähigkeit durch zielorientierte Integration von Prozess- und Bauplanung. wt Werkstattstechnik Online. 2001;91(4):186–91.
8. Wiendahl H-P, Reichardt J, Nyhuis P. Handbuch Fabrikplanung: Konzept, Gestaltung und Umsetzung wandlungsfähiger Produktionsstätten. München: Hanser; 2014.
9. Kampker A, Osebold R, Trautz M, Burggräf P, Krunke M, Meckelnborg A, *et al.* Innovative Fabriken interdisziplinär planen. wt Werkstattstechnik Online. 2012;102(4):186–92.
10. Bougain A. Definition of the needs for enhanced building operation using Building Information Modeling (BIM). In: Achammer CM, Kovacic I, editors. Integrale Planung für Industrie-Bau 4.0: Praxisreport 2015 = Integrated planning for industrial building 4.0. Wien: Klein Publishing GmbH; 2015. pp. 36–51.
11. Kampker A, Meckelnborg A, Burggräf P, Netz T. Factory planning scrum: integrative factory planning with agile project management. Stellenbosch: Stellenbosch Univ.; 2013. Proceedings International Conference on Competitive Manufacturing: COMA '13.
12. Uhlenbruch W. Smart data — Erstellung eines Daten-Tools zur Analyse des Informationsmanagements in Planung, Bau und Bewirtschaftung von Gebäuden. In: Achammer CM, Kovacic I, editors. Integrale Planung für Industrie-Bau 4.0: Praxisreport 2015 = Integrated planning for industrial building 4.0. Wien: Klein Publishing GmbH; 2015. pp. 106–13.
13. Trogisch A. Planerische Grundlagen. In: Albers K-J, editor. Taschenbuch für Heizung und Klimatechnik: Einschließlich Trinkwasser-und Kältetechnik sowie Energiekonzepte. 79. Auflage, 2019/2020. Augsburg: ITM InnoTech Medien GmbH; 2018. pp. 553–72 (Recknagel Edition).
14. Pučko Z. Zeit- und Kostenermittlung bei Bauprojekten mit Building Information Modeling (BIM). In: Achammer CM, Kovacic I, editors. Integrale Planung für Industrie-Bau 4.0: Praxisreport 2015 = Integrated planning for industrial building 4.0. Wien: Klein Publishing GmbH; 2015. pp. 80–95.
15. Loos MN. Daten- und termingesteuerte Entscheidungsmethodik der Fabrikplanung unter Berücksichtigung der Produktentstehung [Zugl.: Karlsruhe, KIT, Diss., 2013]. Hannover, Karlsruhe: Technische Informationsbibliothek u. Universitätsbibliothek; KIT Scientific Publishing; 2013.
16. Vajna S, Weber C, Zeman K, Hehenberger P, Gerhard D, Wartzack S. CAx für Ingenieure: Eine praxisbezogene Einführung. 3., vollständig neu bearbeitete Auflage. Berlin: Springer Vieweg; 2018.
17. DIN Deutsches Institut für Normung. Organisation und Digitalisierung von Informationen zu Bauwerken und Ingenieurleistungen, einschließlich Bauwerksinformationsmodellierung (BIM)—Informationsmanagement mit BIM: Teil 1: Begriffe und Grundsätze. Berlin: Beuth Verlag; 2019. (35.240.67; 91.010.01).
18. Landherr M, Neumann M, Volkmann J, Constantinescu C. Digitale Fabrik. In: Westkämper E, Spath D, Constantinescu C, Lentes J, editors. Digitale Produktion. Berlin, Heidelberg: Springer; 2013. pp. 107–32.
19. Verein deutscher Ingenieure. Simulation von Logistik-, Materialfluss- und Produktionssystemen: Grundlagen. Berlin: Beuth Verlag; 2014. (vol 03.100.10) 2014.
20. Menck N, Yang X, Weidig C, Winkes P, Lauer C, Hagen H, *et al.* Collaborative factory planning in virtual reality. Procedia CIRP. 2012;3:317–22.
21. Shan W, Jian-feng L, Hao Z. The application of augmented reality technologies for factory layout. International Conference on Audio, Language and Image Processing (ICALP). 2010;873–876.

22. Singh S, Weeber M, Birke KP. Advancing digital twin implementation: a toolbox for modelling and simulation. Procedia CIRP. 2021;99:567–572.
23. Uhlemann TH-J, Lehmann C, Steinhilper R. The digital twin: realizing the cyber-physical production system for Industry 4.0. Procedia CIRP. 2017;61:335–40.
24. Rosen R, Wichert G von, Lo G, Bettenhausen KD. About the importance of autonomy and digital twins for the future of manufacturing. IFAC-PapersOnLine. 2015;48(3): 567–72.
25. Kovacic I. BIM for BaMa: Energie- und Ressourceneffiziente Smart Production. In: Achammer CM, Kovacic I, editors. Integrale Planung für Industrie-Bau 4.0: Praxisreport 2015: Integrated planning for industrial building 4.0. Wien: Klein Publishing GmbH; 2015. pp. 126–35.
26. Ungerer M. Industrie 4.0 bei Planung und Errichtung des Rautendachs für den neuen Hauptbahnhof Wien. In: Achammer CM, Kovacic I, editors. Integrale Planung für Industrie-Bau 4.0: Praxisreport 2015 = Integrated planning for industrial building 4.0. Wien: Klein Publishing GmbH; 2015. pp. 194–203.

https://doi.org/10.1142/9789811245626_0006

Chapter 5

Modeling and Simulation in Battery Cell Manufacturing — Concepts and Applications

Soumya Singh[*], Johannes Wanner[†] and Max Weeber[‡]
Fraunhofer Institute for Manufacturing Engineering and Automation IPA
Nobelstraße 12, 70569 Stuttgart, Germany
[*]*soumya.singh@ipa.fraunhofer.de*
[†]*johannes.wanner@ipa.fraunhofer.de*
[‡]*max.weeber@ipa.fraunhofer.de*

Digitalization of battery cell manufacturing requires data processing in modeling and simulation tools in order to obtain a better understanding of the material product, process, process chain and building characteristics. This chapter introduces different modeling and simulation concepts. Also, the coupling of individual models at different scales is explained. In addition, some practical examples from the literature are presented. The advantages and difficulties related to the different modeling concepts, as well as their application in the context of battery cell manufacturing, are shown. It can be recognized that modeling and simulation are widely used for spatial or temporal scales of individual systems or processes, and have different degrees of progress in various domains (such as, material science, biology, and industrial engineering). However, comprehensive model coupling and especially multi-scale simulation are only used in few applications. This is in particular the case for interdisciplinary collaboration due to increased complexity.

1. Multi-Scale Production Simulation

Modeling is the digital representation of a product or system under study, while simulation is the process of executing that model over time. Multi-scale modeling represents a framework to construct mathematical and computational models of a phenomenon by examining connections between models at different scales.[1] The idea behind multi-scale modeling is to describe the behavior of one or multiple systems by considering different scales (e.g. spatial, temporal, and material) using the best-suited methods and tools.[2] The concept of multi-scale modeling and simulation is well established in some fields (material science, applied mathematics), while it is relatively new in others (biomedical, manufacturing engineering).[3]

Analogous to physical or chemical systems, production systems can be considered in different scales from single processes to process chains and technical building services (TBS) up to buildings. Multi-scale modeling makes it possible to significantly reduce empirical assumptions by explicitly considering mechanisms at scales that are neglected in simpler models. However, this increases the complexity of the overall model and increases the modeling effort and the error rate, and risks a lower acceptance of the results. Thus, a balance between model complexity and result accuracy in modeling must be strived for. To this end, it is necessary to consider which models are needed for the defined objectives and their required resolution.[2,3]

A model verification of a multi-scale model deals with the verification of the internal consistency of a model. Validation examines the correspondence between the model and reality. To quote Arsham,[4] validation deals with the question "Are we building the right system" and in verification, "Are we building the system right". Subsequent calibration verifies that the data generated by the simulation match the real data. Due to the complexity of multi-scale modeling, a calculus is necessary for validation, error propagation and model consistency. Formulating mathematical theories of multi-scale modeling can provide formal and sound approaches to error analysis of multi-scale models. A related possibility is "coupled numerical analysis". This idea uses numerical analysis of single models and extracts them in the scale domain to analyze coupled models. Validation, uncertainty quantification and sensitivity analysis of multi-scale models also need to be better understood and put into practice. Benchmark results obtained from fully resolved models are needed to be used to "calibrate" parameters of multi-scale models. Subsequently, uncertainty quantification and sensitivity analyses of the resolved models can estimate the quality of predictions made with non-fully resolved multi-scale models.[3]

This complex modeling in the field of battery cell manufacturing with the accompanying problems has not yet been modeled in a fully resolved model. Up to now, individual and coupled problems have been worked on and answered in interdisciplinary work programs. This approach allows an interdisciplinary working method and makes it possible to approach the most important questions.[3] In the following, the modeling methods and the associated literature from the field of modeling and simulation in battery cell manufacturing and their model coupling are highlighted.

1.1. *Material-level simulation*

The materials used in a battery cell have a major influence on the resulting product properties and are therefore defined as part of the design of a battery cell. Figure 1 describes the different materials of a battery cell.[5]

However, within tolerance limits, these materials have deviations in their physical and chemical properties.[6] The deviations of the material properties result from minimal process inaccuracies in the material processing steps, which in consequence affect the material characteristics of products. In order to quantify these and draw conclusions about the resulting product quality, various simulations exist that deal with the materials at the micro- and mesoscale level of a battery cell (Fig. 2).[7,8] On the contrary, the electrode materials can be examined for their properties, such as pore size, when they are delivered. Due to various mixing and forming processes during the manufacturing process, it is reasonable to characterize the

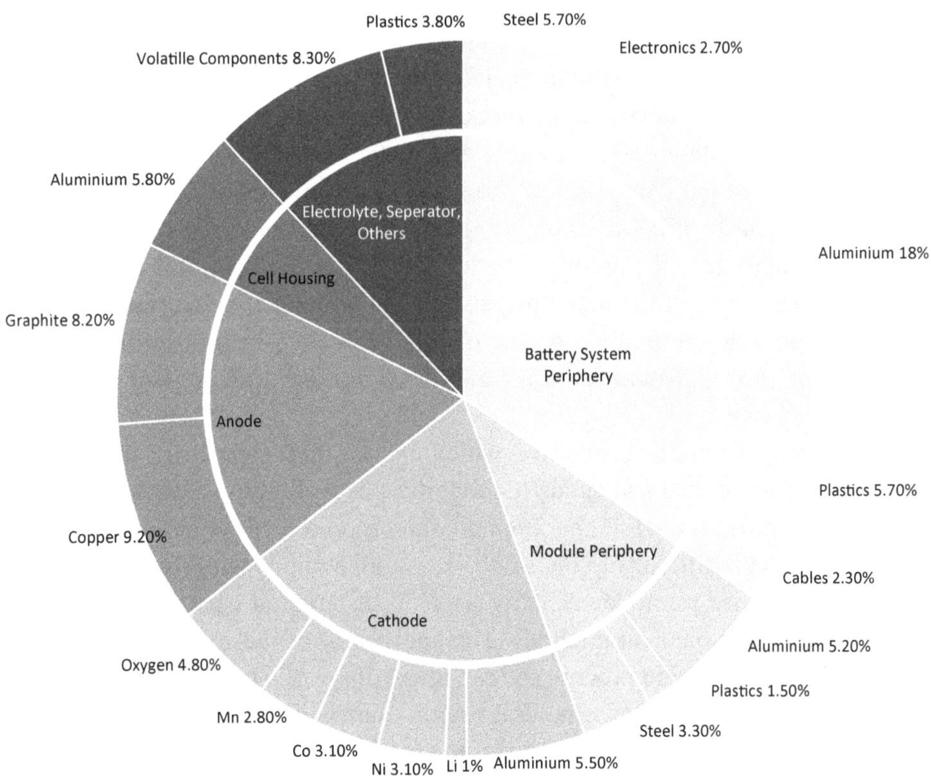

Fig. 1. Materials used in a battery system and their distribution among the individual components. In the style of Diekmann *et al.*[5]

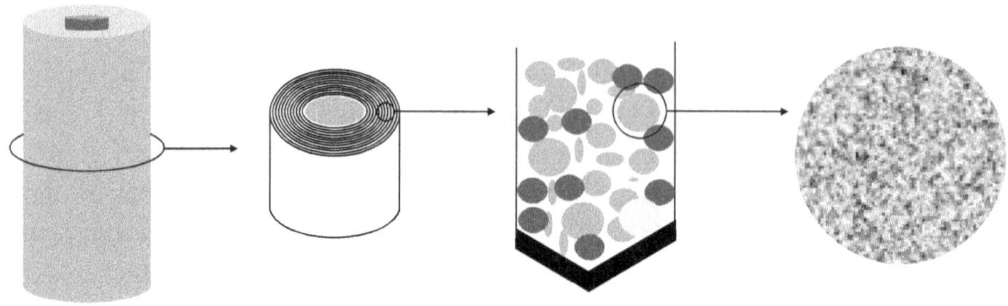

Fig. 2. Length scales of multi-length scale microstructural characterization: full cell, electrode assembly, electrode coating and particles. In the style of Heenan *et al.*[10]

product materials during the manufacturing process.[9] A distinction must be made between in-line and sample-based testing. For this purpose, there are different characterization methods, mostly optical or electrochemical methods, which can show the electrode structure, for example. The relevant information, such as particle sizes, must be extracted from the images.

The recorded values and their standard deviation from the norm/target value can be stored in the database of the production process and serve as input for process or product modeling. Besides, this systematic recording offers the possibility to create an extended data sheet of the battery cell in which the parameters are included. This also offers the opportunity to add additional information on the purchased materials to the produced battery cell.[10] For example, this can be information on the origin of the materials, such as the mining location and time. But, it can also provide information on the material sourcing, its processing methods and whether the used material originates from primary (mining) or secondary (recycling) sources.

The physical material properties, which include thermodynamic, mechanical, electrodynamic, optical and acoustic properties, as well as the chemical material properties, can be recorded experimentally and represented within mathematical models. The main difficulty is the small size of all components.[8] The coupling of different properties in models usually leads to differential equations which can be solved by finite element methods. The knowledge of these properties has relevance for the production process and the usage phase of a battery cell. To predict the microscopic behavior of structural battery composites, the thermal, mechanical and electrical analyses are coupled into a multi-physics finite element simulation. Carlstedt *et al.* incorporate local variation of mass and charge concentrations into refined models with appropriate boundary conditions to capture the microscopic conditions inside the material.[11] Zhu *et al.* review the existing mechanical modeling, for example, the plastic anisotropy and ductile fracture of current collectors.[8] The deformation mechanisms of coated electrode strips at micro- and mesoscale have not been investigated yet.[8] The nanostructure of the electrodes captured by focused ion beam (FIB) SEM images can be modeled analogously to the structure of granular materials such as sand and concrete. The internal structure of the separator and coated materials increases the complexity of the modeling and requires the use of experimental nanotechnology techniques. Based on the behavior of a single particle, Zhao and coworkers have conducted a series of studies on the elasticity, plasticity, fracture and debonding of electrodes and proposed several models to describe these processes.[8] The separator as an important inactive component has already been investigated in numerous studies. Zhang *et al.* investigate and model the average pore size, ultimate tensile stress and tensile elongation of various separators. Further, a finite element model of the separator was developed to predict the load displacement behavior in the test scenarios. The model succeeded in simulating anisotropic deformation during the punch test and shear deformations in the diagonal direction tensile realistically.[12] In addition, the orthotropy of the separator and elasto-viscoplasticity and temperature dependence are investigated.[8] An interaction of adhesion strength between the coating and current collector has also been studied in various publications.[8,13]

1.2. *Cell-level simulation*

A main challenge in the development of Li-ion batteries is the prediction of their behavior in different operating states. Generally, Li-ion battery models can be divided into empirical models and electrochemical models (Fig. 3).[14,15] These usually aim to estimate the state of charge (SOC) and state of health (SOH) of the battery.[16] In addition, some other battery performance criteria can be studied such as the diffusion rates of the cell current. Here, cell-level modeling and simulation are closely related to the material simulations.

Equivalent circuit or neural network based models are commonly used as empirical battery models in the battery management systems (BMSs) of the electronics and automotive industry.[17] These models are based on historical experimental data of the battery, such as the cell potential and current. They are used to estimate their future states. Simplicity and computationally fast execution are the main advantages of the empirical models.[16] The lack of integration of physical parameters in the model is a major disadvantage. Therefore, the battery characteristics are never updated during the aging of the battery. In addition, the empirical model of a battery cannot be applied to other kinds of batteries.[15,18]

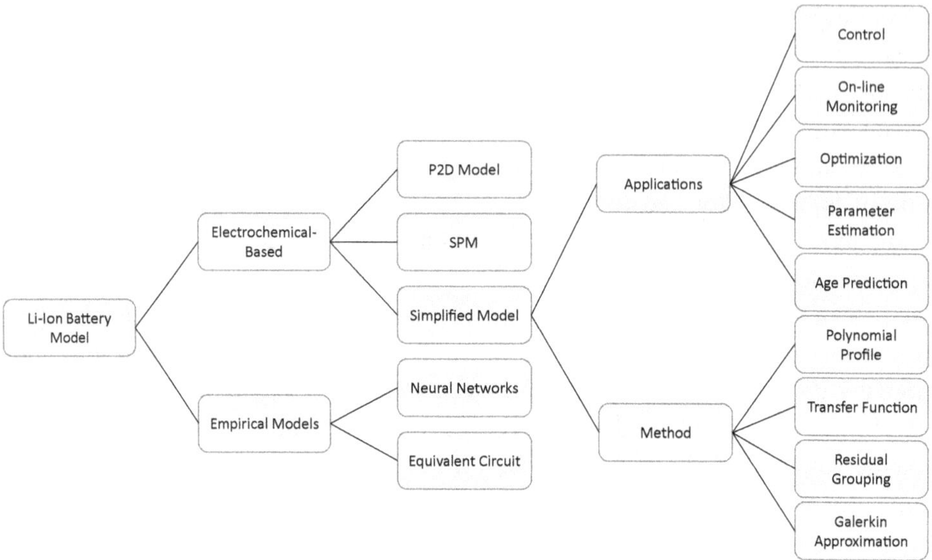

Fig. 3. Classification of Li-ion battery models. In the style of Jokar *et al.*[16]

1.2.1. *Empirical models*

It is also possible to describe the battery states purely electrically by using an electrical equivalent circuit diagram (ECM). The functionality is shown in Fig. 4. Here, empirical measurement results of impedances at different measurement frequencies can be used to parameterize a battery model.[19] It is possible to extend the ECM by several RC elements to increase the accuracy of the modeling. However, this also increases the computational effort when using the EC models. As shown in Fig. 4, the individual RC links describe the electrical properties of the individual components of a battery cell. The depicted resistance R describes the electrolyte and the current collectors, the inductance L the porous structure of the electrode, the RC element of the anode and cathode the double layer effect of each electrode. The Washburn impedance models the diffusion process at low frequencies of the impedance spectroscopy. The amounts of the RC elements increase over the aging process of the cell. However, a heterogeneous development of the RC parameters, analogous to the individual components of the battery cell, can be observed here as a function of the cell loads.[20]

Data-driven methods can be used especially for the determination of SOC and SOH. The input values must be retested for the desired output values for each application. There is usually no standardized procedure for this. Hence, these models require costly development efforts, but the continuous adjustments can

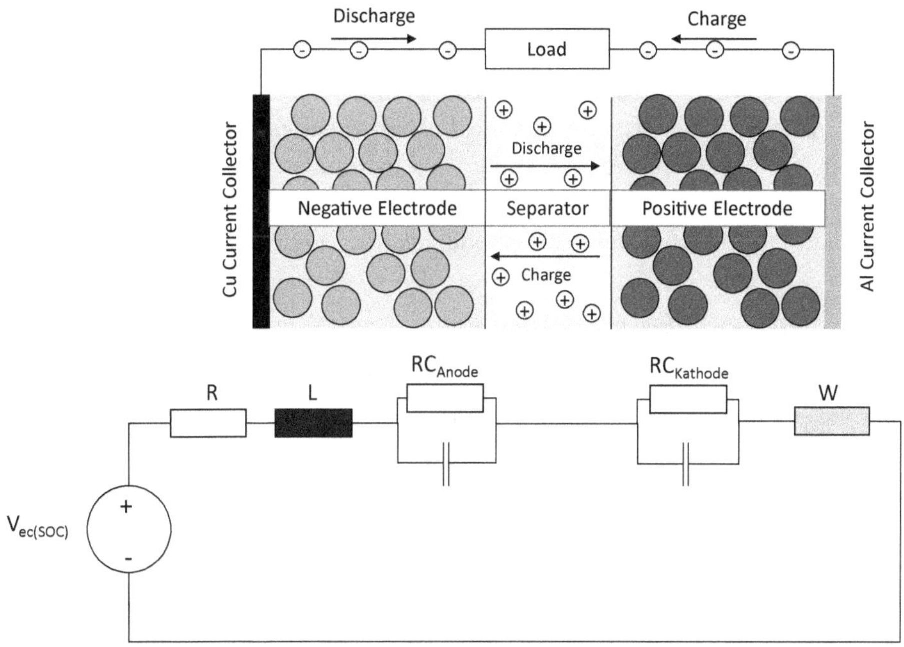

Fig. 4. Structure of an Equivalent circuit battery model (ECM). In the style of Xia *et al.*[20]

also improve the models' quality during operation.[21] For example, Chen *et al.* use a neural network paired with a Kalman filter. This allows the classical ampere-hour counting method to determine the state of charge. The measurement errors due to the different temperatures are then corrected with an artificial neural network. The feed-forward network is given the current battery voltage and the battery temperature as input data. As output data, the network determines an SOC error, which is offset against the measured SOC if necessary. This procedure makes it possible to determine the SOC down to a small error of 2%, especially at low temperatures and small SOCs.[22]

In another publication by Lin *et al.*, artificial neural networks are used to determine the state of health during battery cell aging. The input data of the network are dynamic voltage characteristics during load steps. First, during the constant current–constant voltage (CCCV) full charge, the duration of the constant current phase is measured. In addition, at the start of the upcoming discharge phase, the voltage drop is measured. During the subsequent load change to the charge phase, the positive voltage jump is again measured. These three values change in the course of battery aging. Thus, the charge duration shortens and the voltage jumps increase in magnitude as the battery ages. These data are processed in a probabilistic neural network for each load cycle. This makes it possible to estimate the aging process of a battery cell with an uncertainty of about 1%.[23]

The design of electrochemical models is more sophisticated than data-driven models. In these models, the chemical/electrochemical kinetics and transport equations are considered. They can be used to simulate the properties and reactions of the Li-ion batteries.[14] The most popular electrochemical-based models are the pseudo-two-dimensional (P2D) model and the single particle model (SPM).[16]

1.2.2. *Pseudo-two-dimensional model*

In 1993, Doyle *et al.* introduced the P2D model for Li-ion batteries, which combines the porous electrode theory, the concentrated solution theory and the equations of kinetics. Through numerous tests and validations, this model is the most popular model for Li-ion batteries to date.[16] The P2D model, shown in Fig. 5, predictions are relatively accurate and generally have a good agreement with experimental data.[24]

The behavior of the two electrodes is modeled with spherical particles as a porous matrix surrounded by the electrolyte. The intercalation and de-intercalation processes of the Li ions are carried out over the surface of the particles. Moreover, the predominantly unidirectional transfer processes result in a 1D mathematical model along the x-axis.[24]

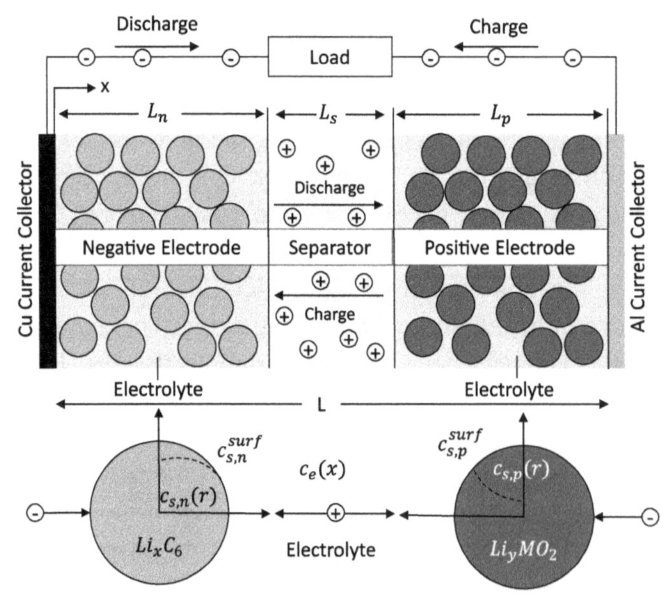

Fig. 5. Functionality of the Li-ion battery P2D model. In the style of Jokar *et al.*[16]

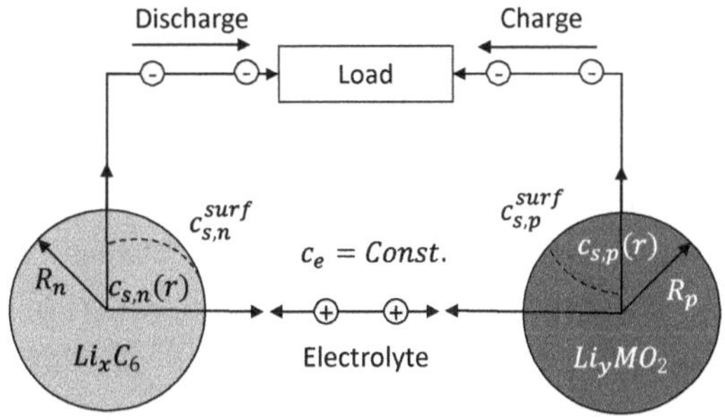

Fig. 6. Functionality of the single particle model (SPM). In the style of Jokar *et al.*[16]

The 1D mathematical models of the P2D model are the Fick's diffusion law for spherical particles to describe the solid-state Li-ion concentration in the electrodes and the Ohm's law to describe the liquid-phase Li-ion concentration in the electrolyte and in the separator. In addition, the liquid-phase potential of the electrolyte and the separator is described by Kirchhoff's and Ohm's laws. The Butler–Volmer kinetics equation describes the pore-wall flux of Li ions in the electrodes.[24] Complete analytical solutions of the governing P2D model equations are not available.[25] With various numerical methods such as the finite difference method (FDM), finite element method (FEM) and finite volume method (FVM), the model parameters are able to be estimated.[16]

1.2.3. *Single Particle Model (SPM)*

In 2000, Zhang *et al.* proposed the Single Particle Model (SPM) to simplify the P2D model. This SPM, shown in Fig. 6, models each electrode as one spherical particle in which intercalation and de-intercalation phenomena arise. In addition, the variations in electrolyte concentration and potential are neglected.[26]

The solid concentration equations and the Butler-Volmer kinetics equations at both negative and positive electrodes are the governing equations of the SPM. Due to the simplicity of the SPM and its minimal computational cost, the model can be used for many applications such as online estimation and lifetime modeling of Li-ion batteries.[16] Its main drawback is that the electrolyte properties must be adjusted for high discharge rates and for thick electrodes.[14] However, improved versions of the SPM that mitigate these problems are available.[27] Multiple particle (MP) models have also been proposed to deal with varying particles with different radii, properties and a range of contact resistances for Li-ion batteries with, for example, LiFePO4 electrodes.[28] This simplified P2D model is mainly developed for applications to control, monitor and optimize the battery parameter estimation and aging prediction.[16]

1.2.4. *Thermal models*

Since the operating temperature plays a key role in the aging of batteries, Bernardi *et al.* developed an energy balance model for battery systems in 1985.[29] To calculate the heat evolution rate of a single battery cell, Pals and Newman used a P2D model coupled with the energy balance model. In order to explain Li-ion batteries' behavior under operating conditions, 2D and 3D models with the relevant thermal and electrochemical equations have been developed.[30] However, these models are computationally intensive and require additional input data such as specific heat capacity, thermal diffusivity and thermal conductivity of the materials. To simplify the complex thermal models, Guo *et al.* coupled the SPM with the heat equations. Empirically based simulations are also able to quantify the heat generation of the Li-ion battery[31] and contribute to updating the battery parameters.[16]

1.2.5. *Product modeling in manufacturing*

In addition, there are product modelings that are not designed for battery cell use but for modeling in a production environment. These models are used, for example, to represent production variants and to evaluate the quality of the product. Reinhardt *et al.* present an object oriented approach in which the cell consists of individual components. These, in turn, have different physical properties and functions which influence the product properties, such as safety or cycle stability from the main battery class. Kornas *et al.* take up the idea and extend the battery model to a multivariate KPI-based battery model.[32,33]

The hierarchical structure shown in Fig. 7 describes the multi-scale model between material, product and process. The lowest level represents the physically measured input characteristics (Cpk). These can represent material properties, process parameters and target values or tolerance values (Cpmk). The second level describes the intermediate product characteristics (MCpk) such as the thickness of the cell body, which is calculated from the characteristics of the first level, such as the length of the electrode strip and properties of the winding process. Here, a loss of quality can be detected with increasing deviation to the allowed tolerance values. The third level represents the final product properties of the cell, such as the Coulomb efficiency or the capacity. The top level summarizes the properties in a single KPI value. Due to the hierarchical structure, all input characteristics can be correlated against each other and related to one KPI value.[33]

1.3. *System-level simulation*

The next scale of modeling is the module or system level. Here, the individual cell models are connected in parallel or series according to the specified arrangement

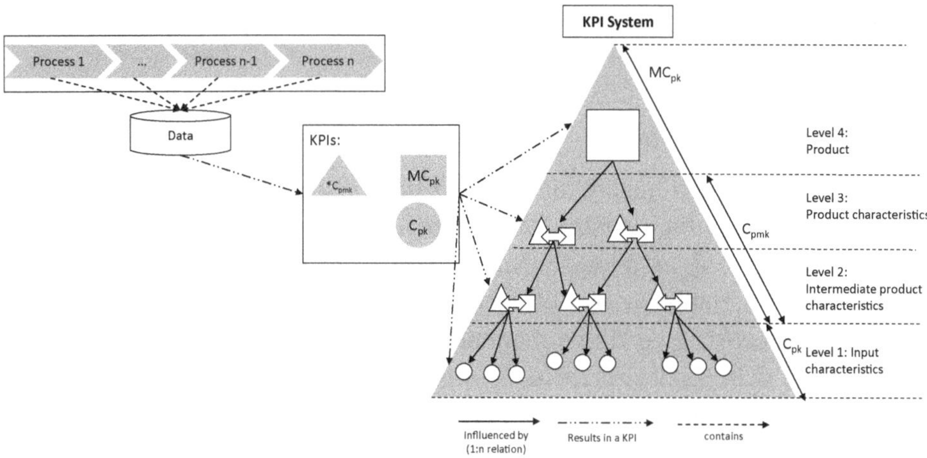

Fig. 7. Structure of a multivariate KPI-based battery model and its use in a manufacturing environment. In the style of Kornas *et al.*[33]

in the module and to represent its electrochemical properties. On the contrary, system-level simulations offer the possibility to reproduce the application conditions better. Thus, cell connectors, the module housing and the battery management system can be added to the simulation. This offers the possibility to perform a variety of simulations that go beyond the classical electrochemical issues. In many cases, thermal simulations are performed.[34,35] Here, questions regarding the optimal module designs, cooling and heating strategy of the thermal battery management system can be answered. In addition, the electrochemical effects of extreme temperatures on the cells can be estimated, and permissible temperature variations among the cells can be simulated. Thermal management strategies such as an SoH-balanced management method are developed to affect the airflow direction. In this strategy, the various SoH inconsistencies of the cells also matter to extend the lifetime of the Li-ion battery system.[35] From this, the thermal management system can provide the energy to control the temperature and maintain the system in optimal operating conditions as well as calculate the thermal runaway risk score (TRR).[34] Mechanical simulations can also be performed on this basis. On the one hand, volume thrusts due to the cells' charge transfer and the resulting forces between the cells in the module can be simulated. On the other hand, the effects of physical forces from outside, such as a crash, can be simulated.[8]

One aspect of system modeling represents the assembly and selection of cells, also considering cell quality variations originating in manufacturing. It can be shown that the capacity of the module is limited by the weakest cell within a module. Thus, the resulting module properties such as SoC distribution and aging appearance both at the individual cell and the overall system level can be concluded based on the variability of cell quality used in the assembly of the respective modules.[6]

1.4. *Process-level simulation*

High demands on cell quality, durability, safety and environmental friendliness require precise and high-quality cell manufacturing. Improved use of process-level simulation tools in the manufacturing environment is one way to improve quality, quality variation, energy efficiency and material usage.[36,37] The simulation tools require real-time condition monitoring, in-line process adjustments and go hand in hand with lifetime predictions of precise cell models considering manufacturing data in addition to data obtained during cell operation. There are many quality-relevant process parameters in a complex production process such as that of a lithium-ion battery. Due to the diversity of the processes, consistent quality management is difficult to implement. Therefore, it makes sense to consider the individual processes steps separately. The chosen simulation method varies depending on the process and the target criterion. The upcoming literature references demonstrate the importance to align the data requirements of a particular cell model with the data provision of the process model.

Kenney *et al.*[6] extended the single particle battery model to investigate the effects of tolerances in the electrode fabrication process on a battery module's performance. The three main electrode fabrication parameters studied were electrode thickness, electrode density, which is associated with the electrode's porosity, and the weight fraction of the active material, which depends on the quality of mixing of the active material. The simulations show that cell quality varies with the extent of manufacturing variations. The simulations also indicate that two manufacturing parameters, electrode thickness and electrode porosity, are of great relevance to maximizing the battery module's useable capacity.[6]

Gimenez *et al.*[38] present a Discrete Element Method (DEM) approach to describe lithium-ion electrodes' behavior during the calendaring production step from micro to macro scale. The model is suitable to describe the macroscopic behavior of the electrode along the calendaring step. In particular, the elastic properties and the elastic recovery after the electrode's calendaring can be described. This effect cannot be determined experimentally and is relevant for better control of the calendaring process. Moreover, a numerical overview of the interactions between the electrode structure and its properties during the calendaring process is given to predict important cell properties and design improved materials.[38,39]

Schreiner *et al.*[40] present a qualitative machine/material process structure model to illustrate the interactions during the calendaring process. Here, the empirically recorded process parameters such as roll temperature and speed, structural parameters of the electrodes (e.g. layer thickness, adhesion strength) and machine parameters of the machine behavior (displacement and bending line) are investigated for their influence on the achieved electrode porosities.[40]

Schmitt and Raatz[41] show an extended FMEA method to optimize the electrode stacking process. The goal here is to rank possible positioning errors and multilayer handling to optimize the manufacturing step. The authors present an algorithm to simulate and predict the positioning error using fuzzy logic for cells with different dimensions and tolerances.[41]

Jeon[42] uses a lattice Boltzmann simulation to show the electrolyte distribution during the electrolyte filling process and understand the cathode and anode's wetting properties. A porous electrode model is implied to identify the influence of wettability on cell performance. Thus, in combination with a cell simulation, it can be shown that wettability is closely related to air entrapment in the electrode and lithium plating occurrence.[42]

1.5. *Production chain simulation*

Achieving high product quality while reducing production costs is the main challenge in battery cell manufacturing.[43] The existing methods of quality assurance in manufacturing have been discussed in publications for years. However, an adaptation of these methods to the context of battery cell manufacturing is not easy due to the complexity inherent to the individual process steps used to manufacture lithium-ion batteries.[44] The respective process chain consists of a combination of batch, continuous and discontinuous processes. Some of these processes also require special ambient and boundary conditions.[45] Composing these single process models to an overall process chain model is very challenging due to missing models for individual process steps. In order to deal with the complexity of the entire process chain, several comprehensive quality management concepts have been introduced in recent years. These quality management concepts differ in their methodology as well as in the identification of the quality relevant parameters.

1.5.1. *Failure Mode and Effects Analysis*

Failure Mode and Effects Analysis (FMEA) has a long tradition in engineering as a bottom-up approach to identify and evaluate potential failures of components, products and processes. At the beginning of an FMEA, a structural analysis is performed in which the (sub)processes are defined, and the interrelationships of the process elements are presented. A subsequent functional analysis identifies the parameters that are involved in the process and potentially influence it. Subsequently, failure modes, effects and causes are identified and defined.[37,41]

1.5.2. *Cross-Industry Standard Process for Data Mining*

The Cross-Industry Standard Process for Data Mining (CRISP-DM) method is used as an analytical method for data mining. It is divided into the phases Business Understanding, Data Understanding, Data Preparation, Modeling, Evaluation and Deployment. The phases are variable and have to be run through loops. As a result, new, usually more targeted questions can emerge through the process. The subsequent data mining processes benefit from the results of the previous runs.[46,47]

Westermeier's publication "Method for Quality Parameter Identification and Classification in Battery Cell Production" from the year 2013 is the first publication that takes a holistic approach for the investigation of quality parameters in battery cell manufacturing. With the help of a modified FMEA analysis, the main influencing factors in battery cell manufacturing are recorded and error chains are identified.[37] Based on this, a Multiple Domain Matrix (MDM) can be created over the entire process chain, which provides parameters with a direct or indirect influence on the quality parameters of the end product. Thus, in further steps, the identified influences can be quantified and target values for the individual processes can be derived.[48] Based on this, a comprehensive quality management concept was proposed by Schnell and Reinhardt. This concept detects process deviations and can minimize the effort for quality control by defining internal decision points (quality gates) and additionally enables the implementation of process control and feedback.[49]

Schnell *et al.*[46] apply the CRISP-DM methodology to process the data captured along the manufacturing process chain. The captured cell parameters are assigned to a cell and correlated with the resulting cell capacity to identify process dependencies. To enable prediction of product quality, various data mining methods are used and evaluated. For example, Generalized Linear Model (GLM), Random Forest (RF) and Gradient Boosted Trees (GBT) methods are better at correlating quality parameters to the resulting cell capacity before forming than Artificial Neural Networks (ANN), Support Vector Regression (SVR) and Decision Trees (DT) methods. Thus, cells with a likely low cell capacity can be ejected from the process chain before the costly and energy-intensive formation process. Thus, electrode fabrication and electrolyte filling can be identified as key quality drivers.[33,46] Turetskyy (2019) extends the data pool by the information of the technical building systems (TBS). The focus here is on the various data resources. Here, automatically and manually acquired data sources are coupled. Analogous to Schnell *et al.*, the resulting cell capacity is attempted to be predicted

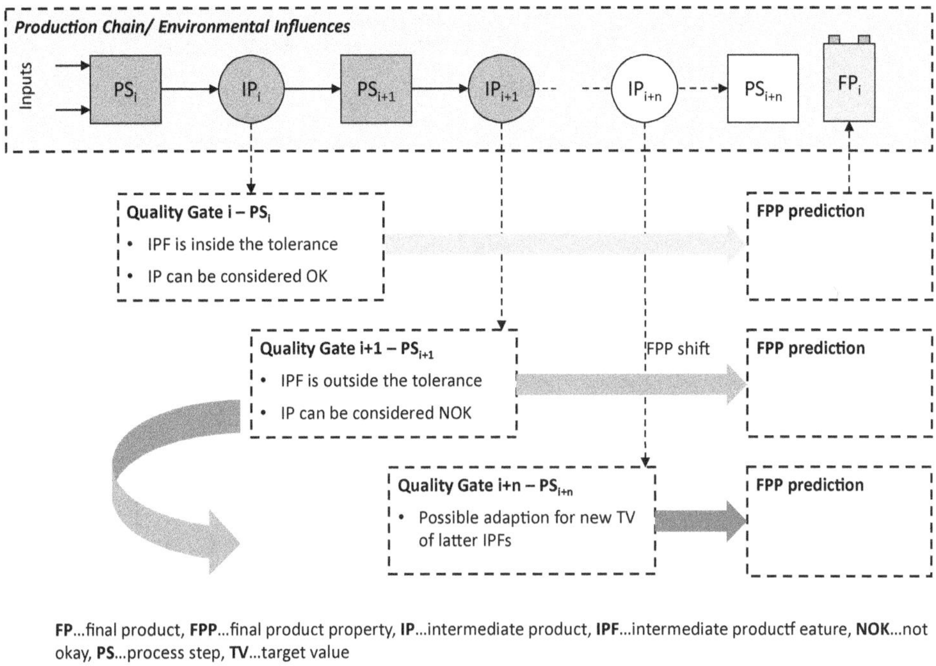

Fig. 8. Concept of the quality gates in manufacturing. In the style of Turetskyy *et al.*[45]

from the process data using data mining methods. In addition, the individual parameters are examined for their influences on the cell capacity. From this, it is recognized that the yield point during dispersion and the weight of the cell before electrolyte filling have the greatest influence on the resulting cell capacity.[50] Thiede *et al.*[47] use these findings to assign and calculate the data sources for intermediate product features and final product features in the production process. Based on the idea of Schnell and Reinhardt, Thiede *et al.* also established so-called quality gates in manufacturing in order to be able to eliminate bad parts prematurely.[47]

In this concept, shown in Fig. 8, the quality of the intermediate product (IP) and the resulting battery cell are evaluated. The intermediate product characteristics (IPF) of the IP have defined target values and tolerances, such as the viscosity of the electrolyte or the particle size of the electrode materials. If the intermediate product characteristics are outside the tolerances, the intermediate product is declared as not okay. In addition, the quality gates estimate the target values of the final product characteristics of the cell. In addition to the resulting cell capacity, these can be extended by the parameters SOH after 400 cycles and formation loss after the first forming cycle. In this process, the target values update during the

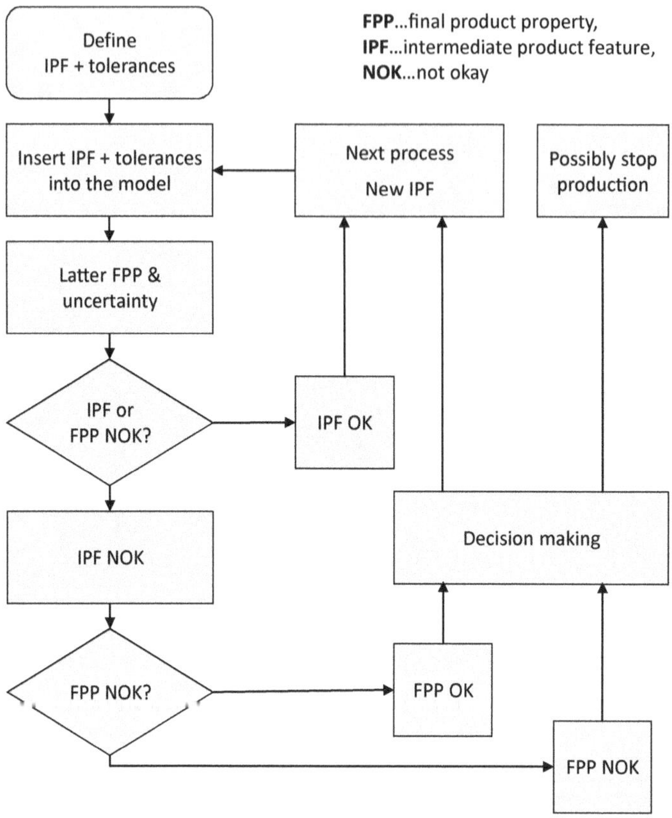

Fig. 9. Flowchart of the logic for adaptive improvement. In the style of Turetskyy *et al.*[45]

manufacturing process as more information is added to the model. Machine learning methods are used to identify, select and model the intermediate product features. Thus, the influence of the respective parameter on the target criterion can be evaluated and selected via the feature importance. In the next step, the future processes can use the information of the intermediate product to readjust the processes. This also includes not continuing the processing of the intermediate products and, if necessary, ejecting them in order to save energy and costs. This concept is also based on an empirical machine learning model, which uses data from already produced cells and relates the intermediate products and final product properties.[45]

The flowchart in Fig. 9 shows the adaptive improvement logic. After the relevant intermediate product characteristics have been selected and their relationship to the end product characteristics modeled, process experts define the target values and tolerances of the intermediate product characteristics. From this, the

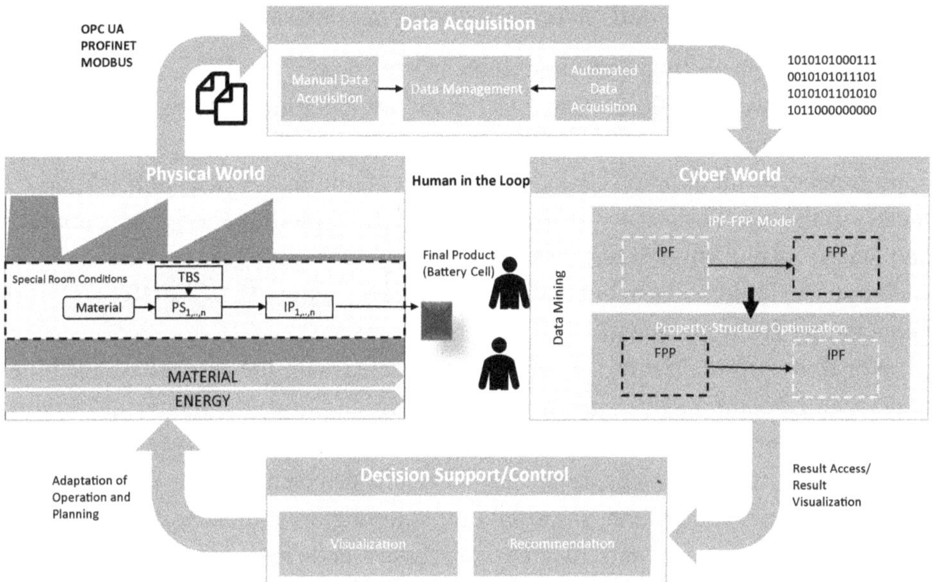

Fig. 10. Framework of the cyber-physical systems for the deployment of the quality gates concept. In the style of Turetskyy *et al.*[45]

target values and tolerances of the end product characteristics are calculated. Now, the intermediate product characteristics are calculated in loops. If the characteristics meet the specifications, the next process step is calculated. If the characteristics are outside the tolerances, it is checked whether and in which way this has negative influences on the end product characteristics. A decision is then made whether to continue production of the cell and a new target value is determined for further process settings. At this point, it is important to decide whether it is advantageous to stop production and perform damage limitation.[45]

This can be extended to a cyber-physical production system as shown in Fig. 10. This consists of the elements of physical world, data acquisition, cyber world and a control element. The physical world represents battery cell production. The data acquisition is performed in a manual or automated way, as already described. The cyber world contains the described single process simulations as well as the cross-process simulation methods and their data. The control element visualizes and derives recommendations from the models of the cyber world. Thus, the current process can be improved and future process planning can be positively influenced.[45] Thus, this cyber-physical production system represents the target stage of battery cell manufacturing modeling.

1.6. *Factory building and technical building service simulation*

The presented modeling and simulation of the product quality can be supplemented to answer questions related to the factory operation. Here, the utilization rates of individual manufacturing processes can be examined alongside the energy consumption of the process chain, the technical building services (TBSs) and the factory building. This provides the possibility to show the energy consumption and the associated costs. It also provides the basis for further assessment methods such as Life Cycle Assessment (LCA), Life Cycle Costing (LCC) or Total Cost of Ownership (TCO) methods.[51]

For this purpose, the process chain is modeled with the associated machines and logistics components, shown in Fig. 11. In addition, the technical building services are included, which have to maintain ambient conditions (e.g. of the dry room) within narrow bounds, especially in electrode production. In order to simulate the dynamic energy consumption, the modeling must include different material flows, continuous and discrete processes and machines, organizational framework conditions, discrete and batch processing, specific machine states and various models for the technical building services.[51] If, in addition, different factory locations are considered, the climatic framework conditions, the electrical emission factors and energy prices of the factory location must also be taken into account.[52] Thomitzek *et al.* and Weeber *et al.* investigate the effects of innovation leaps on the process, such as temperature reduction of the calendaring machine or reduced throughput time of the drying process due to the implementation of dry coating. In both cases, the influence of machine availability is also investigated. Thus, a strongly varying energy demand for the produced battery capacity

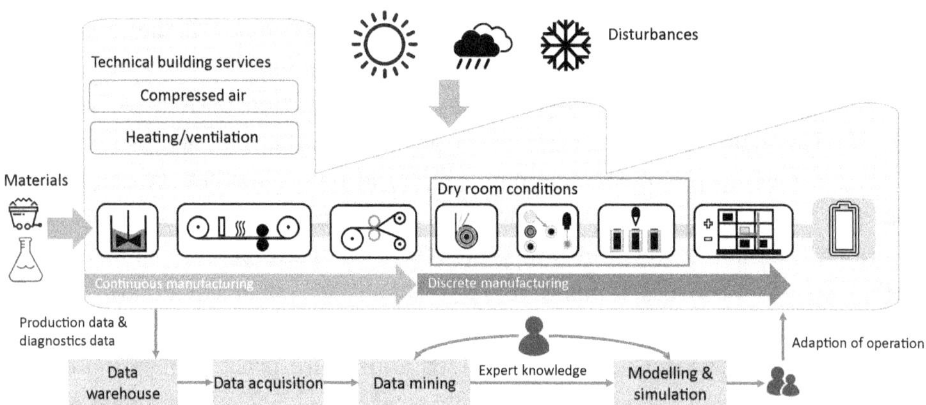

Fig. 11. A Lithium-ion battery cell manufacturing factory — A complex system composed of the process chain, the technical building services and the building shell. In the style of Turetskyy *et al.*[50]

can be recognized in both cases. The simulations of Thomitzek *et al.* are in the range of 69 Wh/Wh_{prod} to 213 Wh/Wh_{prod} and of Weeber *et al.* in the range of 93 Wh/Wh_{prod} to 328 Wh/Wh_{prod}. The modeling of Weeber *et al.* in comparison to Thomitzek *et al.* includes the building simulation and the energy source gas, which can explain the higher energy demand. Furthermore, additional energy sources such as compressed air or long distance heating can be considered in future simulations. Here, specific energy efficiency measures can be tested for their effectiveness. For example, the installation of a photovoltaic system for self-supply or the implementation of various energy flexibilization methods can be examined. This simulation environment can then be connected to the cyber-physical production system to additionally investigate the influences on product quality.[51,52]

Real data of machine conditions and their energy consumption as well as the data of the TBS can be used to realize a more realistic simulation. In this way, the buildup of these models can also provide valuable information for the planning of new production facilities or for the retrofitting and ramp-up of existing sites.[50,53]

2. Classification of Modeling and Simulation

On a software level, there exist a wide array of approaches and techniques for performing modeling and simulations. Choosing a simulation method that can enhance the range of information derived from a model is crucial as well as highly sought-after. There is no generally accepted classification of models and methods of simulation, as they have been classified with respect of varying factors in the past, such as the degree of automated decisions, level of competitiveness and application of simulation results. This section sheds light on the wide range of literature, especially those from the recent years that depict the topic of modeling and simulation in the context of industry 4.0, industrial engineering and system engineering.

It is essential to differentiate between modeling and simulation in order to derive a justifiable classification.[54] The purpose of a **model** is to represent a system of interest by incorporating most of its salient features, which can enable an analyst to predict the effect of changes to the system.[55] Complementarily, **simulation** is the imitation of the operation of an existing or planned system and its behavior over time by using a specific approach. In a broader sense, simulation is a tool to evaluate the performance of a system, existing or proposed, under different configurations of interest and over long periods of real time.[56] The function of simulation is to investigate dynamic models. A model is designed to imitate the time evolution of a real system, while a simulation results when the equations of the underlying dynamic model are solved.[57] Simulation models have rich and complex structures that make them a special type of scientific model. Additionally, a simulation model can be distinguished from other forms of models by various algorithmic structures, representation relations and new semantic connections involved in their architecture[58] identified by "computer simulation" as the entire process of constructing, using and justifying a model involves analytically intractable mathematics.

Thus understood, these definitions treat mathematical models and simulation models on an equal methodological and epistemological footing. Indeed, a simulation could be used to evaluate a behavioral model, data-driven model, descriptive model, conceptual model or an analytical model. It is rather more pivotal to shift from the strategic level of modeling or domain-specific model to an operational and a lower level of abstraction. The classification of models is hence done on a structural level. Different models and different simulation approaches serve different purposes. Its classification can be useful to select the right type of model for the intended scope. Figure 12 provides a consolidated overview of types of models and simulation approaches.[59–61]

Model selection, model fidelity and the simulation approach have a direct impact on the evaluation results of a simulation.[62] There are trade-offs between

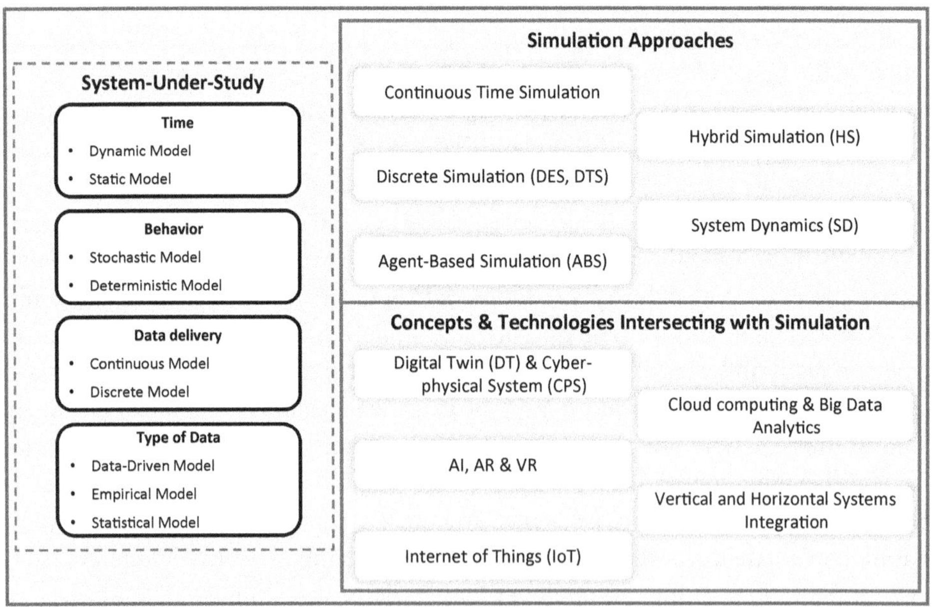

Fig. 12. Type of models; Simulation approaches; Concepts and technologies intersecting with simulation in Industry 4.0.

Notes: AI: Artificial Intelligence, AR: Augmented Reality, VR: Virtual Reality.

model development cost, simulation effort and prediction validity of the model that have to be considered when developing models for operational performance evaluations. This in turn generates the question of the impact of model type on the simulation approach, e.g. a static model should not be simulated using a system dynamics approach. It rather depends on the state of the system. Additionally, the system under study greatly influences the modeling and simulation decisions.[63,64] In some systems, the state changes all the time, not just at a specific time as a discrete event. Here, a "continuous simulation" is more appropriate, although discrete event simulation can also serve as an approximation. In order to identify the type of model and the simulation approach that need to be undertaken, we will describe the system characteristics in more detail.

2.1. *Classification of models*

The literature consists of classifications such as linear or nonlinear model, static/ steady-state or dynamic model, deterministic or stochastic model, continuous or discrete model, local or distributed. To understand model classification, this section focuses on the properties of the type of systems that can be modeled and simulated. The structural description and implementation methods of these systems are provided next.

Time is the influencing factor that defines a static or dynamic simulation model. A system is called **static**, whenever the relations between all relevant elements (parameters, variables) of the system do not depend on time, i.e. a time-independent view of the system leads to a static model.[65] One example is the number of students in a class in a year. Static models do not contain an internal history of input values previously applied, values of internal variables or output values. The mathematical representation of static models is a set of algebraic equations:

$$y_i = f_i(u_1, u_2, \ldots, u_n)$$

Each output variable y_i is dependent on a function f_i of the inputs u_i. The output variables can also be dependent on other output variables and the result has no simple solution. The function extends from algebraic expressions, i.e. physics-based equations, lookup and interpolation tables as well as static regression models.

Systems that change with time are called **dynamic** systems. The variable time is either a sequence of instants (discrete time, discrete event) or values in an interval (continuous time). For example, during weather forecasting, the state of the system changes with respect to time but the previous states of the system also serve as inputs to the simulation. Dynamic models provide a means to simulate the time-dependent behavior of the systems. Unlike a static model, here an internal memory is maintained, which can be some combination of prior inputs, internal variables and outputs. The sample mathematical representation of dynamics models involving a combination of algebraic and differential equations is shown as follows:

$$\frac{dx_i(t)}{dt} = f_i((u_1(t), u_2(t), \ldots, u_m(t)), x_1(t), x_2(t), \ldots, x_n(t)), t)$$

$$y_i(t) = g_i((u_1(t), u_2(t), \ldots, u_m(t)), x_1(t), x_2(t), \ldots, x_n(t)), t)$$

u_i and y_i are still the input and output variables, respectively, but they are represented as a function of time. The primary addition is the series of state variables x_i that are dependent on both input variables and themselves. The output variables y_i(t) are a set of functions, "g", dependent on the current state of the state variables and the input variables. The internal memory of the dynamic model is provided by the state variables. In the above example, only the rate of change of state variables $\left(\frac{dx(t)}{dt}\right)$ is provided instead of values of the state variables, and its relationship is an example of 1st order differential equations. If the rate of change is zero, the value of state variables would remain constant. Also, it is necessary to define the initial conditions in this case, at $t = 0$ or any other point of time when the simulation should start. This shows the dynamic model characteristic that *time is the independent variable that drives the model.* Within a simulation model, "time" is often referred to as the simulation clock and is mostly managed by the software.

In a static model, providing the same input values always results in the same set of output values. In the case of the dynamic model, the output values at any point in time depend not only on the input values at the current point in time but may also depend on all input values provided to the model at previous points in time.

The factor *behavior* defines a stochastic or deterministic simulation model. A **deterministic** system is one whose behavior can be predicted and it includes no randomization in the development of future events of the simulation. This implies that the output of a deterministic model is determined by initial conditions and parameter values. Ideally, deterministic functions return the same results whenever they are called with particular input value sets and the same parameter states. In contrast, a non-deterministic function returns varying results every time with specific input value sets and the same parameter state. Figure 13 represents a deterministic system vs. a stochastic system.

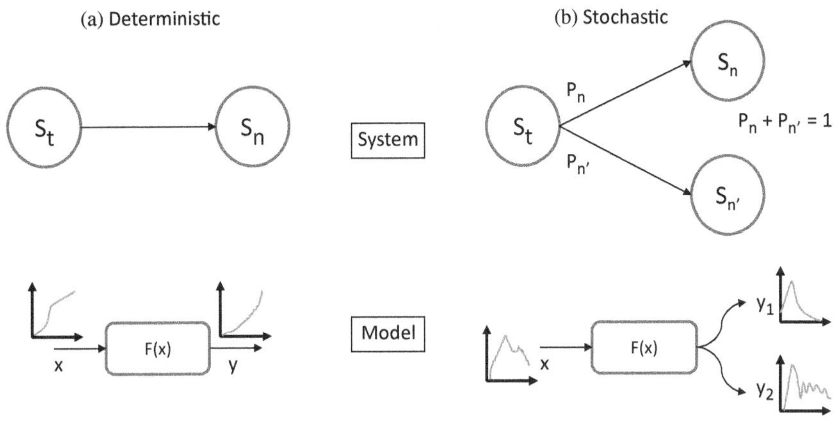

Fig. 13. Deterministic vs. Stochastic.[68,69]

A deterministic model is characterized by a set of equations that describe exactly how the system will progress over time. Deterministic models are commonly easier to analyze than stochastic models, because they are not significantly constrained by processing complexity, while probabilistic modeling easily becomes computationally intractable in the case of time-invariant, nonlinear processing.[66] In many cases, however, stochastic models are more realistic, especially for "small samples".

The distinction between deterministic and stochastic models has been blurred slightly by the development of chaotic models in the last few years. A chaotic model is a special deterministic model that is highly sensitive to the values of the parameters in the model. A very small variation of the values of these parameters can result in completely different outcomes of the model.[60] There is a famous quote which exemplifies the distinction as follows[67]: *A mountain stream, a beating heart, a smallpox epidemic, and a column of rising smoke are all examples of dynamic phenomena that sometimes seem to behave randomly. In actuality, such processes exhibit a special order that scientists and engineers are only just beginning to understand. This special order is 'deterministic chaos', or chaos, for short.*[70]

The word stochastic (Greek roots — "pertaining to chance") is jargon for random. A **stochastic** system possesses inherent randomness so that the same set of initial values and parameters produces different outputs at different points of time. A stochastic model accounts for uncertainties caused due to the varying behavioral characteristics of one or more fluctuating parts of the system, such as random time delays, noisy disturbances and even stochastic dynamic processes. Stochastic models tend to be more realistic and informative that deterministic models. For example, simulation of random walk process, traffic control and rare species management should be represented as stochastic systems.

By incorporating a little randomness of the real world, a stochastic model brings simulation results far closer to understanding the real world than could possibly be achieved with models which neglect such random behavior.[70] Stochastic process must include a family of random variables $\{X_t\}$, where t is a point in a space T called the *parameter space*, and where, for each t T, X_t is a point in a space S called the *state space*. The family $\{X_t\}$ may represent the path of a particle moving "randomly" in space S, its position at time t being X_t. A record of one of these paths is called a realization of the process. The basic steps to create a stochastic model are as follows:

- Create the sample space, i.e. a list of all possible outcomes X_t
- Assign probabilities to the elements of $\{X_t\}$
- Identify the events of interest, i.e. relations between the X_t for different fixed values of *t*
- Compute the probabilities for the events of interest, i.e. apply the theory of probability to determine the relationships identified.

The factor *data delivery* defines a discrete or continuous simulation model. The models may be continuous or discrete with respect to *time* or *space*. A **continuous** system is a system in which the state variable(s) changes continuously over time, e.g. fluid aspects and heat transfer equations across an electrical component. The input and output are both continuous signals in this case. A **discrete** system is a system where the state variable(s) changes only at a discrete set of time points, e.g. sales and production lines. In discrete systems, both input and output signals are discrete signals.

Time advancement mechanism of simulation is divided into next event and time step, called discrete event simulation (DES) and discrete time simulation (DTS), respectively. The simulations are explained in more detail in the next section (Simulation Methods cross-reference). However, it is necessary to differentiate between discrete and continuous models with respect to space and time, as shown in Fig. 14.

Dynamic models have dependent and independent variables and one, several or all may change continuously or in a discrete way. Similarly, variables of both deterministic and stochastic systems may change continuously or at discrete instants of time.

Lastly, the factor *type of data* drives a statistical, empirical and data-driven model, as shown in Fig. 15. A **data-driven model** is based on the analysis of the

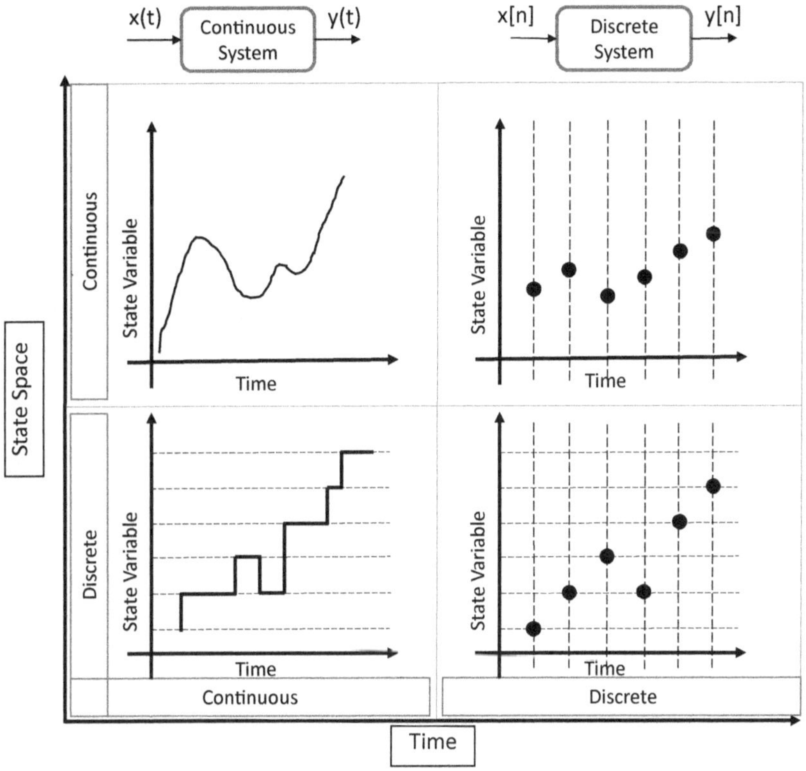

Fig. 14. Discrete vs. continuous time and discrete vs. continuous state space models. In the style of Holzer *et al.*[71]

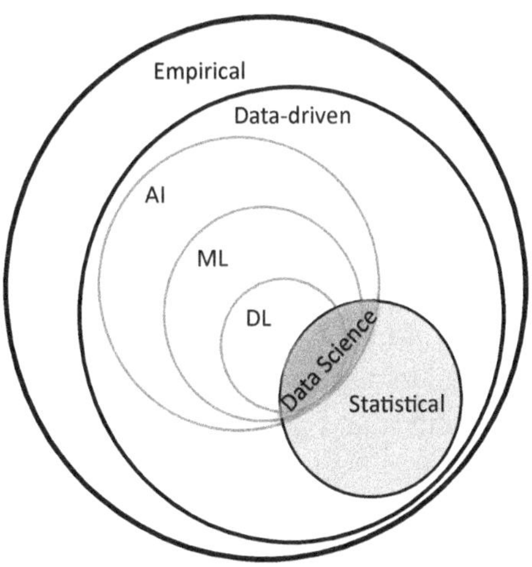

Fig. 15. Differentiating empirical, data-driven and statistical models.

Notes: AI: Artificial Intelligence, ML: Machine Learning, DL: Deep Learning.

data about a specific system. It assumes the presence of a considerable and sufficient amount of data describing the underlying system. Data are used to perform tasks of classification, pattern recognition, and associative and predictive analysis. Data-driven models find relationships between the system state variables (input and output) without explicit knowledge of the physical behavior of the system.[72] They may complement or replace the "knowledge-driven" physical models. There are a number of (overlapping) areas contributing to data-driven modeling: data mining, knowledge discovery in databases, computational intelligence, machine learning, intelligent data analysis, and soft computing and pattern recognition. Examples of the most common data-driven modeling methods are statistical methods, artificial neural networks, machine learning and fuzzy rule-based systems. Manufacturing process controls, resource management, weather and climate prediction, traffic management, systems engineering, geo-exploration and bio-sensoring are examples of areas which can benefit from data-driven application systems.[73]

Statistical models are subsets of data-driven models that mathematically approximate real-world scenarios. A statistical model uses statistics to build a representation of data and then conducts analysis to infer relationships between variables or discover insights. Methods used in statistical modeling include hypothesis testing, confidence intervals, sampling and probability spaces. Regression is a classic example which determined how a single or a number of variables are affected by one or more independent variables.

Statistical and mathematical models both involve mathematical formulas and equations, but they are not necessarily the same. Mathematical models are mostly static and represent a natural phenomenon in mathematical form; the models once formulated do not need to change the form after they are formed. However, statistical or data-based models are flexible enough to change as per arrival of new data as they can incorporate new and emerging patterns and trends. Statistical models are composed of statistical relationships which are developed from experimental data or historical studies.

Empirical models are based on observations and supported by experimental data.[74] Although they are at a higher abstraction level than the models we have discussed so far, it is essential to discuss the term as it is often used to describe data-drive models. The fundamentals and physical mechanisms underlying the system are not considered. Rather, data obtained from experiments are correlated to obtain some derived quantities. Functions of an empirical model capture the trend of data. Data-driven models and statistical models are subsets of empirical models.

A common drawback of the empirical modeling principle is that experimental data are always needed and in most cases, they are applicable only for the modeled experiment under a particular operating condition. Therefore, empirical models are not able to make predictions beyond a particular operating range or designed experiment.

2.2. *Simulation approaches in Industry 4.0*

The penetration of simulation into various aspects of business processes as a decision support tool has allowed solution development, validation and testing for systems and individual elements of systems.[75] Simulation has entered a new era with the advent of Industry 4.0. With increase in digitalization, new uses and needs of simulation will emerge in manufacturing.[61] The Industry 4.0 paradigm requires modeling of manufacturing and other systems via the virtual factory concept, cognitive process control, self-organization and various concepts which will be further discussed in this section.

High costs are associated with the development of experiments with the actual system to observe the behavior of processes in the real world. The advantage of using simulation is that it is often the time-effective and cost-effective means of detecting causal effects, estimating critical parameter and determining the changes in a process over time. It allows people to analyze systems optimization prior to implementation, especially when the problem cannot be formulated in mathematical terms.[76] Figure 12 provides an overview of the simulation-based approaches and the concepts/technologies of Industry 4.0 that intersect with simulation. A description of each of them is provided in the following section.

In **Continuous Time Simulation**, the underlying space–time structure as well as the set of possible states of the system is assumed to be continuous. In the Industry 4.0 simulation approaches, continuous time simulation is often used in combination with other methods. This is because in a simulation environment, the empirical data might not be available continuously or the state variables might not change continuously with respect to time. The simulation of a continuous time model is equivalent to the numerical integration of differential equations. There are a large number of methodologies to accomplish accurate, fast, efficient numerical integrations such as the Forward/Backward Euler method and the Runge-Kutta method.[77]

In **Discrete Time Simulation** (DTS), there is a fixed time quantity Δt with which the time of simulation advances and the value of the state variable is updated accordingly. It corresponds to the numerical methods for solving differential equations. However, in **Discrete Event Simulations** (DES), the events can occur rather arbitrarily as adaptive time steps. The time of the next "event" is calculated and the clock is directly advanced to that time. Often, time discretization is accompanied by state discretization.[78] Discrete event systems are also dynamic systems which evolve in time by the appearance of events at possibly irregular time intervals.[4] Examples include traffic systems, flexible manufacturing systems, computer communications systems, production lines, coherent lifetime systems and flow networks. Most of these systems can be modeled in terms of discrete events whose occurrence causes the system to switch from one state to another. DES is less helpful in simulations involving forces such as gravity, since a force is felt at every infinitesimal point in time, like a continuous, ongoing event. In such cases, stepping is required.

Agent-based modeling and simulation (ABMS) is a relatively new approach to model dynamic, autonomous and interacting entities or agents to achieve a specific goal. The state of an agent at a given time is defined by a collection of constraints that describe the agent's communication with other agents. The evolution of states of the agent may be deterministic or stochastic.[76,79,80] The rules of interaction of the agents result in an emergent system behavior as time and space evolve. An agent is defined as a software unit able to operate autonomously, pursuing a set of specific goals.[81] An agent can represent manufacturing systems' elements such as material, sensors, machines, products or people.[61] The model consists of three components: a set of agents, the agents' environment and a set of agents' relationships. The model should be capable of capturing the behavior of individuals within an environment.[82] ABMS plays an important role in I4.0 as a modeling paradigm for CPS and the simulation method.[83] Abar *et al.* (2017) provided a wide range of software tools for ABMS.[81]

System Dynamics (SD) is an analytical approach that complements systems thinking, based on feedback systems theory. It involves the discovery and representation of the feedback processes that, together with stock and flow structures, time delays and nonlinearities, define the dynamics of a system.[84] Wolstenholme *et al.* defined system dynamics as "A rigorous method for qualitative description, exploration and analysis of complex systems in terms of their processes, information, organizational boundaries and strategies; which facilitates quantitative simulation modeling and analysis for the design of system structure and control".[85] The technique for constructing SD model focuses on interdependencies, feedback effects, time dependencies and causality in the object that is being represented. Its modeling approach has three characteristics: First, the information feedback loops involve the collection of information about the state of the system and influencing factors which change the state of the system.[86] Second, computer simulation is used to deduce the behavioral consequences over time of the hypothesized causal network. Lastly, quantitative information and more subjective aspects of a given situation are held as mental models.[84]

Hybrid Simulation (HS) is a combination of two or more simulation methods leading to a multi-method/hybrid simulation model.[76] HS allows physical testing of only a part of the system that is of interest, while the rest of the system is simultaneously simulated with numerical models with good accuracy and confidence. This substructure technique provides cost savings, insights into detailed local behavior of the physical subsystem and better understanding of the complex structural systems (including multiple components and complex interactions).[87] Mustafee *et al.* proposed four types of hybrid models. Each model type aligned with different simulation methods such as SD-ABS, ABS-DES and ABS-DES-SD.[88]

Digital Twin (DT) is a key initiative for driving the digital transformation in smart manufacturing. It is identified as the new simulation modeling paradigm. DT is a virtual dynamic model of a system, process or service, with real-time (or near real-time) data interactions that facilitate improved system analysis and comprehensive representation of a physical system. Combining the real-life data with the simulation models from design enables accurate productivity and maintenance predictions based on the realistic data, and hence, continuous support for the decision-making process.

According to,[89] a DT is intended to answer the following questions:

- What are the necessary parameters for updating the simulation model, in order to reflect the physical system?
- What information should the simulation model return to the physical environment?
- How is the connection and interface between the simulation model and the physical process established to enable the exchange of data and information between them?
- During analysis, which scenarios should be tested considering the desired model responses already defined and how often should the model be executed given the decision-making needs?

Cyber Physical Systems (CPSs) are multidimensional and complex systems that integrate the cyber world and the dynamic physical world. It is the blend of software with mechanical and electronic devices which can communicate through a data exchange medium. In a broad sense, CPSs and DTs have similar features, and both describe the fusion of the cyber and physical worlds. However, CPSs and DTs are not identical.[90] CPSs primarily emphasize sensors and actuators, while DTs concentrate on models, data and their integration. CPSs include smart machines, processes, factories and storage systems which can autonomously exchange information through standardized service interfaces and Plug-and-Play principles, transforming the factory into a highly modular and flexible setup.[91]

Artificial Intelligence (**AI**) is a broad area of computer science that deals with the simulation of intelligent behavior in computers. Its subfields include machine learning, deep learning, natural language processing, computer vision and cognitive computing.[92] AI can also be further characterized as a range of techniques for modeling and simulation of environmental systems, which include artificial neural networks, fuzzy models, reinforcement learning, cellular automata and meta-heuristics.[93] AI developers are working on a wide range of business applications in areas such as manufacturing, logistics, supply chains, urban transportation, healthcare, business processes and asset management. Successful application of AI in these areas requires simulations in order to generate synthetic data, train a neural network and test the trained network.

Virtual Reality (VR) is a computer-generated environment with objects that appear to be real, making the user immersed in a responsive virtual environment.[94] **Augmented Reality (AR)** is the process of overlaying animations and graphics on actual scenes in real time through a set of technologies (e.g. capturing device, visualization devices, interaction device and tracking system). Both VR and AR engineering uses 3D modeling tools and visualization techniques as part of the design process. They have a wide range is applications in the manufacturing industry as they can support the assembly and maintenance of products.

Internet of Things (IoT) refers to a network of interconnected physical devices that are embedded with sensors and software in order to apply analytics and exchange valuable information between the devices over the Internet. Manufacturing machines, transporters, storage systems and even products can communicate and exchange information with IoT technology. When devices can sense and communicate via the Internet, they can go beyond local embedded processing to access and take advantage of remote computing nodes.[95]

Cloud Computing and Big Data have experienced an explosive growth in the recent times. Cloud computing means accessing and storing data as well as programs over the Internet rather than on the computer's hard drive.[96] Big Data technologies characterize a new generation of technologies and architectures, designed to extract economic value from very large volumes of a diverse data by enabling rapid collection, discovery and/or analysis.[97] Both cloud computing and big data facilitated *simulation optimization algorithms* to execute a large number of time-consuming simulations and provide datasets to specify simulation input processes and make simulation modeling and optimization a reality in a variety of application contexts.[98]

Vertical and Horizontal Systems Integration: The Smart Factory in Industry 4.0 requires two types of system integration, vertical and horizontal.

Vertical integration means that the systems within a smart factory must be aware of each other, and manufacturing systems and products must be hierarchically organized. Horizontal integration means that smart factories and businesses must be networked and must cooperate.[61] The concept is at quite a high abstraction level, but its realization is closely related to simulation methods CPS, IoT and DTs.[99]

3. Multi-Scale Model Coupling and Interfaces

Multi-scale Modeling and Simulation run on the principle of modeling of system elements on different temporal and spatial scales and the exchange of generated results between scales during simulation runs.[100] Coupling of processes at different scales and transforming information from one scale and its transfer to another scale are the key requirements in multi-scale modeling and simulation. The **goal** is an integrated analysis of system elements and their interrelations by coupling of multiple simulation models or tools for systems and subsystems. This *scale-bridging* technique or *co-simulation* is strongly dependent on specific application domains and is motivated by an increasing complexity of systems that requires an interdisciplinary collaboration of experts.[2,3] In many cases, developing multi-scale simulation amounts to developing software to combine existing pieces of software that implement single-scale models and scale-bridging methods.

Each model at different scales represents a part of the system at a specific temporal or spatial scale. Data exchange between different models is a great challenge with respect to accuracy and stability of the transfer and the results. This section is a discussion of product- and process-level multi-scale simulations, focusing on information exchange and synchronization methods of both processes.

3.1. *Product-level multi-scale simulation*

Multi-scale models used to investigate battery materials involve modeling the behavior of active materials, electrolyte and interfaces. Generally speaking, the computational methods used for material-scale simulations are quantum mechanics, molecular dynamics, mesoscale and continuum medium. They provide a better understanding of the fundamental physical and chemical characteristics of the materials and therefore help to identify ways to modify and design the materials from an atomic level.[101] Atomic, continuum and particle-scale simulations provide computational techniques or tools to describe the interactions among the atoms and electrons, calculate the charge distribution and charge transfer, simulate materials' kinetic behaviors and collect all necessary information on the materials' physical and chemical properties at the atomic level, either from a kinetics or thermodynamics point of view.

Multi-scale models used to investigate battery electrodes are supported on abstract representations of the electrode structure aimed at capturing the main features, such as their porosity, tortuosity, constituting particles' shapes, effect of the particle arrangement on the state of charge/discharge, conductivity and material damage. Models are built through various methods such as electrode images by X-ray tomography and 3D focused ion beam/scanning electron microscopy.[102]

Multi-scale models for battery cells and modules are used to investigate the electrochemical performance, thermal behavior and aging (degradation) process. Typically, studies include coupling of Pseudo-Two Dimensional (P2D) electrochemical models to 2D or 3D thermal models. Moreover, integration of physical models, i.e. electrical, thermal and electrochemical and their corresponding software into a computational framework, is often practiced.[103] At the component and cell level, software such as Matlab, Ansys Fluent or COMSOL or software combinations are used. Some of these utilize finite element solvers, making it possible to model complex geometries.

Multi-scale models for battery applications are mainly used to investigate the effect of operating conditions and environmental conditions on the state and aging process of the battery. Studies by Ahmed *et al.* and Venugopal *et al.* provide SOH and SOC estimation methods using data-driven and physical models by adopting the dynamic load profile condition of EVs.[104,105]

Figure 16 represents the battery-specific, temporal and spatial scales of model development. The time scales vary from sub-nanoseconds (electrochemical reactions) to seconds (transport) up to hours (electrode compositional changes) and days or months (structural and chemical degradation). The length scales vary from sub-nanometers (material), to micrometers (electrode), millimeters (cell) and meters (module and pack).

Multi-scale modeling of Li-ion batteries used for EVs includes material-level analysis, modeling of electrodes and cell/module/pack models followed by system-level models, i.e. battery pack in EVs. Simulation of each of these model scales serves various purposes and certain examples for each of them are indicated in Fig. 16.[102] Understanding the relationship between these multiple-scale mechanisms plays a key role in research and innovation of materials, components, cells and/or battery pack operation strategies. With increased complexity, the system under study cannot be understood using simplified approaches. For example, cell performance is not necessarily estimated by analyzing structural and transport mechanisms of the materials; local evaluations using only one level may lead to

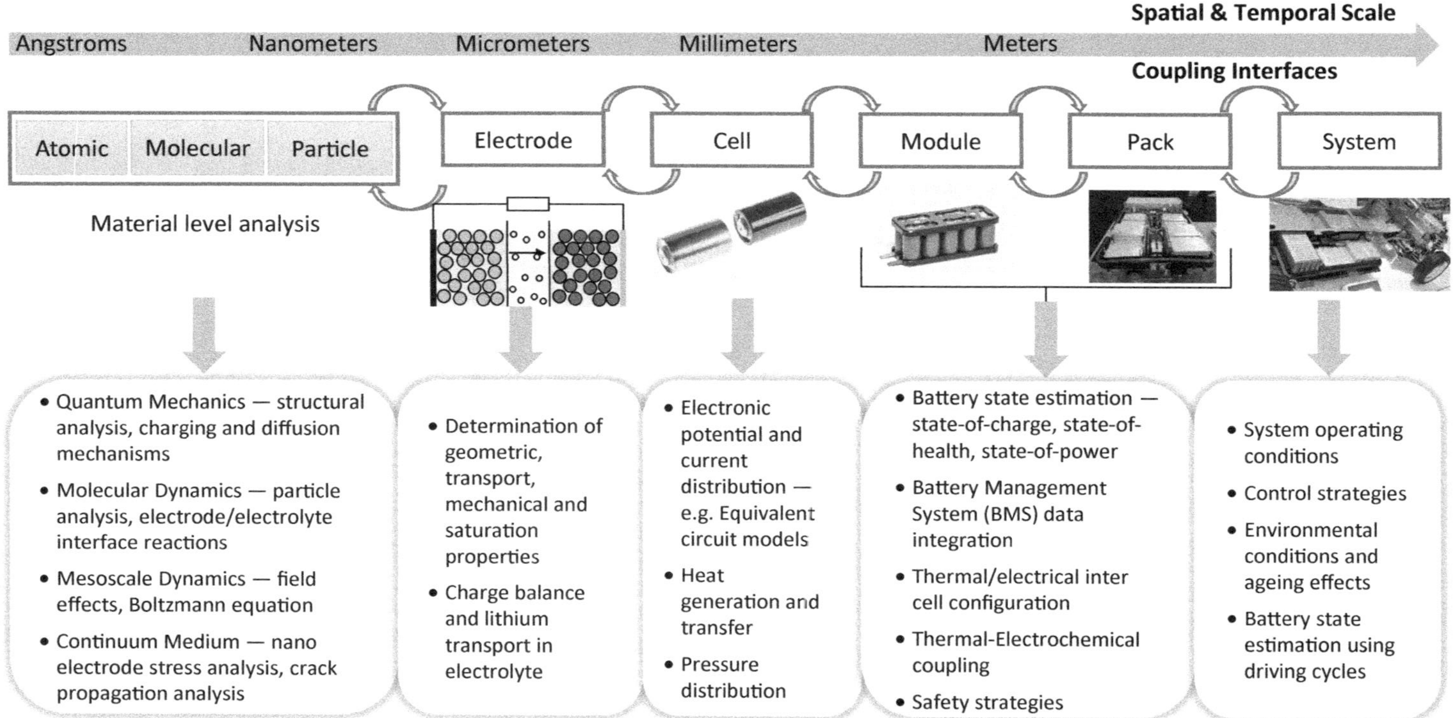

Fig. 16. Product-level multi-scale simulation and coupling interfaces.[101,106]

Sources: https://ecomento.de/2019/12/24/vw-erklaert-meb-elektroauto-batteriesystem/" and "https://www.greencarreports.com/news/1118974_heres-the-battery-pack-behind-vws-global-electric-vehicle-push".

erroneous or at least inaccurate predictions. A complete understanding only results from viewing the system as a whole, where effects are correlated through a holistic viewpoint. In order to capture the interplay of mechanisms occurring at multiple spatiotemporal scales, the coupling interfaces including the synchronization and information exchange methods from the literature are hereby further discussed.

Veen *et al.* identified that a significant amount of time is spent in solving technical issues of multi-scale simulation such as experimenting with different model formulations or performing efficient validation and uncertainty quantification or ensuring continued interoperability between model components.[107] A **model coupling software framework** (MultiScale Coupling Library and Environment version 3 (MUSCLE 3)) was proposed which allows coupling in a simple configuration file, which specifies the components of the simulation and how they should be connected together. At runtime, a simulation manager would coordinate the submodels, while data would be exchanged over the network via the MUSCLE library. MUSCLE 3 is developed openly on GitHub and is available as Open Source software under the Apache 2.0 license.

Models in various spatiotemporal scales can be coupled by direct connections between simulation tools or via **centralized middleware software**. The task of the middleware software is systematic coupling of the different simulation programs and synchronized data communication.[108] A commercial middleware solution, TISC, consists of a server and different clients for integration of simulation tools and data acquisition software. TISC integration is available in MATLAB/Simulink, LabVIEW, Modelica, etc. Additionally, the MATLAB/Simulink software, often used for battery modeling, offers data exchange interfaces and data stores across multiple models in order to facilitate multi-scale model coupling. Likewise, Wetter and Wetter *et al.*[109,110] developed an open-source middleware interface Building Control Virtual Test Bed (BCVTB) that allows coupling of different tools such as EnergyPlus, Dymola (Modelica), Matlab and Simulink in order to exchange data during time integration. Some common applications of the BCVTB are in the context of building energy and control system evaluation.[109,110]

Brocke *et al.* presented a multirate algorithm for co-simulation of electric-chemical systems with the goal of minimizing the communication between the components. Co-simulation often imposes the problem of the variables' availability in the mutually dependent components, i.e. not all required variables are available at a point of time due to the independent discretization of the components. The multirate algorithm allowed handling each component of the large system with a step size appropriate to its time scale.[111]

3.2. *Process-level multi-scale simulation*

Multi-scale modeling and simulation of a complex system are established methods in diverse disciplines and enable the replication of the behavior by modeling different scales separately. Within a scale, production systems can be observed at different spatial and temporal scales, each using the most appropriate methods and tools. The spatial scales range from product and single processes to process chains and TBS to the building. Spatial scales can be analyzed in temporal scales such as seconds, minutes and hours or up to several years.[2] As described in the previous chapters, the single elements of a production system can be simulated in different ways. The spatial and temporal scales form the basis for interactions of relevant simulation models within a multi-scale simulation framework. This framework clusters models of the individual production system elements, defines the necessary interfaces and enables a lean information flow between the simulation models. Due to the generic approach, the framework can be used for the production of cells as well as for the assembly of modules or battery systems.[100]

Figure 17 illustrates the framework which consists of different layers and model instances. First, the product is considered with its material components and

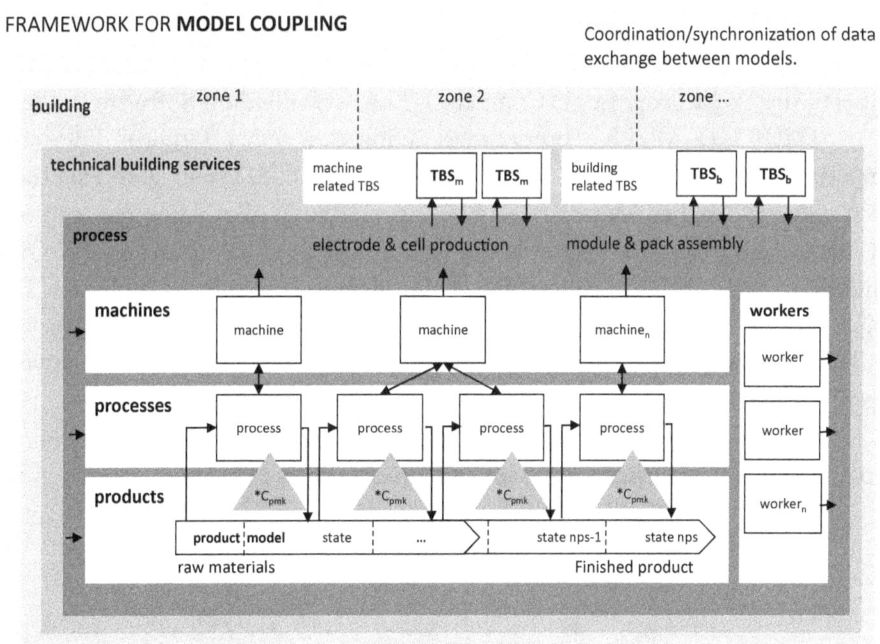

Fig. 17. Framework for multi-scale simulation of the manufacturing process. In the style of Schönemann and Kornas *et al.*[2,33]

properties. This represents the smallest spatial and temporal scale of the production system. The next layer contains models for machines, processes and intermediate or final products. This embeds with the workers in the process chain. Here, the spatial and temporal scale difference is particularly apparent. The process chain model is the core scale of the model structure and coordinates the material, energy and information flow between the models and scales. In addition, this layer provides the transition to the TBS and the building at the lowest layer. This represents the infrastructure for model coupling, with which the modeling described in the previous chapter can be combined and information exchanged by means of interfaces and data sources.[2,100]

Here, the process chain model acts as a coordinator of product-specific production activities in the multi-machine environment. It collects and aggregates information from all machines to calculate the TBS supply requirements.[2] An object-oriented approach is used to model the machines, buffer stations or workers. Production configuration as sequential or agile production can enable agent-based modeling at this level. Here, the individual stakeholders such as machines and workers can be modeled as agents. This enables a realistic mapping of a process chain and the planning of new process chain configurations.[2,100] Based on the process chain, the energy flows of the machines and processes are also modeled. Here, the focus is on the exchange with the building and the TBS. Here, the software tools are Tecnomatix Plant Simulation (Siemens AG); Dosimis (SDZ GmbH); Any Logic (Any Logic Company); FlexSim (FlexSim Software Products Inc.); SIMUL8 (SIMUL8 Corporation); Witness (Lanner Group); Enterprise Dynamics (INCONTROL Simulation Solutions); ExtendSIM (Imagine That Inc.); and AutoMod (Applied Materials).[112] For the simulation of building energy flows and their TBS, the most suitable tools are Energy Plus, IDA ICE (EQUA Simulation AB), IES-VE (Integrated Environmental Solutions) and TRNSYS (University of Wisconsin-Madison).[113]

The focus is on the exchange of information between the models, as shown in Fig. 18. There are four different concepts presented in Thiede *et al.*[114]

Offline coupled models are executed individually and one by one. Here, the model information is usually exchanged manually or via text files. However, the

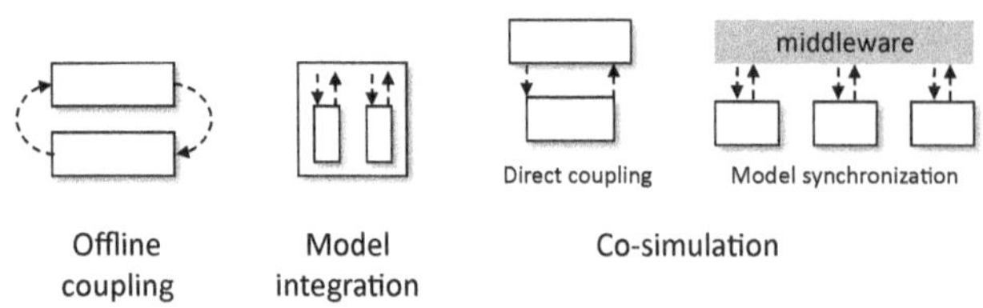

Fig. 18. Concepts of coupling simulation models. In the style of Thiede *et al.*[109]

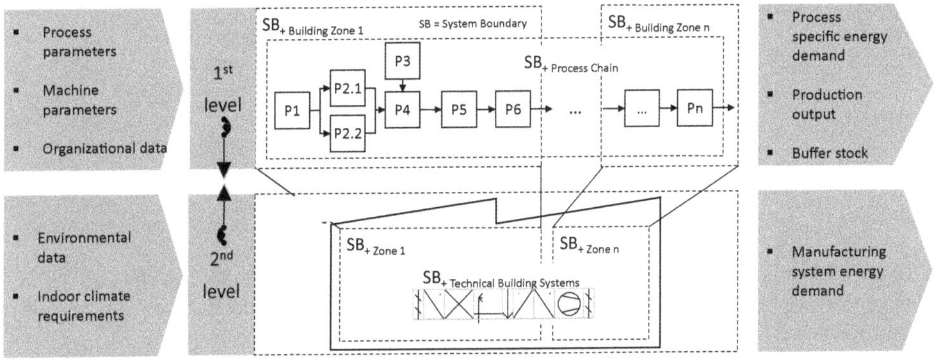

Fig. 19. Macrostructure of a battery cell manufacturing system model. In the style of Weeber *et al.*[52]

individual models can also be integrated into a so-called functional mock-up interface (FMI). This forms a standardized interface in which data can be exchanged continuously or at certain events.[109] The use of co-simulations enables close interlocking and simultaneous modeling. This requires a synchronous data exchange. If two software tools are used, the exchange can be done by direct coupling; if several simulation tools are used, middleware software has to be coupled. This enables data exchange and synchronization. The middleware software enables data exchange and synchronization. Building Controls Virtual Test Bed (BCVTB) is a co-simulation tool that was presented to simulate the energy flows in manufacturing systems[115] and the middleware — "TISC software suite" was used, which has also been successfully applied to multi-scale simulation of a manufacturing system[2,108]

A manual model coupling to determine the energy demand of the processes, TBS and the building was performed in Weeber *et al.* and is shown in Fig. 19.

In the process, the machines and process chain model were created on the first level, which determines the process energy demand with the organizational data using the Tecnomatix Plant Simulation software. These machines and process data are divided into the different rooms and integrated offline in a text file exchange as internal heat sources in a building model. Here, the environment-specific data like climate and solar radiation can be integrated into the software model in EQUA IDA ICE. In this way, the energy consumption of the technical building services can be determined in addition to the process energy. This approach provides an elegant and straightforward method to perform a multi-scale simulation in the context of battery cell manufacturing.[52]

Another example is shown in Fig. 20 from Schönemann *et al.* Here, the multi-scale simulation is performed with the middleware TISC Suite.[100]

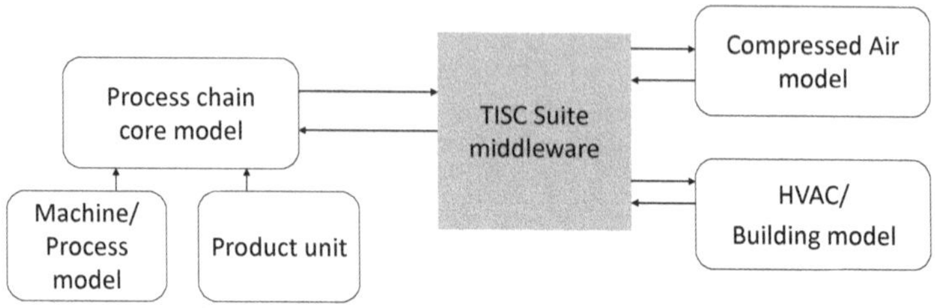

Fig. 20. Modeling structure of Schönemann *et al.*: a middleware software connects the process chain core model with a compressed air generation plant model and a building model. In the style of Schönemann *et al.*[97]

Analogous to Weeber *et al.*, the process chain is simulated in AnyLogic and the software provides the required compressed air quantities and power requirements to the middleware. This middleware transmits the data to an HVAC and building model in Simulink. Weather information is also included here. The middleware enables time-efficient execution of co-simulation runs in the respective software environment.[100] Thus, in addition to the evaluation of the production systems, the multi-scale simulation can also be used around the production planning and control. Real-time capability and prediction of operation are enabled in principle in both implementations. If the software is included in a production environment with the necessary computing resources and interfaces, it offers the possibility of intelligent decision-making and production control, taking into account various scenarios and stochastic effects.[52,100]

References

1. Weinan E. Principles of multiscale modeling. Cambridge: Cambridge Univ. Press; 2011.
2. Schönemann M. Multiscale simulation approach for battery production systems. Cham: Springer International Publishing; 2017.
3. Hoekstra A, Chopard B, Coveney P. Multiscale modelling and simulation: A position paper. Philos Trans A Math Phys Eng Sci. 2014. 372 (2021):1–8. doi:10.1098/rsta.2013.0377.
4. Arsham H. Systems simulation: The shortest route to applications. National Science Foundation. 1995. http://home.ubalt.edu/ntsbarsh/Business-stat/simulation/sim.htm.
5. Diekmann J, Hanisch C, Froböse L, Schälicke G, Loellhoeffel T, Fölster A-S, Kwade A. Ecological recycling of lithium-ion batteries from electric vehicles with focus on mechanical processes. J Electrochem Soc. 2017;164:A6184–91. doi:10.1149/2.0271701jes.
6. Kenney B, Darcovich K, MacNeil DD, Davidson IJ. Modelling the impact of variations in electrode manufacturing on lithium-ion battery modules. J Power Sources. 2012;213:391–401. doi:10.1016/j.jpowsour.2012.03.065.
7. Thomitzek M, Schmidt O, Röder F, Krewer U, Herrmann C, Thiede S. Simulating process-product interdependencies in battery production systems. Procedia CIRP. 2018;72:346–51. doi:10.1016/j.procir.2018.03.056.

8. Zhu J, Wierzbicki T, Li W. A review of safety-focused mechanical modeling of commercial lithium-ion batteries. J Power Sources. 2018;378:153–68. doi:10.1016/j.jpowsour.2017.12.034.
9. Meyer C, Bockholt H, Haselrieder W, Kwade A. Characterization of the calendering process for compaction of electrodes for lithium-ion batteries. J Mater Process Technol. 2017;249:172–8. doi:10.1016/j.jmatprotec.2017.05.031.
10. Heenan TMM, Jnawali A, Kok MDR, Tranter TG, Tan C, Dimitrijevic A, *et al.* An advanced microstructural and electrochemical datasheet on 18650 Li-Ion batteries with Nickel-Rich NMC811 cathodes and graphite-silicon anodes. J Electrochem Soc. 2020;167:140530. doi:10.1149/1945-7111/abc4c1.
11. Carlstedt D, Xu J, Runesson K, Larsson F, Asp L. Unit cells for multiphysics modelling of structural battery composites. Undefined. 2019.
12. Zhang X, Sahraei E, Wang K. Deformation and failure characteristics of four types of lithium-ion battery separators. J Power Sources. 2016;327:693–701. doi:10.1016/j.jpowsour.2016.07.078.
13. Haselrieder W, Westphal B, Bockholt H, Diener A, Höft S, Kwade A. Measuring the coating adhesion strength of electrodes for lithium-ion batteries. Int J Adhes Adhes. 2015;60:1–8. doi:10.1016/j.ijadhadh.2015.03.002.
14. Ramadesigan V, Northrop PWC, De S, Santhanagopalan S, Braatz RD, Subramanian VR. Modeling and simulation of lithium-ion batteries from a systems engineering perspective. J Electrochem Soc. 2012;159:R31–R45. doi:10.1149/2.018203jes.
15. Seaman A, Dao T-S, McPhee J. A survey of mathematics-based equivalent-circuit and electrochemical battery models for hybrid and electric vehicle simulation. J Power Sources. 2014;256:410–23. doi:10.1016/j.jpowsour.2014.01.057.
16. Jokar A, Rajabloo B, Désilets M, Lacroix M. Review of simplified pseudo-two-dimensional models of lithium-ion batteries. J Power Sources. 2016;327:44–55. doi:10.1016/j.jpowsour.2016.07.036.
17. Salkind AJ, Fennie C, Singh P, Atwater T, Reisner DE. Determination of state-of-charge and state-of-health of batteries by fuzzy logic methodology. J Power Sources. 1999;80:293–300. doi:10.1016/S0378-7753(99)00079-8.
18. Northrop PWC, Suthar B, Ramadesigan V, Santhanagopalan S, Braatz RD, Subramanian VR. Efficient simulation and reformulation of lithium-ion battery models for enabling electric transportation. J Electrochem Soc. 2014;161:E3149–57. doi:10.1149/2.018408jes.
19. Yannlia WB. Modeling of lithium ion cells?A simple equivalent-circuit model approach. Solid State Ionics. 2004;175:835–839. doi:10.1016/j.ssi.2004.09.049.
20. Xia Z, Abu Qahouq JA. Evaluation of parameter variations of equivalent circuit model of lithium-ion battery under different SOH conditions. In: 2020 IEEE Energy Conversion Congress and Exposition (ECCE); 2020 Oct 11–15; Detroit, MI, USA: IEEE; 2020. pp. 1519–23. doi:10.1109/ECCE44975.2020.9236339.
21. Charkhgard M, Farrokhi M. State-of-charge estimation for lithium-ion batteries using neural networks and EKF. IEEE Trans Ind Electron. 2010;57:4178–4187. doi:10.1109/TIE.2010.2043035.
22. Chen C, Xiong R, Yang R, Shen W, Sun F. State-of-charge estimation of lithium-ion battery using an improved neural network model and extended Kalman filter. J Cleaner Prod. 2019;234:1153–64. doi:10.1016/j.jclepro.2019.06.273.
23. Lin H-T, Liang T-J, Chen S-M. Estimation of battery state of health using probabilistic neural network. IEEE Trans Ind Inf. 2013;9:679–85. doi:10.1109/TII.2012.2222650.
24. Doyle M, Newman J, Gozdz AS, Schmutz CN, Tarascon J-M. Comparison of modeling predictions with experimental data from plastic lithium ion cells. J Electrochem Soc. 1996;143:1890–903. doi:10.1149/1.1836921.

25. Botte GG, Subramanian VR, White RE. Mathematical modeling of secondary lithium batteries. Electrochimica Acta. 2000;45:2595–609. doi:10.1016/S0013-4686(00)00340-6.
26. Zhang D, Popov BN, White RE. Modeling lithium intercalation of a single spinel particle under potentiodynamic control. J Electrochem Soc. 2000;147:831. doi:10.1149/1.1393279.
27. Khaleghi Rahimian S, Rayman S, White RE. Extension of physics-based single particle model for higher charge–discharge rates. J Power Sources. 2013;224:180–94. doi:10.1016/j.jpowsour.2012.09.084.
28. Farkhondeh M, Safari M, Pritzker M, Fowler M, Han T, Wang J, Delacourt C. Full-range simulation of a commercial LiFePO 4 electrode accounting for bulk and surface effects: a comparative analysis. J Electrochem Soc. 2014;161:A201–12. doi:10.1149/2.094401jes.
29. Bernardi D, Pawlikowski E, Newman J. A general energy balance for battery systems. J Electrochem Soc. 1985;132:5–12. doi:10.1149/1.2113792.
30. Chen SC, Wan CC, Wang YY. Thermal analysis of lithium-ion batteries. J Power Sources. 2005;140:111–24. doi:10.1016/j.jpowsour.2004.05.064.
31. Chen K, Unsworth G, Li X. Measurements of heat generation in prismatic Li-ion batteries. J Power Sources. 2014;261:28–37. doi:10.1016/j.jpowsour.2014.03.037.
32. Reinhart G, Kurfer J, Westermeier M, Zeilinger T. Integrated product and process model for production system design and quality assurance for EV battery cells. AMR. 2014;907:365–78. doi:10.4028/www.scientific.net/AMR.907.365.
33. Kornas T, Knak E, Daub R, Bührer U, Lienemann C, Heimes H, *et al.* A multivariate KPI-based method for quality assurance in lithium-ion-battery production. Procedia CIRP. 2019;81:75–80. doi:10.1016/j.procir.2019.03.014.
34. Yang S. A review of lithium-ion battery thermal management system strategies and the evaluate criteria. Int. J. Electrochem. Sci. 2019:14:6077–6107. doi:10.20964/2019.07.06.
35. Xia Q, Yang D, Wang Z, Ren Y, Sun B, Feng Q, Qian C. Multiphysical modeling for life analysis of lithium-ion battery pack in electric vehicles. Renewable Sustainable Energy Rev. 2020;131:109993. doi:10.1016/j.rser.2020.109993.
36. Rumpf K, Naumann M, Jossen A. Experimental investigation of parametric cell-to-cell variation and correlation based on 1100 commercial lithium-ion cells. J Energy Storage. 2017;14:224–243. doi:10.1016/j.est.2017.09.010.
37. Westermeier M, Reinhart G, Zeilinger T. Method for quality parameter identification and classification in battery cell production quality planning of complex production chains for battery cells. In: 2013 3rd International Electric Drives Production Conference (EDPC); 2013 Oct 29–30; Nuremberg, Germany: IEEE; 2013. pp. 1–10. doi:10.1109/EDPC.2013.6689742.
38. Sangrós Giménez C, Finke B, Schilde C, Froböse L, Kwade A. Numerical simulation of the behavior of lithium-ion battery electrodes during the calendaring process via the discrete element method. Powder Technol. 2019;349:1–11. doi:10.1016/j.powtec.2019.03.020.
39. Meyer C, Kosfeld M, Haselrieder W, Kwade A. Process modeling of the electrode calendering of lithium-ion batteries regarding variation of cathode active materials and mass loadings. J Energy Storage. 2018;18:371–9. doi:10.1016/j.est.2018.05.018.
40. Schreiner D, Oguntke M, Günther T, Reinhart G. Modelling of the calendering process of NMC-622 cathodes in battery production analyzing machine/material–process–structure correlations. Energy Technol. 2019;7:1900840. doi:10.1002/ente.201900840.
41. Schmitt J, Raatz A. Failure mode based design and optimization of the electrode packaging process for large scale battery cells. AMR. 2014;907:309–19. doi:10.4028/www.scientific.net/AMR.907.309.

42. Jeon DH. Wettability in electrodes and its impact on the performance of lithium-ion batteries. Energy Storage Mater. 2019;18:139–47. doi:10.1016/j.ensm.2019.01.002.
43. Kwade A, Haselrieder W, Leithoff R, Modlinger A, Dietrich F, Droeder K. Current status and challenges for automotive battery production technologies. Nat Energy. 2018;3:290–300. doi:10.1038/s41560-018-0130-3.
44. Wanner J, Weeber M, Birke KP, Sauer A. Quality modelling in battery cell manufacturing using soft sensoring and sensor fusion — A review. In: 2019 9th International Electric Drives Production Conference (EDPC); 2019 Dec 3–4; Esslingen, Germany: IEEE; 2019. pp. 1–9. doi:10.1109/EDPC48408.2019.9011847.
45. Turetskyy A, Wessel J, Herrmann C, Thiede S. Data-driven cyber-physical system for quality gates in lithium-ion battery cell manufacturing. Procedia CIRP. 2020;93:168–73. doi:10.1016/j.procir.2020.03.077.
46. Schnell J, Nentwich C, Endres F, Kollenda A, Distel F, Knoche T, Reinhart G. Data mining in lithium-ion battery cell production. J Power Sources. 2019;413:360–6. doi:10.1016/j.jpowsour.2018.12.062.
47. Thiede S, Turetskyy A, Kwade A, Kara S, Herrmann C. Data mining in battery production chains towards multi-criterial quality prediction. CIRP Annals. 2019;68:463–6. doi:10.1016/j.cirp.2019.04.066.
48. Westermeier M, Reinhart G, Steber M. Complexity management for the start-up in lithium-ion cell production. Procedia CIRP. 2014;20:13–19. doi:10.1016/j.procir.2014.05.026.
49. Schnell J, Reinhart G. Quality management for battery production: A quality gate concept. Procedia CIRP. 2016;57:568–73. doi:10.1016/j.procir.2016.11.098.
50. Turetskyy A, Thiede S, Thomitzek M, Drachenfels N von, Pape T, Herrmann C. Toward data-driven applications in lithium-ion battery cell manufacturing. Energy Technol. 2020;8:1900136. doi:10.1002/ente.201900136.
51. Thomitzek M, Drachenfels N von, Cerdas F, Herrmann C, Thiede S. Simulation-based assessment of the energy demand in battery cell manufacturing. Procedia CIRP. 2019;80:126–31. doi:10.1016/j.procir.2019.01.097.
52. Weeber M, Wanner J, Schlegel P, Birke KP, Sauer A. Methodology for the simulation based energy efficiency assessment of battery cell manufacturing systems. Procedia Manuf. 2020;43:32–9. doi:10.1016/j.promfg.2020.02.179.
53. Leiden A, Herrmann C, Thiede S. Cyber-physical production system approach for energy and resource efficient planning and operation of plating process chains. J Cleaner Prod. 2021;280:125160. doi:10.1016/j.jclepro.2020.125160.
54. Maria A. Introduction to modeling and simulation. In: Andradóttir S, Healy KJ, Withers DH, Nelson BL, editors. The 29th Conference; 1997 Dec 7–10; Atlanta, Georgia, USA. Piscataway: IEEE; 1997 Nov. pp. 7–13. doi:10.1145/268437.268440.
55. Banks J. Discrete-event system simulation. 5th ed. Upper Saddle River NJ u.a.: Pearson Prentice Hall; 2010.
56. Hartmann S. The world as a process. In: Hegselmann R, Mueller U, Troitzsch KG, editors. Modelling and simulation in the social sciences from the philosophy of science point of view. Dordrecht: Springer; 1996. pp. 77–100. doi:10.1007/978-94-015-8686-3_5.
57. Durán JM. What is a simulation model? Minds Mach. 2020;30:301–323. doi:10.1007/s11023-020-09520-z.
58. Frigg R, Reiss J. The philosophy of simulation: Hot new issues or same old stew? Synthese. 2009;169:593–613. doi:10.1007/s11229-008-9438-z.
59. Paula Ferreira W de, Armellini F, Santa-Eulalia LA de. Simulation in Industry 4.0: A state-of-the-art review. Comput Ind Eng. 2020;149:106868. doi:10.1016/j.cie.2020.106868.

60. Seel NM, editor. Encyclopedia of the sciences of learning. Boston, MA: Springer US; 2012.
61. Gunal MM. Simulation and the fourth industrial revolution. In: Gunal MM, editor. Simulation for Industry 4.0. Cham: Springer International Publishing; 2019. pp. 1–17. doi:10.1007/978-3-030-04137-3_1.
62. Nielsen JB, Sandvik E, Pedersen E, Asbjørnslett BE, Fagerholt K. Impact of simulation model fidelity and simulation method on ship operational performance evaluation in sea passage scenarios. Ocean Engineering. 2019;188:106268. doi:10.1016/j.oceaneng.2019.106268.
63. Arsham H. Performance extrapolation in discrete-event systems simulation. Int J Syst Sci. 1996;27:863–9. doi:10.1080/00207729608929287.
64. Adlakha VG, Arsham H. A simulation technique for estimation in perturbed stochastic activity networks. Simulation. 1992;58:258–67. doi:10.1177/003754979205800406.
65. Troch I. Proceedings / MATHMOD 09: 6th Vienna Conference on Mathematical Modelling, Vienna; 2009 Feb 11–13, Vienna University of Technology, Austria; full papers CD volume. Vienna: ARGESIM Publ. House; 2009.
66. Sarkar TK, Schwarzlander H, Choi S, Palma MS, Wicks MC. Stochastic versus deterministic models in the analysis of communication systems. IEEE Antennas Propag Mag. 2002;44:40–50. doi:10.1109/MAP.2002.1043146.
67. Krause C. Oak ridge national laboratory. Review. 1992;25(3,4):1992.
68. Platon V, Constantinescu A. Monte Carlo method in risk analysis for investment projects. Procedia Econ Finance. 2014;15:393–400. doi:10.1016/S2212-5671(14)00463-8.
69. Pooch UW, Wall JA. Discrete event simulation: A practical approach. Boca Raton: CRC Press; 1993
70. Coleman R. Stochastic processes. Dordrecht: Springer; 1974.
71. Holzer R, Wüchner P, Meer H de. Modeling of self-organizing systems: An overview 2010. Electronic Communications of the EASST, 2010;27:1–12. doi:10.14279/TUJ.ECEASST.27.385.
72. Solomatine DP, Ostfeld A. Data-driven modelling: Some past experiences and new approaches. J Hydroinf. 2008;10:3–22. doi:10.2166/hydro.2008.015.
73. Darema F. Dynamic data driven applications systems: A new paradigm for application simulations and measurements. In: Kanade T, Kittler J, Kleinberg JM, Mattern F, Mitchell JC, Naor M, *et al.*, editors. Computational science–ICCS 2004. Berlin, Heidelberg: Springer Berlin Heidelberg; 2004. p. 662–669. doi:10.1007/978-3-540-24688-6_86.
74. Ashoor BB, Giwa A, Hasan SW. Full-scale membrane distillation systems and performance improvement through modeling. In: Current trends and future developments on (Bio-) membranes. Elsevier, San Diego; 2019. pp. 105–140. doi:10.1016/b978-0-12-813551-8.00005-x.
75. Blaž R. Industry 4.0 and the new simulation modelling paradigm. Organizacija. 2017;50: 193–207.
76. Galvão Scheidegger AP, Fernandes Pereira T, Moura de Oliveira ML, Banerjee A, Barra Montevechi JA. An introductory guide for hybrid simulation modelers on the primary simulation methods in industrial engineering identified through a systematic review of the literature. Comput Ind Eng. 2018;124:474–92. doi:10.1016/j.cie.2018.07.046.
77. Register A. Continuous time simulation. In: Loper ML, editor. Modeling and simulation in the systems engineering life cycle. London: Springer London; 2015. pp. 111–137. doi:10.1007/978-1-4471-5634-5_11.
78. Buss A, Al Rowaei A. A comparison of the accuracy of discrete event and discrete time. In: Johansson B, editor. 2010 Winter Simulation Conference — (WSC 2010); 2010 Dec 5–8; Baltimore, MD, USA. Piscataway, NJ: IEEE; 2010. pp. 1468–1477. doi:10.1109/WSC.2010.5679045.

79. Laubenbacher R, Jarrah AS, Mortveit HS, Ravi SS. Agent based modeling, mathematical formalism for. In: Meyers RA, editor. Computational complexity: Theory, techniques, and applications. New York, NY: Springer Science+Business Media LLC; 2012. pp. 88–104. doi:10.1007/978-1-4614-1800-9_6.
80. Macal CM, North MJ. Agent-based modeling and simulation. In: Rossetti MD, editor. 2009 Winter Simulation Conference — (WSC 2009); 2009 Dec 13–16; Austin, TX, USA. Piscataway, NJ: IEEE; 2009. pp. 86–98. doi:10.1109/WSC.2009.5429318.
81. Abar S, Theodoropoulos GK, Lemarinier P, O'Hare GM. Agent based modelling and simulation tools: A review of the state-of-art software. Comput Sci Rev. 2017;24:13–33. doi:10.1016/j.cosrev.2017.03.001.
82. Buth L, Broderius N, Herrmann C, Thiede S. Introducing agent-based simulation of manufacturing systems to industrial discrete-event simulation tools. In: 2017 IEEE 15th International Conference on Industrial Informatics (INDIN); 2017 Jul 24–26; Emden. Piscataway, NJ: IEEE; 2017. pp. 1141–1146. doi:10.1109/INDIN.2017.8104934.
83. Houston C, Gooberman-Hill S, Mathie R, Kennedy A, Li Y, Baiz P. Case study for the return on investment of internet of things using agent-based modelling and data science. Systems. 2017;5:4. doi:10.3390/systems5010004.
84. Kunc M. System dynamics: a soft and hard approach to modelling. In: Chan WK, D'Ambrogio A, Zacharewicz G, Mustafee N, Wainer G, Page EH, editors. 2017 Winter Simulation Conference (WSC); 2017 Dec 3–6; Las Vegas, NV. Piscataway, NJ: IEEE; 2017. pp. 597–606. doi:10.1109/WSC.2017.8247818.
85. Wolstenholme EF. System enquiry: A system dynamics approach. Chichester: Wiley; 1994.
86. Oyarbide A, Baines TS, Kay Jm, Ladbrook J. Manufacturing systems modelling using system dynamics: Forming a dedicated modelling tool. J Adv Manuf Syst. 2003;02:71–87. doi:10.1142/S0219686703000228.
87. Song W, Chang C-M, Dertimanis VK. Recent advances and applications of hybrid simulation: Frontiers Media SA. 2021.
88. Mustafee N, Brailsford S, Djanatliev A, Eldabi T, Kunc M, Tolk A. Purpose and benefits of hybrid simulation: Contributing to the convergence of its definition. In: Chan WK, D'Ambrogio A, Zacharewicz G, Mustafee N, Wainer G, Page EH, editors. 2017 Winter Simulation Conference (WSC); 2017 Dec 3–6; Las Vegas, NV. Piscataway, NJ: IEEE; 2017. pp. 1631–1645. doi:10.1109/WSC.2017.8247903.
89. Santos CHd, Queiroz JA de, Leal F, Montevechi JAB. Use of simulation in the Industry 4.0 context: Creation of a digital twin to optimise decision making on non-automated process. Journal of Simulation. 2020:1–14. doi:10.1080/17477778.2020.1811172.
90. Tao F, Qi Q, Wang L, Nee A. Digital twins and cyber–physical systems toward smart manufacturing and Industry 4.0: Correlation and comparison. Engineering. 2019;5:653–61. doi:10.1016/j.eng.2019.01.014.
91. Gorecky D, Weyer S. SmartFactoryKL Systemarchitektur für Industrie 4.0-Produktionsanlagen: Whitepaper SF-1.1: 04/2016. Kaiserslautern; 2016.
92. Carvalho TP, Soares FAAMN, Vita R, Da Francisco RP, Basto JP, Alcalá SGS. A systematic literature review of machine learning methods applied to predictive maintenance. Comput Ind Eng. 2019;137:106024. doi:10.1016/j.cie.2019.106024.
93. Chen SH, Jakeman AJ, Norton JP. Artificial Intelligence techniques: An introduction to their use for modelling environmental systems. Math Comput Simul. 2008;78: 379–400. doi:10.1016/j.matcom.2008.01.028.
94. Turner CJ, Hutabarat W, Oyekan J, Tiwari A. Discrete event simulation and virtual reality use in industry: New opportunities and future trends. IEEE Trans Human-Mach Syst. 2016; 46:882–94. doi:10.1109/THMS.2016.2596099.

95. Karimi K, Atkinson G. What the Internet of Things (IoT) needs to become a reality. White Paper, FreeScale and ARM. 2013;1–6.
96. Griffith E. What is cloud computing. Retrieved from PC Mag: http://au.pcmag.com/networking-communications-software-products/29902/feature/what-is-cloud-computing.2016 May.
97. Villars RL, Olofson CW, Eastwood M. Big data: What it is and why you should care. White paper, IDC. 2011;14:1–4. http://www.tracemyflows.com/uploads/big_data/IDC_AMD_Big_Data_Whitepaper.pdf.
98. Xu J, Huang E, Chen C-H, Lee LH. Simulation optimization: A review and exploration in the new era of cloud computing and big data. Asia Pac J Oper Res. 2015;32:1550019. doi:10.1142/s0217595915500190.
99. Haag S, Simon C. Simulation of horizontal and vertical integration in digital twins. In: Iacono M, Palmieri FAN, Gribaudo M, Ficco M, editors. 33rd International ECMS Conference on Modelling and Simulation; 2019 Jun 11–14. UK: European Council for Modelling and Simulation; 2019. pp. 284–289. doi:10.7148/2019-0284.
100. Schönemann M, Bockholt H, Thiede S, Kwade A, Herrmann C. Multiscale simulation approach for production systems. Int J Adv Manuf Technol. 2019;102:1373–90. doi:10.1007/s00170-018-3054-y.
101. Shi S, Gao J, Liu Y, Zhao Y, Wu Q, Ju W, *et al.* Multi-scale computation methods: Their applications in lithium-ion battery research and development. Chinese Phys B. 2016;25:18212. doi:10.1088/1674-1056/25/1/018212.
102. Franco AA, Rucci A, Brandell D, Frayret C, Gaberscek M, Jankowski P, Johansson P. Boosting rechargeable batteries R&D by multiscale modeling: Myth or reality? Chem Rev. 2019;119:4569–4627. doi:10.1021/acs.chemrev.8b00239.
103. Allu S, Kalnaus S, Elwasif W, Simunovic S, Turner JA, Pannala S. A new open computational framework for highly-resolved coupled three-dimensional multiphysics simulations of Li-ion cells. J Power Sources. 2014;246:876–86. doi:10.1016/j.jpowsour.2013.08.040.
104. Ahmed R, Gazzarri J, Onori S, Habibi S, Jackey R, Rzemien K, *et al.* Model-based parameter identification of healthy and aged Li-ion batteries for electric vehicle applications. SAE Int J Alt Power. 2015;4:233–47. doi:10.4271/2015-01-0252.
105. Venugopal P. State-of-health estimation of Li-ion batteries in electric vehicle using IndRNN under variable load condition. Energies. 2019;12:4338. doi:10.3390/en12224338.
106. Qi W, Shapter JG, Wu Q, Yin T, Gao G, Cui D. Nanostructured anode materials for lithium-ion batteries: Principle, recent progress and future perspectives. J Mater Chem A. 2017;5:19521–40. doi:10.1039/C7TA05283A.
107. Veen LE, Hoekstra AG. Easing multiscale model design and coupling with MUSCLE 3. In: Krzhizhanovskaya VV, Závodszky G, Lees MH, Dongarra JJ, Sloot PMA, Brissos S, Teixeira J, editors. Computational science — ICCS 2020. Cham: Springer International Publishing; 2020. pp. 425–438. doi:10.1007/978-3-030-50433-5_33.
108. Kossel R. Simulation of complex systems using Modelica and tool coupling. 5th Modelica Conf. 2006;485–90.
109. Wetter M. Co-simulation of building energy and control systems with the Building Controls Virtual Test Bed. J Build Perform Simul. 2011;4:185–203. doi:10.1080/19401493.2010.518631.
110. Wetter M, Zuo W, Nouidui TS. Recent developments of the Modelica "Buildings" library for building energy and control systems. In: The 8th International Modelica Conference, Technical University, Dresden, Germany; 2011 Mar 20–22; Linköping University Electronic Press, Dresden, Germany; 2011. pp. 266–275. doi:10.3384/ecp11063266.
111. Brocke E, Djurfeldt M, Bhalla US, Kotaleski JH, Hanke M. Multirate method for co-simulation of electrical-chemical systems in multiscale modeling. J Comput Neurosci. 2017;42:245–56. doi:10.1007/s10827-017-0639-7.

112. Fuss, E. und Beißert, U. Energie- und Materialflusssimulation in Produktion und Logistik. In: 19. Magdeburger Logistiktage — Sichere und nachhaltige Logistik.
113. Svetel I, Jaric M, Budimir N. BIM: promises and reality. Spatium, SPATIUM International Review, 2014;32:34–38.
114. Thiede S, Schönemann M, Kurle D, Herrmann C. Multi-level simulation in manufacturing companies: The water-energy nexus case. J Cleaner Prod. 2016;139:1118–27. doi:10.1016/j.jclepro.2016.08.144.
115. Leobner I, Ponweiser K, Neugschwandtner G, Kastner W. Energy efficient production — A holistic modeling approach. In: 2011 World Congress on Sustainable Technologies (WCST); 2011 Nov 7–10; London. Piscataway, NJ: IEEE; 2011. pp. 62–67. doi:10.1109/WCST19361.2011.6114239.

Part III

Operating Battery Cell Manufacturing Lines with Digital Tools

https://doi.org/10.1142/9789811245626_0007

Chapter 6

Concepts for Full Traceability — Establishing the Foundation for Smart Manufacturing within the Framework of the Digital Shadow

Anna Buss[*,‡], Joachim Peter Doppler[*,§], Ira Effenberger[*,¶], David Ensling[†], Christoph Haar[*,|], Silke Hartleif[*,**], Günther Riexinger[*,††], Klaus Schöbel[†], Michael Trierweiler[*,‡‡], and Thomas Bauernhansl[*,§§]

[*]*Fraunhofer-Institute for Manufacturing Engineering and Automation IPA Nobelstr. 12, 70569 Stuttgart, Germany*

[†]*VARTA Microbattery GmbH, VARTA-Platz 1, 73479 Ellwangen, Germany*

[‡]*anna.buss@ipa.fraunhofer.de*

[§]*doppler.joachim@bcg.com*

[¶]*ira.effenberger@ipa.fraunhofer.de*

[|]*christoph.haar@ipa.fraunhofer.de*

[**]*silke.hartleif@ipa.fraunhofer.de*

[††]*guenther.riexinger@ipa.fraunhofer.de*

[‡‡]*michael.trierweiler@ipa.fraunhofer.de*

[§§]*thomas.bauernhansl@ipa.fraunhofer.de*

This chapter describes the concept and integration approaches of the Digital Shadow including an underlying traceability system in lithium-ion battery production. Specific challenges are addressed, such as technical and process-related changes in the batch structures. Thus, integration concepts for full traceability, including solution morphology with focus on identification technologies in lithium-ion battery production along with the challenges but also the advantages, like the reduction of costs, have been elaborated and described. The established concepts form the basis for Smart Manufacturing within the framework of the Digital Shadow.

1. Introduction

In the context of connecting technical systems in real time, the economic coordination of the resource information plays a central role. The digital capture and management of a growing flood of information makes this task a challenge, because socio-technical and technical information systems are becoming increasingly complex due to the growing number of system elements and their relationships to one another. Due to the demand of the market for more dynamic value-adding processes, the demand for changeability and dynamics of control systems is also increasing.[1] Current information systems are often characterized by static and stable information processes that manage the supply of information to all actors.[2] However, this status is no longer given in highly networked value creation systems.[2] The development in information technology also contributes to the described problem. Due to proprietary infrastructure components, interoperability is often prevented. In the future, cyber-physical systems have to be able to process near-real-time information. This results in a closer interlocking of the digital and physical worlds.[3]

The digital shadow as a framework including traceability is understood as a field of action of Industry 4.0.[3] Its purpose is to create an information base that enables actors to derive recommended actions for production.[4] However, the challenges in the information supply described above are not addressed by an information base as such. The information itself does not provide any benefit to a value creation system. This benefit is only generated if the information is made available at the right time, at the right place and in the right quality. Otherwise, waste in the sense of Lean production occurs in both information logistical processing and in its provision. This leads to an increased time requirement or even to an incorrect processing of a task.

A sector, in which process improvement via digital tools is aimed at, is the cell production of lithium-ion batteries. Since its introduction in 1991, the importance of lithium-ion batteries for society has been considerably significant and is still increasing.[5] At the beginning, this battery technology was used in the consumer industry, e.g. for the energy supply of camcorders, notebooks or mobile phones.[5] That is why these batteries are seen as the enabler for the portability of the Internet. In the last few years, lithium-ion batteries have also more and more been used in applications with much higher energy demands, as in vehicles and energy storage systems. Currently, lithium-ion battery production is characterized by a high scrap rate.[6] While this was kind of negligible in the production of smaller batteries, the scrap rate for larger batteries has a more decisive influence on the resource efficiency and also the profitability of battery production. Here, the use of the digital shadow to enable a traceability system is considered as a promising approach to reduce the scrap rate and to enable more efficient production.

Especially in quality assurance, sufficient information is needed to detect scrap and to identify errors at an early production stage. In particular, if the individual cells are assembled into large packs, detecting scrap is mandatory. One aspect of quality assurance is the realization of traceability of individual cells in the production process. This use case requires the availability of a large amount of information, some of which is near real time and often very granular. In order to achieve a demand-oriented and dynamic information supply, the concept of the Digital Shadow is adapted and presented in the following for the described use case.

2. The Digital Shadow — Theoretical Background

To understand the adopted concept of the Digital Shadow, different theoretical components have to be defined. The Digital Shadow, as it is presented in,[7,8] is an information system consisting of various services. The Digital Shadow provides the right information at the right time, at the right place and in the right quality for all actors involved in a value creation process. The effective supply of information is a task of information management. To make the best possible use of the resource information, information management involves the use of the three subtasks of management of information economy, management of information systems and management of information and communication technologies.[9] It can be seen that the tasks of information management include technical and organizational aspects. The management of the information economy tries to find a balance between information supply and information demand. This balance is mostly dynamic and requires regular readjustment. The information system management specifies requirements for the information technology and organizes the life cycles of different information (sub)systems. At the level of information and communication technology, it is all about the organization of the technical infrastructure and architecture.[9] In managing the information economy, the demarcation or the relationship between information supply, information demand and information need plays an important role. According to Augustin, a need for information only includes information that is necessary for an optimal decision. Information demand, on the contrary, is information that is requested by an actor in a formalized form. Information offer comprises all information that is available in the system.[10]

2.1. *Theoretical concept, differentiation and adoption*

Many different terminologies are currently being used for similar topics in the context of digitalization of value creation. In particular, the terms Digital Shadow and Digital Twin are often confused with each other. Siemens and Daimler use the model of the so-called digital twin to collect all-encompassing data of a product both during production and in later use. This offers the possibility of an improved life cycle management.[11–13]

Therefore, the Digital Twin describes the digital image of a product or extension of an object in production. As described in the first section of this chapter, a digital image of the production alone does not provide any benefit, because the database can only be used if it is made available as needed. However, the Digital Twin as a digital image has no functionalities in this respect. The concept of the Digital Shadow uses the Digital Twins of different objects to provide information according to specific needs. To fulfill this functionality, the Digital Shadow combines four sub-functionalities in its final development stage[7]:

(1) The Digital Shadow links information supply and demand.
(2) The Digital Shadow controls the flow of operationally relevant information (→ operational information).
(3) The Digital Shadow guarantees application-specific information quality.
(4) The Digital Shadow optimizes independently the used data and information base.

The individual expansion stages of the Digital Shadow and its four sub-functionalities are described on the basis of development stages that build on each other. In particular, the semantic description of information sources and information sinks plays a major role. This is based on a profound knowledge of the production process and the activities taking place within it. A detailed examination of the various dimensions of information quality is also addressed within the Digital Shadow concept.[8]

2.2. *General architectural overview*

For the functionalities described, the Digital Shadow requires several components (services). One important component of the digital shadow, which is first of all regarded generically, is the component of directory services. These comprise data models, data self-descriptions and elements of a production plant. Depending on the specific application or the specific use case, the respective directories differ.

The Digital Shadow as a generic, tailor-made information supplier might or even must be instantiated for each production system or rather for each use case, because both the components of the Digital Shadow and their combination alter from case to case.

The single subsystems of the Digital Shadow should be designed for high availability since they form the backbone of production in a highly automated factory. A total failure of the system might have significant, negative effects on the production. The management of the single services is therefore indispensable.

One option is the implementation of the Digital Shadow as a lose compound consisting of several (micro-) services which complement one another and are designed redundantly. External D-, O- and DO-components, attached to the Digital Shadow, which demand (D), offer (O) or both demand and offer (DO) information, are able to find the central directory service for components via so-called discovery protocols, for example, UPnP or Bonjour/Zeroconf. Discovery protocols enable finding services via a network connection without the need for attaching a manual configuration.

The directory service offers entry points for each component, by which they are included in the complete system, and provides interfaces, so that they can register in the Digital Shadow. Equally, they can make enquiries to the Shadow with respect to own information requests. The Shadow is then instructed to meet these requirements at its best by keeping the quality demands.

All D-, O- and DO-components which are attached using a different transport technology have to use particular transfer points by means of gateways. The included components of the Digital Shadow can be easily implemented in a container virtualization. Thus, server/services are provided in single virtual isolated containers. Consequently, the system's security and the portability of the application increase. Hence, a largely automated management of internal structures with already existing and approved technology is enabled, see Fig. 1.

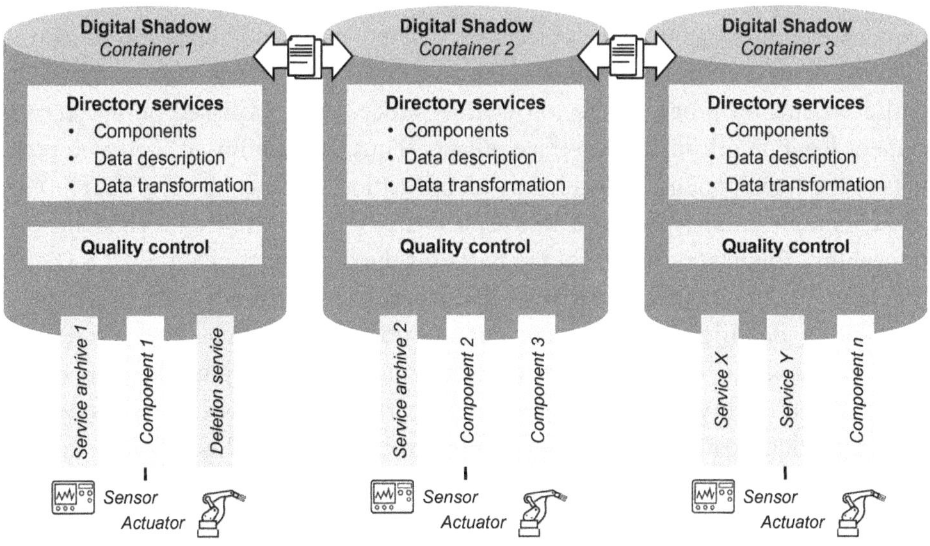

Fig. 1. Distributed containers comprising the main components of the Digital Shadow.

3. Challenges in the Context of Lithium-Ion Battery Production

Battery-powered systems and devices are generally seen as ecological or more sustainable than systems which rely on fossil fuels as energy sources. Nevertheless, there is a considerable impact on the environmental aspects concerning battery production. In order to reduce this impact, research work is focused on advanced technologies and more efficient production concepts.[14] The utilization of traceability systems is one approach to enable a more efficient production of batteries. Thus, production parameters can be related to final product characteristics to gain a holistic system understanding, depending on the available production data of the individual batch.[15] In order to further evaluate and assess production processes, production parameters and their effect on performance or quality indicators can be considered using a traceability system which connects production data and real performance indicators with the resolution of single product or battery cells.

A traceability concept for lithium-ion batteries needs to overcome two main challenges: At first, several process-related changes in the batch structure occur.[16] Identification markers need to be preserved or new identifiers need to be applied during a batch changeover. The second challenge is to mark tracing objects without impairing them in relation to process and product performance or quality. Especially the electrode foil with coated functional layers must not be negatively affected by the application of physical markers. Besides, the sensor data acquisition write and read rates must correspond to the respective process speeds.[17]

3.1. *Traceability system*

Information from trace objects along phases of the product life cycle is collected by a traceability system. Trace objects are discrete units, individual pieces as well as whole batches that are tracked from a specific processing step or during an entire production process. The retrieval of process data is based on the current state or location information of trace objects. Thus, traceability of produced products or their components is enabled by identification, data acquisition, data linking and communication aspects.[18] These aspects also comprise the main elements of a traceability framework and form the basis for the core functionalities of a Digital Shadow. Within the presented research work, the focus is on the description (semantic), identification and serialization of trace objects. In general, a traceability system enables forward and backward tracing within the value chain, as shown in Fig. 2.[19] If quality issues are detected in a particular production process or trace object, products can be identified and rejected if the traceability system is able to forward track objects following the timeline. Besides, relevant batches and process can be identified to investigate the cause of a product failure by tracing parts and quality issues backward.

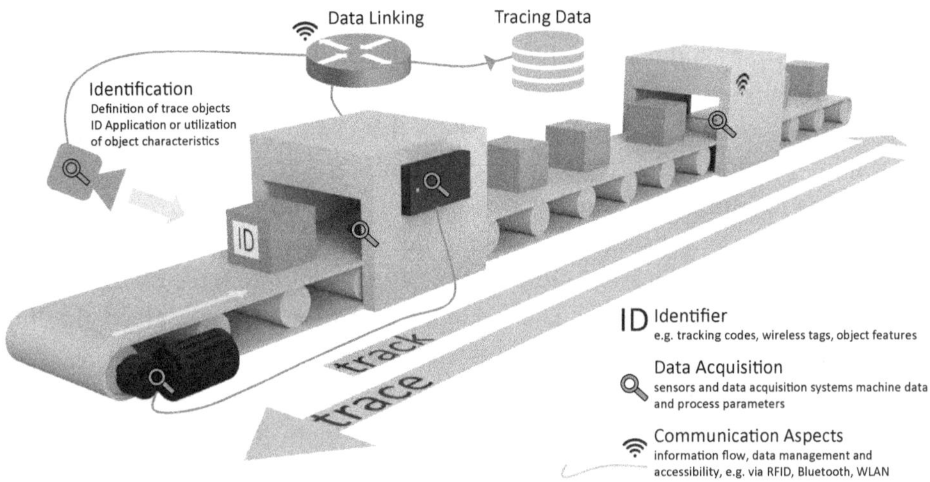

Fig. 2. Traceability system overview.

In order to identify an object or production element, its inherent object properties such as their characteristic features or applied identification markers can be tracked accordingly. The changing structures of production batches are a challenge for traceability by means of applied markers as they have to be removed or are hidden. It can also be necessary to adjust the resolution or the granularity of the tractability system. Thus, it may be necessary to clarify whether, where and which identification features have to be newly applied to ensure continuous identification chains.

A traceability system is also extracting and merging data from different sources (machine data, logistic information, etc.). Thus, different interfaces of machines and databases are often a challenge when connecting existing devices. When a traceability system is set up, it must be determined which data from existing plants or control systems can be utilized and where missing information must be obtained by installing new sensors or manual interfaces. This also serves to optimize the data and information base used by the Digital Shadow, see also sub-functionality four of the Digital Shadow concept. The data acquisition, communication and data linking of the traceability system need to be designed to comply with process requirements such as cycle times and required data acquisition rates.

3.2. *Cell-specific traceability of lithium-ion batteries*

The state of the art of cell-specific traceability of lithium-ion batteries with a focus on identification out of the four main elements of a traceability framework as mentioned in Section 3.1 is described within this section. The identification of certain units from its raw materials to the finished foil as well as the data acquisition of certain points inside the manufacturing chain can be seen as problem specific and is assessed accordingly. One of the challenges in implementing a traceability system in electrode foil production is the changes of manufacturing units during certain manufacturing processes.

The manufacturing starts with mixing the electrode paste or slurry preparation. Then, a carrier film is coated with the slurry, resulting in the electrode foil.[5] After coating, the electrode foil is dried in a continuous furnace. Calendaring follows, where the electrode foil is compressed to reduce the porosity of the electrode material. During the following slitting process, the electrode foil is cut into several subsidiary coils. These coils are dried in a drying room. At this point, the manufacturing of the electrode foil is completed and the assembly steps to create the battery cells can follow. In the case of pouch cells, the electrode foil is usually separated into single sheets and formed to stacks of cathode, anode and separator sheets. The stack is contacted and packaged to a battery cell. Round cells are produced by winding the anode, cathode and separator foils. Then, they also get packed in a case.[17] After this point, the identification of the cells is possible through, e.g. identifiers on the cell cases. From here on, the traceability of cells can be seen as state of the art. To enable traceability in the before-mentioned manufacturing process, options of identification for the different manufacturing units and the changes in batch structure need to be defined, whereby each individual production object must still be clearly identifiable.

To identify assembled battery cells, markers on the surface of the cell cases can be used. Current research deals with cell-specific tracking during manufacturing of the electrode foil, from the raw materials to finished cells. First, ideas exist for cell-specific tracking of the electrode foil using optical features or applied markers by lasers or inkjet printers.[20] Apart from this, no approaches to enable identification at the cell level during manufacturing of the electrode foil can be found, either in industry or in research. Another approach could be the application of marker-free identification techniques by using existing surface microstructures, such as the "Track & Trace Fingerprint" system by Fraunhofer IPM.[21] Industries with similar requirements on traceability and comparable process characteristics are, for example, the food,[22] paper,[23] process[24] or steel industries.[25] Kvarnström *et al.* describe generic methods for traceability solutions in continuous processes. Here, ways of identification and data acquisition are described.[26] However, the described methods are not completely transferable to the needs within the production of battery cells. Westermeier *et al.* describe which process data in the several production steps need to be recorded and assigned via data links to each other.[16]

In data mining research projects inside cell manufacturing processes, data acquisitions and identification of manufacturing units are also needed. Thiede *et al.* describe an approach inside the so-called Battery LabFactory Braunschweig.[27] Here, quality characteristics of intermediate and the final product are recorded, however, mostly manually and batch-wise instead of per cell.[15] In summary, no concept for traceability is currently available which can be simply adapted to the cell production of lithium-ion batteries. Accordingly, further developments based on the described approaches are necessary.

3.3. *Reduction of costs by using traceability systems in lithium-ion battery production*

Establishing traceability systems can reduce two main cost drivers regarding the manufacturing process itself: costs of a poor-quality rate as well as process costs (e.g. speed losses). The application of traceability systems offers the possibility to improve product quality by merging process and quality data, which is the basis for process modeling. Process models give a better understanding of dependencies between process parameters and product quality. They are used for optimizing the product quality as well as the overall equipment effectiveness (OEE). Additional benefit can be gained by having more targeted recall campaigns, if necessary.

Figure 3 shows an example of cost savings and expenses for a traceability system depending on the degree of part identification. The degree of part identification can differ from batch-wise to single-piece tracking.

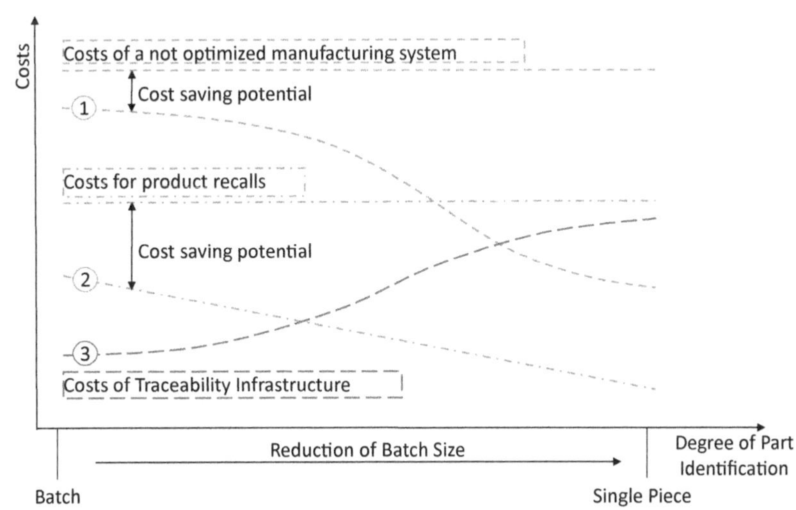

Fig. 3. Costs savings and expenses for traceability systems.

Graph 3 in Fig. 3 shows a typical cost trend of the expenditures for the required infrastructure. Identification systems as well as IT infrastructure, e.g. server capacity, are needed. The initial costs for the infrastructure components and their installation are only slightly increasing by downscaling the tracked batch size. If the batch size is becoming too small for the used infrastructure, for example, standard tags cannot be applied on a batch carrier any longer, different tracking methods are necessary. This can lead to a sharp rise of investment costs, as shown in the chart.

Saving potentials regarding process optimizations are often hard to estimate due to the complexity of the process and the high variation of its parameters. The current quality rate can give an indication of how much costs can be saved by increasing the quality at this process step. Graph 1 shows a cost saving trend where most benefit is gained by linking single piece quality and process data. If process parameters are very stable over a batch, the results may vary. Furthermore, the integration of tracking solutions will gain extra benefits if product recalls are necessary. The higher the degree of part identification, the more precisely a recall campaign can be conducted. This is shown in the chart with graph 2 by a linear trend.

The actual economic benefit as well as the integration costs of a traceability system depends on the initial situation of the production system and its infrastructure.

4. Traceability Concept for Battery Production

To fulfill the described requirements of a traceability system in lithium-ion battery production, one of the main challenges is the selection of the trace objects and their identification methods.[28]

Besides the physical shape (e.g. object and carrier), the resolution of the observed components is important for the development of the traceability system. Therefore, two main classification levels, single object tracing and group object tracing, can be identified. Implementing single object tracing gives the possibility of tracing each object individually compared to using object group tracing, in which only one entire object group can be traced. In many cases, object groups or their carriers are easier to trace than single objects, but the information of their tracing path is less specific. Along the process chain of lithium-ion battery cell production, no stable batch size exists. This means that units are divided into subunits, or components are joined to connected batches. If the batch size and structure are stable over several process steps, these steps are defined as one process cluster.

Not only the tracing objects but also the identification methods can be classified into two main groups: On the one hand, an identifier can be directly attached

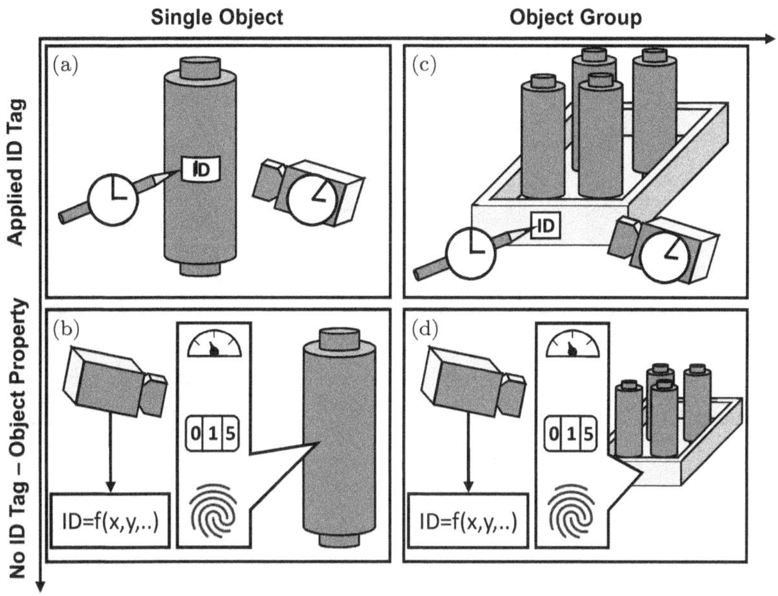

Fig. 4. Solution matrix for tracing methods.[28]

on the surface of the trace object. On the other hand, different types of systems can trace objects without a physical marker by using inherent information like logic, mechanical data or object properties. Together, identification methods and tracing objects can be merged into a method matrix, which shows the field of possible concepts for these tracing tasks as shown in Fig. 4.

For each process cluster, the effort and the benefit of single tracing compared to group tracing must be weighed against each other to identify a resolution, which allows both a cost-effective and useful solution. Besides choosing one matrix field, a combination is also possible, like applying an ID Tag on a group carrier and identifying each single object inside this group by using a logical approach. Figure 5 shows different possible combinations of single object tracing depending on the chosen technical implementation.

4.1. *Trace objects in battery production*

Excluding pouch cell production, battery cell production in general can be divided into five process clusters, see in Table 1. Each process cluster is defined by its process-related batch structure. Every change of the structure of a trace object results in a new cluster. The trace objects can be grouped into discrete (dis) and continuous (con) trace objects.

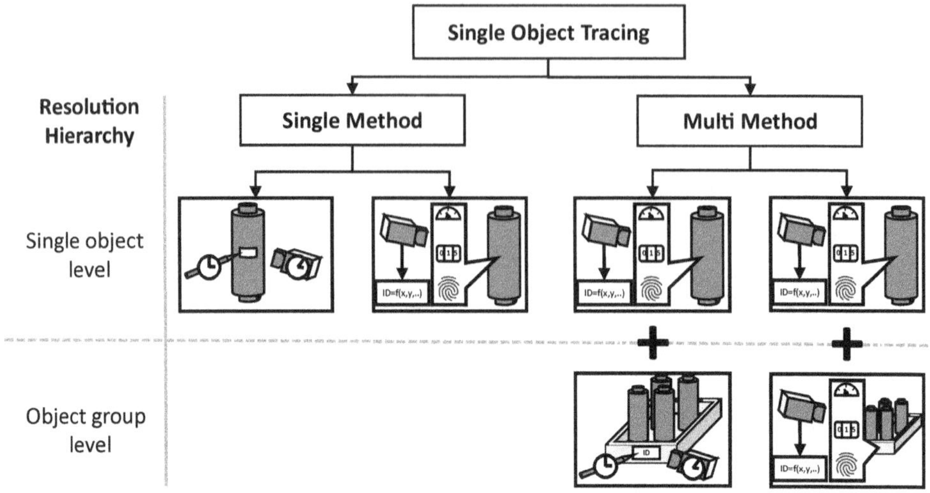

Fig. 5. Combinations for single object tracing.[28]

Table 1. Main process cluster and trace objects.[28]

Process Cluster	Content	Trace Objects
Slurry production	Mixing, storage	Containers (dis) Paste material (con)
Electrode coating	Coating, drying, calendaring	Coils, carrier coil batches (dis) Electrode foil (con)
Electrode cutting	Electrode foil separation/ slitting	Sub-coils, carrier sub-coil batches (dis) Electrode foil segments (con)
Cell assembly	Winding/stacking cell conductor welding	Stacks, coils, container, batches (dis)
Finalization	Encasing, welding, electrolyte induction, formation and aging, packaging	Battery batches, transport container (dis)

4.2. *Identification techniques and morphological analysis*

For the unique identification of trace objects, an applied identifier can be used. The information obtained using an identifier is stored on the workpiece, on the workpiece carriers or freight carriers[29–31] and can be read out either in a contactless or tactile manner. For transferring, contactless optical systems, like cameras, are suitable.[32] Identifiers like a clear print, symbols or an n-dimensional tracking code (barcode, QR code, color code) can be captured with them. Also, an object can be identified by its inherent features by measuring surface characteristics or roughness.[21,33,34]

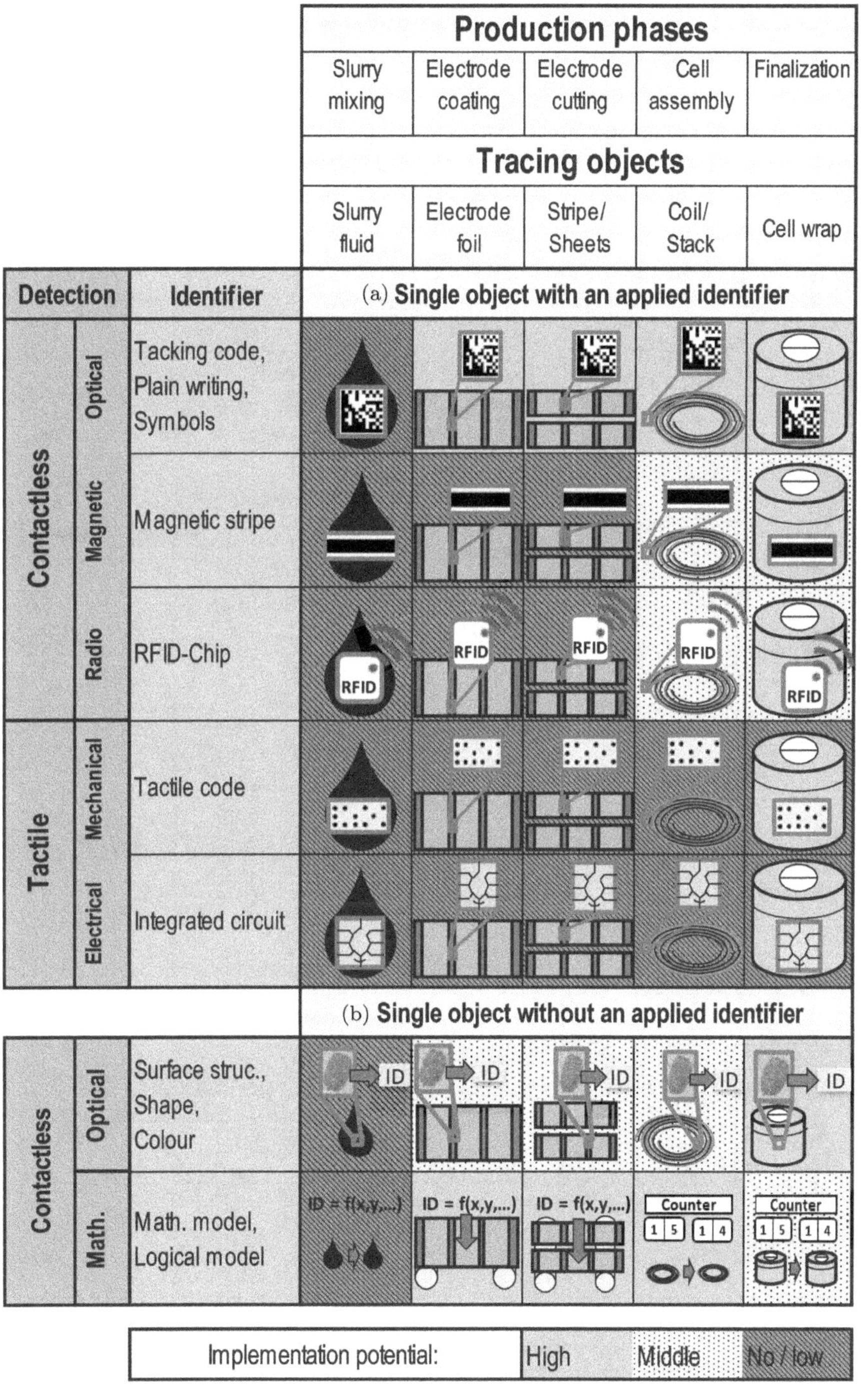

Fig. 6. Morphological analysis: Solution combinations for single trace objects (a) with applied identifiers or (b) without applied identifiers.[28]

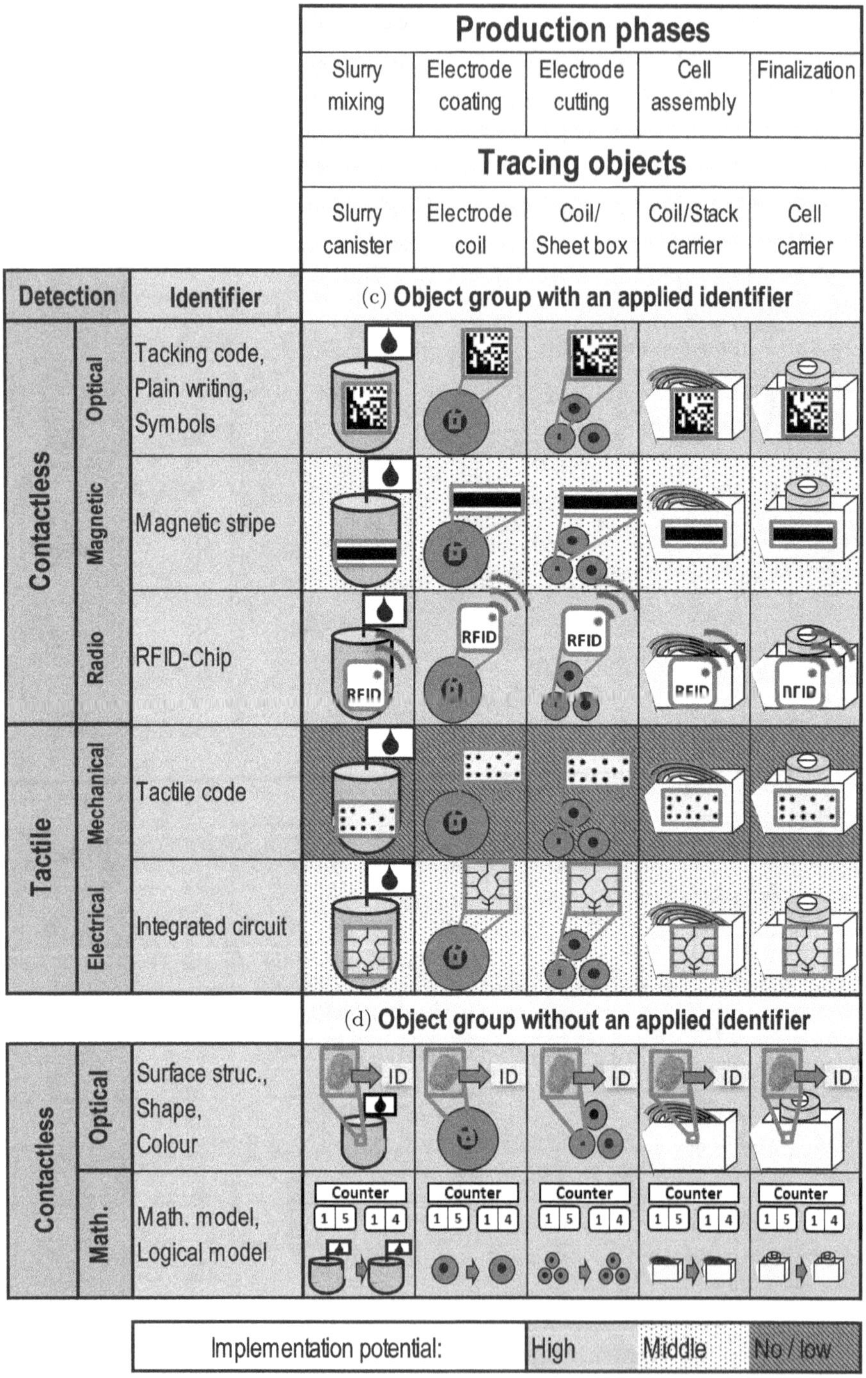

Fig. 7. Morphological analysis: Solution combinations for object groups (c) with applied identifiers and (d) without applied identifiers.[28]

Besides optical systems, magnet stripes could also be used for identification. Furthermore, wireless approaches, like using active or passive wireless modules RFID, Bluetooth or WLAN, are a possibility to transmit information about trace objects.

Tactile methods include approaches like readers which read electronic circuits, e.g. within chip cards. Moreover, trace objects can also be tracked without having an applied identifier by its inherent features. This can be performed optically by measuring surface characteristics or roughness.[21,33,34] Another possibility is to calculate its identity by physical-logical models.[19] An example can be the calculation of the localization of a foil section based on its process parameters.[24,35]

By matching identification techniques with process clusters and their tracing objects of lithium-ion battery production, a solution space can be defined. Figure 6 shows possibilities for single object tracing, and in Fig. 7 solutions for tracing object groups are developed and evaluated.

Each possible solution is evaluated by its technical, process-specific requirements and costs. The result of the evaluation is as follows:

Solutions with a high potential for implementation are illustrated in patternless cells, whereas combinations in hatched cells do not show a potential for successful implementation. The dotted cells show solutions which are either technically less feasible or whose development state is at a low technology readiness level.

5. Adopting the Digital Shadow for the Concept of Traceability

To ensure consistent traceability of individual cells throughout the entire production process, different digital images (twins) must be linked together as required. Currently, single images are acquired along the production process, each of which refers to a specific trace object. These include images of individual batches of paste, electrode tape, batches of bulk material and individual battery cells. By linking and processing these images for different information users, the collected data can be leveraged for optimization and analytical tasks.

In the following, the four sub-functionalities of the Digital Shadow are discussed with respect to the concept of full traceability in battery cell production (see Fig. 8). The discussion is based on two application examples, which are briefly presented below. In the first application case, a machine operator is considered. This operator needs real-time data on various process parameters with a high granularity in order to set up machine parameters in line with current production line goals (e.g. high quality or high throughput). In addition to the process data of individual trace objects, environmental data are important for his work. For example, high humidity can have a negative influence on the drying process of the coils.

The second application case involves a product engineer who uses process and scrap data to create an improved product design. The data he needs do not have to be real time and should be provided to him in a consolidated form. In addition to the process data of the trace objects, usage data such as user charging behavior and corresponding cell lifetime can positively influence his work.

(1) **The Digital Shadow Links Information Supply and Demand:** The concept of full traceability is well defined, as is the amount of information resulting from the production process (information offer). In particular, it includes various digital images of individual trace objects. In order to further optimize the information offer for different user groups, the collected database must be prepared. New, i.e. semantically new, information can be created by data fusion and data analysis (see Fig. 8). To make this process target-oriented, it is necessary to describe the information needs of the users in detail. The specific information needs of different employees or functions depend on their different tasks. Only if the purpose of the information use is clearly defined can a need-oriented information offer be produced. As described above, two use case examples are given: The machine operator needs process data and associated parameter settings as well as environmental conditions. These data should be made available to him. The product engineer needs process data (images of trace objects), product usage data and ideally information on the materials used. Once the specific information requirements have been determined, they and the extended information offer including data such as process, environmental and product usage data can be described semantically and thus linked to each other.

(2) **The Digital Shadow Controls the Flow of Operationally Relevant Information:** To fulfill this functionality, it is also necessary to specify the information needs of different employees. In this case, however, the right time for the provision of information bundles and their specific composition must be defined in particular. Information can be made available by using different control logics. In computer science, these are known under the term interprocess communication. Information bundles on demand are usually required in exceptional situations, such as a machine downtime (often occurs, for example, when winding or stacking the electrodes) or an unusually high scrap rate. These situations are exceptional situations for which a standardized information flow control can hardly be designed.

The situation is different in the two use cases considered. Here, the need for a specific information bundle usually occurs when the required information changes or at a certain point in time. The machine operator only has to become

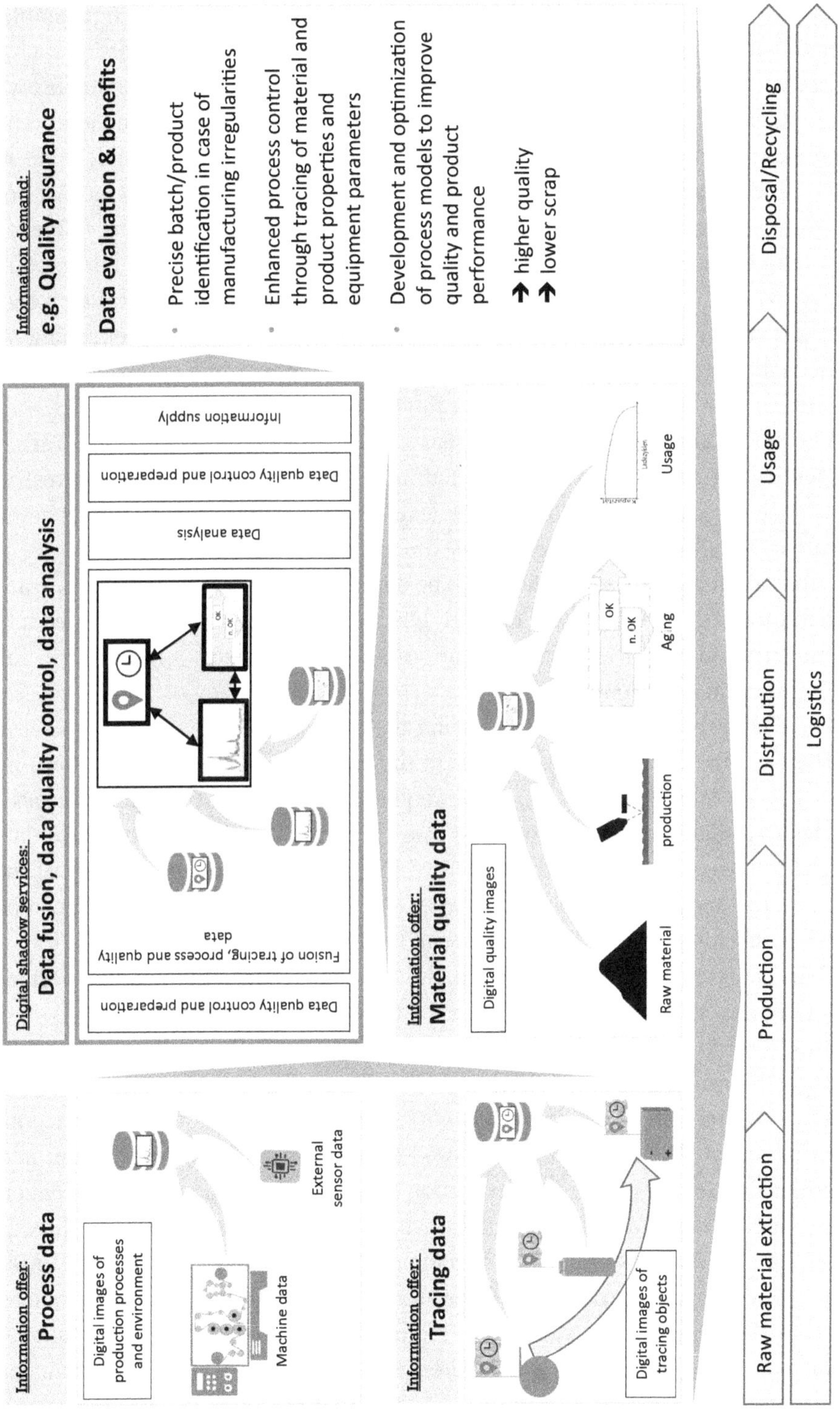

Fig. 8. Overview — concepts for full traceability — establishing the foundation for smart manufacturing.

active if something changes in the process or in the environment. For example, he only needs information on humidity when a critical change occurs. If such a case occurs, he can also use the collected near-real-time process data to observe how a change in the environment affects the product to be manufactured. The product engineer needs bundles of information at different, describable times in the product development process, for example, individual cell usage data of existing products for defining the product design goals and later the influence of temperature on cells for designing the right cell housing. The information flow control for a bundle of information, which is needed in case of a change or at a certain point in time, can be described and implemented. Depending on the right time for the provision of a bundle of information, a manual, semi-automatic or automated information flow control can be designed.

(3) **The Digital Shadow Guarantees Application-Specific Information Quality:** The quality of the collected information is of particular interest in this use case. Granularity, accuracy, level of detail, timeliness, completeness, comprehensibility and accessibility are only some of the 15 dimensions of information quality (see[36]) that must be determined for smooth use of the trace data of a cell. Based on this detailed concept, services can be developed and implemented which evaluate the information quality and improve it in the long term in an automated way.
The challenge can be illustrated using the example of the machine operator. The machine operator needs access to the data at his workplace. If he has no access, he must first search for the required information or organize access. The up-to-dateness of the data must also be guaranteed, because outdated data lead to errors in machine operation, which in turn reduces the quality of the cells. The same applies if the data available to him are incorrect. Both can be ruled out, for example, by an automated plausibility check.
With the results of the services which check the information quality, various actions can be performed. In order not to force any errors, an enterprise can decide not to pass on information with insufficient quality or to send the result of the information quality check as meta-information.

(4) **The Digital Shadow Independently Optimizes the Used Data and Information Base:** The database used by the Digital Shadow must be optimized frequently. Besides the extension of the database, the deletion of data is also a challenge. Whether and if so, which data are deleted depends directly on the IT strategy and data management of the company in question. In the area of battery cell production, a very cautious approach should be taken here, as the data-based knowledge about the production process and the product as such is not yet mature. Data therefore have a high value. In particular,

questions regarding the influence of environmental parameters on the process and product are still partly unresolved (e.g. which electrode materials require a dew point environment of –60°C in the cell assembly and what influence does a fluctuation of these environmental conditions have on the quality of the cells?). Such relationships can be explained with the help of trace data and environmental data. Nevertheless, it has to be decided whether data are archived in pure form or in processed and aggregated form. In the exclusive case of the machine operator, for example, all data that are no longer current could be deleted. The aggregated datasets can only be used for training new machine operators. However, since the data are not exclusively used by machine operators (e.g. by engineers evolving the manufacturing process), deleting the data is not recommended.

One way to expand the information base used by the Digital Shadow is to link current and historical data. Another source of information is the environmental data mentioned above (room temperature, humidity, air pressure), which can have a considerable influence on the quality of the battery cells. For the case of the product developer, it is also useful to add product usage data (e.g. product lifetime) or material data.

6. Implementation Concepts of Identification Techniques

Within this chapter, potential implementation concepts of identification techniques for tracking and tracing are described and evaluated for battery production process clusters. The basis for the implementation concepts is an exemplary coin cell production line consisting of the following process steps: mixing, coating, drying, calendaring, cutting or slitting, cell conductor welding, winding, encasing, welding, electrolyte induction, closing, formation and packaging. For each main process cluster, a possible implementation concept is presented. This concept was developed in cooperation with industry experts and is built up on the morphological analysis and presented research findings illustrated in Section 4.2.

6.1. *Cluster 1: Slurry mixing*

There are tracing concepts that are well tested and can be implemented in a cost-effective manner. For slurry mixing, the identification of a batch container by RFID tags or other n-dimensional tracking codes is a valid solution. Other alternatives do not generate any additional benefits or exceed costs limits.

6.2. *Cluster 2 & 3: Electrode production — Coating and cutting (coil and sub-coil sections)*

Within current production lines, information gaps are occurring between the different batch structures and processes of electrode production (coating, drying and calendaring), electrode foil slitting and the subsequent processes of cell assembly and finalization. To address this information gap, two basic solution alternatives for identification at the cell level have been identified as follows:

- Application of printed or laser-marked n-dimensional tracking codes to the uncoated intermediate areas or sections of the electrode foil — Here, altering of the battery cell or contamination with substances, modifying the cell function and cell, has to be excluded. Thus, further tests and long-term studies are to be carried out in order to avoid the effects of the marker process on cell quality. Possible problems with the cell arrester contact or a possible particle entry should be prevented.
- Equipment of (partial) coils with n-dimensional tracking codes in combination with calculation and localization of foil sections and the assignment of parameters by the use of mechanical-mathematical models — for example, belt speed can be used to localize electrode foil sections and to connect the data generated in the process. As no tracking markers are applied, the surface of the electrode remains unaffected. Besides, further detection and documentation of the removed sections are required for clear traceability as faulty strip sections are removed from the electrode foil.

6.2.1. *Traceability-based data matching*

Based on the possibilities for marking all process steps of battery production described in the previous chapter, it is possible to identify the individual components of the round cells at any time. However, since the possibility of identification alone does not allow any conclusion to be drawn about a process, this section describes the integration and data matching using the coating process as an example.

The sensors installed in the process are dimensioned starting from the DMC printer or sensor (Fig. 9). By means of the belt speed and the distance to the DMC printer/sensor, the time of the operation can thus be assigned to a specific DMC. It is important to note that the operations have a defined length. For example, temperature measurement during drying of the entire drying section plus the entry and exit of the electrode section is relevant.

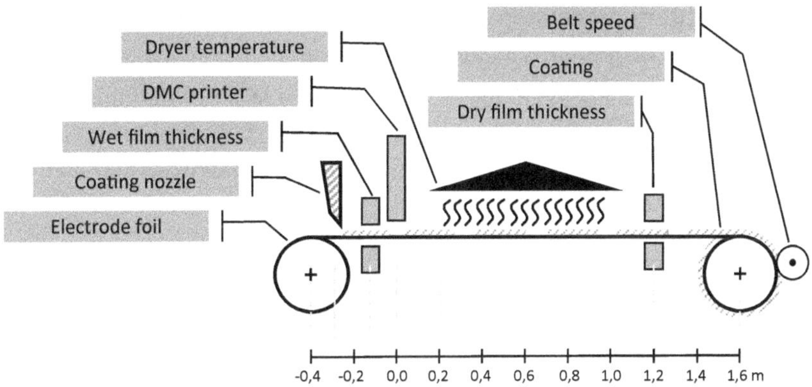

Fig. 9. Reference for data matching in coating process.

As a result, the relevant start and end time stamps of the individual sensors and processes can be calculated for each individual belt section using the following formulas:

$$\text{TS Process start} = \text{TSDMC} + \frac{\text{DDMC}}{\text{BSTI}}$$

$$\text{TS Process end} = \text{TS Process start} + \frac{\text{PL} + \text{IL}}{\text{BSTI}}$$

TS: Time stamp
BSTI: Belt speed in time interval
DDMC: Distance to data matrix code printer
TSDMC: Time stamp of DMC printer
PL: Process length
IL: Intermittent length (Length of one coating section).

When defining the spacing, it must also be considered whether the DMC is printed at the beginning or the end of the tape section.

The data can be matched without gaps over a precisely determined time interval.

6.3. *Cluster 4: Cell assembly*

The transfer of cell information or cell identification markers from continuous electrode foil production to the discrete manufacturing of single cells is one challenge for battery production. Thus, information of the electrode foil must be linked to the case of the individual cell to guarantee the traceability of the individual components and process steps from the start of production to the finished battery cell. A minimum-sized tracking code on the battery cell is currently the best solution, due to the limited space on the cell surface, cheap application methods and sophisticated technology available.

6.4. *Cluster 5: Finalization*

For battery finalization, the identification of individual battery cell cases with tracking codes is already a state of the art. Tracking codes, such as the DataMatrix code, are also used in subsequent phases of the product life cycle. The information link to earlier production processes should be addressed via continuous implementation of traceability and integration of the described tracking possibilities. The individual tracking and tracing solutions from the process clusters have to be combined to form a consistent implementation and information link in order to be able to trace individual components and process steps of the final battery as well as to connect information from previous processes at the cell level.

7. Conclusion

Concepts for full traceability and integration approaches concerning the Digital Shadow provide a basis for Smart Manufacturing of lithium-ion batteries. Thus, the theoretical concept of the Digital Shadow has been presented including its different components. As a prerequisite for the Digital Shadow, a traceability system needs to be applied which can fulfill the core functionalities. Regarding the state of the art in the field of battery production, no consistent solution for traceability, but only individual approaches for some production steps, has been implemented so far.

The main challenges are process-related changes in the batch structures, fast process speeds and the necessity to avoid an influence on the product quality. Possible traceability solutions have been described within a morphological evaluation for each process cluster and trace object. Moreover, not only the technological challenges concerning implementation have been introduced, but also advantages, like the reduction of costs, by using a traceability system in lithium-ion battery production.

As an example, how a traceability system can be used as the basis for the Digital Shadow, the integration of a traceability system and data matching has been presented. The implementation solutions have to be combined to form a consistent implementation and information link within the framework of the Digital Shadow, in order to be able to trace individual components and process steps of the final battery as well as to connect information from previous processes. From a technical point of view, the semantic description and automated linking of information supply and demand (sub-function 1 of the Digital Shadow framework) are already possible in many places, whereby the fourth development stage of this sub-functionality, which involves the trading of trace data between information consumer (demand-side) and information supplier (supply-side), is almost unexplored in the industrial environment. In order to be able to realize this trading of trace data, the effort for the acquisition, generation, archiving and distribution of information must become assessable.

The automatic optimization of some quality defects in information is already possible today. For example, incomplete time series in the measurement of viscosity can be filled with the help of various algorithms. For other information quality dimensions, this is not yet possible today and requires further research and development. From a technical point of view, an information system is able to make suggestions for the deletion of individual data records, which are based, for example, on the retrieval rate of the data. However, the development of a "deletion strategy" and the derivation of machine-processable rules are not yet common practice. Further, basic research in the field of information systems and information and communication technology is required for the automated and independent expansion of the information base used by the Digital Shadow.

References

1. Roy D, Mittag P, Baumeister M. Industrie 4.0–Einfluss der Digitalisierung auf die fünf Lean-Prinzipien. Schlank vs. Intelligent. Productivity Manage. 2015;20:27–30.
2. Knauer D. Act Big–Neue Ansätze für das Informationsmanagement: Informationsstrategie im Zeitalter von Big Data und digitaler Transformation. Wiesbaden: Springer Fachmedien Wiesbaden; 2015.
3. Bauernhansl T, Krüger J, Reinhart G, Schuh G. WGP–Standpunkt Industrie 4.0. [cited 2017 Mar 20]. Available from: https://wgp.de/wp-content/uploads/WGP-Standpunkt_Industrie_4-0.pdf
4. Schuh G, Walendzik P, Luckert M, Birkmeier M, *et al.* Keine Industrie 4.0 ohne den Digitalen Schatten: Wie Unternehmen die notwendige Datenbasis schaffen. ZWF Zeitschrift für wirtschaftlichen Fabrikbetrieb. 2016;111:745–748.
5. Wöhrle T. Lithium-Ionen-Zelle. In: Handbuch Lithium-Ionen-Batterien. Berlin, Heidelberg: Springer Berlin Heidelberg, Reiner Korthauer; 2013, pp. 107–17.
6. Maiser E. *et al.* Roadmap Batterie-Produktionsmittel 2030, VDMA Verlag GmbH, Frankfurt am Main, 2016.
7. Bauernhansl T, Hartleif S, Felix T. Der Digitale Schatten: Gestaltung eines Informationssystems für die Informationsversorgung in wertschöpfenden Systemen. wt Werkstattstechnik online.- Düsseldorf Springer-VDI Verlag. 2018;108:132–6.

8. Bauernhansl T, Hartleif S, Felix T. The Digital Shadow of production — A concept for the effective and efficient information supply in dynamic industrial environments. Procedia CIRP. 2018;72:69–74.
9. Krcmar H. Informationsmanagement. 6th ed. Wiesbaden: Springer Gabler; 2015.
10. Augustin S. Information als Wettbewerbsfaktor: Informationslogistik–Herausforderung an das Management. Zürich: Verl. Industrielle Organisation [u.a.]; 1990.
11. Uhlemann T, Lehmann C, Steinhilper R. The digital twin: realizing the cyber-physical production system for Industry 4.0. Procedia CIRP. 2017;61, 335–340.
12. Daimler AG. Die smart factory: Die komplett vernetzte Wertschöpfungskette. [cited 2017 Nov 6]. Available from: http://media.daimler.com/marsMediaSite/de/instance/ko/Die-Smart-Factory-Die-komplett-vernetzte-Wertschoepfungskette.xhtml?oid=9905147
13. Mundo C. Digitale Werte schaffen. [cited 2017 Oct 26]; pp. 8–9. Available from: https://www.digital-manufacturing-magazin.de/cloud-fuer-manufacturing-und-engineering-digitale-werte-schaffen/
14. Väyrynen A, Salminen J. Lithium ion battery production. J Chem Thermodyn. 2012;46:80–85.
15. Turetskyy A, Thiede S, Thomitzek M, von Drachenfels N, *et al.* Toward data-driven applications in lithium-ion battery cell manufacturing. Energy Technol. 2019;12:1–11.
16. Westermeier M, Reinhart G, Zeilinger T. Method for quality parameter identification and classification in battery cell production quality planning of complex production chains for battery cells. 3rd Int Electr Drives Prod Conf (EDPC). 2013;3:308–17.
17. Heimes H, Kampker A, Lienemann C, Locke M, Offermanns C, Michaelis S, Rahimzei E. Produktionsprozess einer Lithium-Ionenbatteriezelle. [cited 2020 Jan 28]. Available from: https://www.pem.rwth-aachen.de/global/show_document.asp?id=aaaaaaaaaaoqixv
18. Wegner-Hambloch S, van Betteray K, Springob K, Editors. Rückverfolgbarkeit in der Praxis: Artikel 18 und 19 der VO (EG) Nr. 178/2002 schnell und einfach umgesetzt. 1st ed. Behr, B. Behr's Verlag GmbH & Co. KG, Hamburg; 2004.
19. Feldmann K, Schöppner V, Spur G, Editors. Handbuch Fügen, Handhaben, Montieren: Edition Handbuch der Fertigungstechnik. München: Hanser; 2014.
20. Fraunhofer IPA. Abschlussbericht DigiBattPro BW: Machbarkeitsuntersuchung für die digitalisierte Batteriezellenproduktion. 2018b.
21. Hofmann A. Track & Trace per "Fingerabdruck". [cited 2019 Jun 22]. Available from: https://www.ipm.fraunhofer.de/content/dam/ipm/de/PDFs/produktblaetter/PK/IVS/Track-trace-FINGERPRINT-de.pdf
22. Laboratoire national de métrologie et d'essais. Traceability applied to the paper and board food packaging chain. [cited 2019 Jul 17]. Available from: http://www.contactalimentaire.com/
23. Industry guideline: for the compliance of paper & board materials and articles for food contract. 2012.
24. Freier H. Immer auf der richtigen Fährte: Lückenlose Rückverfolgbarkeit von Qualitäts- und Produktionsdaten in der Prozessindustrie. In: Chemie Technik. Hüthig, Armin Scheuermann Heidelberg; 2007, pp. 26–28.
25. Kvarnström B. Traceability methods for continuous processes. Luleå. 2008.
26. Kvarnström B, Oghazi P. Methods for traceability in continuous processes–Experience from an iron ore refinement process. Miner Eng. 2008;21:720–30.
27. Thiede S, Turetskyy A, Kwade A, Kara S, *et al.* Data mining in battery production chains towards multi-criterial quality prediction. CIRP Ann. 2019;68:463–6.
28. Riexinger G, Doppler JP, Haar C, Trierweiler M, *et al.* Integration of traceability systems in battery production. Procedia CIRP. 2020;93:125–30.
29. Abramovici M, Bellalouna F, Flohr M. Durchgängige Rückverfolgbarkeit entlang der Supply-Chain. Rahmenkonzept und Anforderungen. ZWF Zeitschrift für wirtschaftlichen Fabrikbetrieb. 2007;102:367–70.

30. Günther T, Schreiner D, Metkar A, Meyer C, *et al.* Classification of calendering-induced electrode defects and their influence on subsequent processes of lithium-ion battery production. Energy Technol. 2019;5:1–9.
31. Knoche T, Surek F, Reinhart G. A process model for the electrolyte filling of lithium-ion batteries. Procedia CIRP. 2016;41:405–10.
32. Datalogic Communication Division, Strichcode-Fibel. [cited 2020 Jan 28]. Available from: https://www.logitogo.com/assets/Strichcodefibel_Datalogic_2008_deutsch.pdf
33. Hofmann A. Markierungsfreie Identifizierung und intrinsischer Fälschungsschutz in einem Erfassungsschritt: Track & Trace ohne Markierung. [cited 2020 Jan 27]. Available from: https://www.wileyindustrynews.com/topstories/automation/markierungsfreie-identifizierung-und-intrinsischer-faelschungsschutz-einem-erf
34. Bertz A. Rückverfolgen bis zur letzten Schraube. [cited 2020 Jan 27]. Available from: https://www.ipm.fraunhofer.de/content/dam/ipm/de/PDFs/produktblaetter/PK/IVS/Track-trace-jb-2015-2016.pdf
35. Tulupov ON, Moller AB, Kinzin DI, Levandovskiy SA, Ruchinskaya NA, Nalivaiko AV. Structural matrix models for long product rolling processes modeling production traceability and forming consumer properties of products. Vestnik of Nosov Magnitogorsk State Technical University, 2013;5:46–50.
36. Wang R, Strong D. Beyond accuracy: what data quality means to data consumers. J Manage Inf Syst. 1996;12:5–33.

https://doi.org/10.1142/9789811245626_0008

Chapter 7

Exemplary IT Landscape Enabling the Digitalization of the Battery Cell Assembly Process

Ozan Yesilyurt[*], Andreas Schlereth[†], Florian Maier[‡], David Brandt[§] and Michael Oberle[¶]
Fraunhofer IPA, Nobelstraße 12, Stuttgart, Germany
[*]*ozan.yesilyurt@ipa.fraunhofer.de*
[†]*andreas.schlereth@ipa.fraunhofer.de*
[‡]*florian.maier@ipa.fraunhofer.de*
[§]*david.brandt@ipa.fraunhofer.de*
[¶]*michael.oberle@ipa.fraunhofer.de*

Besides a potential for improving battery cell quality by using new materials, there is also potential for the optimization of battery manufacturing processes. The identification of a suitable process setup requires several iterations of manufacturing with alternating process parameters. In order to streamline the identification procedure, flexibility in the physical manufacturing processes, a high degree of connectivity and end-to-end integration of software services with equipment as well as sensors are required. This chapter presents an exemplary IT landscape providing flexibility required for the ramp-up of the cell assembly process. Full transparency and control of the manufacturing process are enabled by tracking and parameterizing the manufacturing process on the level of single cells. Context from the production processes is stored in a semantically understandable and analyzable form to build a solid foundation for further developments of AI-based use cases. The principles of this IT landscape can be adopted for other process steps beyond cell assembly.

1. Introduction

Today's electrical world is characterized by the use of batteries. Mobile devices as well as the transformation of the mobility sector toward electric vehicles lead to a higher global demand for battery cells.

Production of lithium-ion battery cells is a complex process due to converging and diverging material flows, single unit and batch processes, continuous and discontinuous processes.[1] Different data mining approaches[2] show promising notions on how to optimize battery manufacturing processes by utilizing data. Such data-driven approaches require a reliable and encompassing IT landscape, enabling fully software-defined control of the manufacturing process and seamless collection of data.

The well-known functional model of ISA-95[3] thereby tried to describe the different functions of an IT Landscape consisting of Enterprise-Control Systems which was realized in the industry under the names Manufacturing Execution System (MES) and Enterprise Resource Planning (ERP). This article deals with the use of this model in the context of the future battery manufacturing to show an exemplary IT Landscape based on former project results to enable cell assembly of smart battery manufacturing.

The following Section 2 describes shortly the evolvement of IT system architectures based on the automation pyramid. Section 3 presents the cell manufacturing process with a focus on cell assembly and describes the center for battery cell manufacturing. The main work results follow in Section 4, covering the approach, concept and implementation of the IT system landscape. Section 5, the last section, deals with adoption of the principles of the IT landscape for other process steps beyond cell assembly and a short conclusion.

2. Evolvement of Production IT System Architectures

The historical evolvement of IT system architectures is described in this section. First, the concept of the automation pyramid is presented. Then, the evolvement toward Industry 4.0 and Service-Oriented Architecture is illustrated.

2.1. *Automation pyramid*

As early as the 1980s, attempts were made to link parts of production IT by means of Computer Integrated Manufacturing (CIM), but this ultimately failed due to the complexity, the early status of the technology and the individuality of a few suppliers who were using their own proprietary interfaces. At the same time, however, the automation pyramid was developed, which is based on the CIM Pyramid, developed in the 70s and 80s.[4] It describes the integration of IT systems relevant to business processes with IT systems for controlling production technology and was implemented primarily by large companies in their

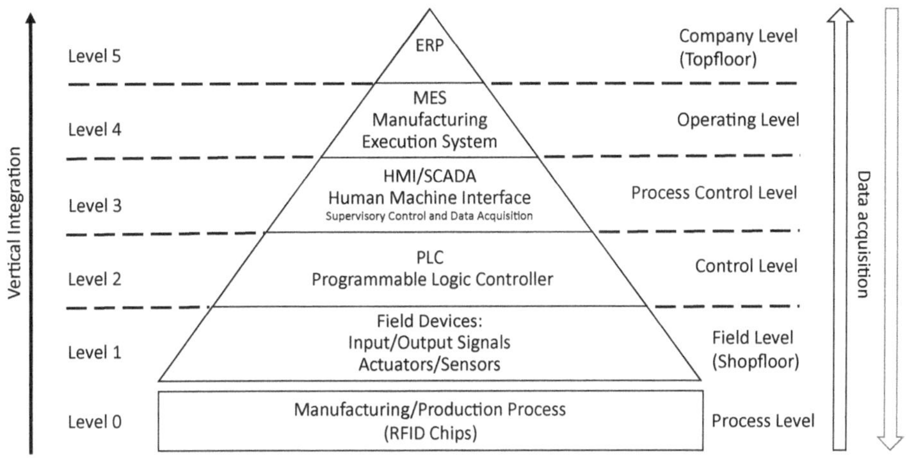

Fig. 1. Automation pyramid.[4]

production.[5] Over the years, many variants of this pyramid have evolved.[6] One of them is shown in Fig. 1 and modeled according to[4] and based on DIN EN 62264. This type of automation pyramid is a detailed example of 6 levels starting from the company level to the process level and encompasses a high number of systems and hardware.

On level 5, the enterprise resource planning system (ERP) is located with tasks such as production planning (type and volume of orders to be produced by the production program and the corresponding planning) and order processing of industrial production.[4,7] At the next lower level (level 4), the manufacturing execution system (MES) is primarily responsible for controlling, steering and monitoring production. It links machine control with the corporate level.[4] Thus, it exchanges data with the ERP system and receives data from the manufacturing level. On level 3, supervisory control and visualization of production-relevant processes take place. Human–machine interfaces allow supervisory operators to make decisions for adjusting controls. Data enabling these decisions are collected and adjustment of process parameters is transmitted to the next lower level. On this level 2, programmable logic controllers (PLCs) evaluate input signals (sensor data) and send the corresponding result data (output signals) back to the field. These electrical signals are processed by actuators and converted into mechanical movements. Level 1 encompasses the whole shop floor including sensors and actuators. The field level supplies and processes production-relevant data in the form of input and output signals. On the lowest level 0, intelligent products and radio frequency identification chips (RFID chips) provide information on product properties and production steps from the process level.[4]

The traditional automation pyramid has a long history and as a model enforces strong hierarchical structure with clear roles. However, it lacks flexibility as information always needs to be passed through several levels for enabling business processes and centralized decision-making. The vision of Industry 4.0 includes more flexible and adaptable manufacturing systems, which renders the pyramid structure too rigid, requiring a new architectural model, which is going to be presented in the following section.

2.2. *Industry 4.0 service-oriented architecture*

Industry 4.0 refers to the intelligent networking of industrial machines and processes with the help of information and communication technology.[8] One of its important features is decentralized decision-making.[9,10] With the Internet of Things (IoT), today's capabilities are much more extensive as industrial processes do not only take place in one manufacturing plant but can span multiple organizations. Therefore, some changes occur on the traditional automation pyramid toward the use of edge computing for local process control and cloud computing for higher-level applications (see Fig. 2). When the automation pyramid was introduced, the technical conditions available at the time did not allow for any other structures. Today's capabilities are much more extensive and involve IoT and information technologies such as micro services.[11]

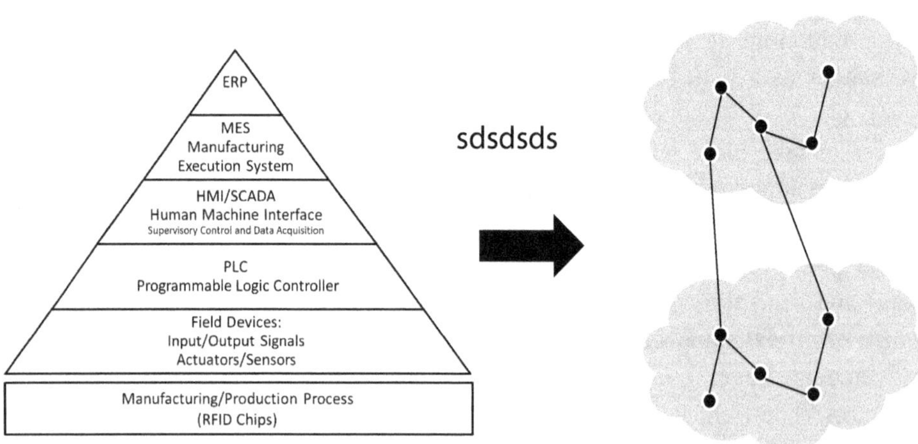

Fig. 2. Changes toward vision of Industry 4.0.

Micro services serve as an implementation variant for a service-oriented architecture (SOA), and therefore are also called micro service architecture.[12] SOA attempts to encapsulate the functionalities of an application in separate services. The micro services go even further by encapsulating single functionalities in smaller independent interconnectable services.[13] This enables engineering these micro services around business capabilities. Thus, the focus is on key capabilities of the business system and not the capabilities of predefined software packages.[12] This micro services architecture style serves as a model for the IT landscape for supporting the ramp-up of the battery cell assembly process as it incorporates one of the key requirements: flexibility.

The following section examines the cell assembly process as a part of the battery manufacturing process in general and more specifically in the context of the laboratory environment for which the IT landscape was designed and implemented.

3. Battery Cell Assembly

3.1. *Cell assembly in the general battery manufacturing process*

The production process of Lithium-Ion-Cells can be separated into three parts: the production of the electrode, cell assembly and cell finishing. The first and last ones are independent of the used cell type. With cell assembly, it is important to consider if it is about manufacturing of pouch cells, cylindrical cells or prismatic cells.[14] Since this chapter is concerned with the manufacturing of cylindrical cells, the following cell assembly description focuses on this type of battery.

The first part, electrode production, encompasses mixing of active material to produce the coating slurry. This slurry is then applied to copper and aluminum foils. A desiccation process dries the coated coils since the adhesion is stopped by a high level of moisture in the active material. A roll compaction happens afterward through a calendar to compensate expansion during the previous drying process by compressing the material.[15]

In the following assembly of the cell, a winding process is applied. The electrode coils as well as two separator coils are therefore wound around a center pin. Afterward, the resulting "jelly roll" is packed into a metal housing in the packaging process. As a final step, the electrolyte is filled into the cell under vacuum and the cell is sealed.[14]

The third and last part begins with the formation. During formation, the battery cell is charged and discharged forming the Solid Electrolyte Interphase (SEI) layer. Gas builds up during this process in pockets next to the cell. These are removed after the cells are finally sealed.[15]

3.2. *ZDB — A laboratory environment for battery manufacturing process*

The center for battery cell manufacturing (ZDB) is an industry-oriented research and development base for sustainable production and life cycle-optimized use of modern energy storage systems using the tools of digitization, modeling and simulation. For this purpose, the ZDB has a laboratory infrastructure with networked individual processes in order to manufacture small series of battery cells.[16] ZDB aims to build an IT infrastructure for digitalizing the entire battery cell manufacturing process. The main goal of the ZDB laboratory manufacturing line is to carry out process experiments. Different parameter variants need to be tested and analyzed to optimize battery cell quality and to improve the stability and efficiency of the processes itself.

Conducting these experiments requires small batches and frequent changes of recipes and machine parameters. This means that in addition to flexibility in physical manufacturing processes, a high degree of connectivity and end-to-end integration of software services with equipment as well as sensors is required. For ZDB this was addressed by a service-oriented IT system landscape that aims at increasing transparency of the manufacturing processes by providing contextual data at the cell level and thus laying the data foundation for sophisticated AI-based used cases. In order to validate this concept the IT system landscape has been implemented for a sub-part of the ZDB: the battery cell assembly process.

3.3. *The ZDB battery cell assembly sequence of operations*

The battery cell assembly process in ZDB consists of several sub-processes, which are conducted manually or automatically on different stations. The first process starts with the delivery of the jelly-rolls, lids, cups, electrolyte and arresters from different suppliers. An input control (inspection) of these materials is performed in the second process. Then, the cups are labeled and placed on trays of a smart workpiece carrier manually. These labeled cups are re-sorted with lids into conveyor systems. After that, they are transferred to a clean room manually. In this room, first the labeled cups are weighted. Then, the jelly-rolls are automatically inserted and welded into cups in a single process step. The next process is again a quality control process in which the weight of the cups with welded-on jelly-rolls is measured individually. After this process, the cups are beaded manually assisted by a special machine. The beaded battery cells are weighed again in a quality control process. Then, a transfer process occurs in which the beaded cells are discharged from the clean room. In the next process, the arrester is welded to the beaded cell in the laser room. Then, the beaded cell with the arrester is transferred to the clean room again, where the weights of the cells are measured in a quality control process. After the measurement, the cells are vacuum dried in an industrial oven. The weight of the dried cells is measured one more time. Then, the cells are moved to the glove box. In the glove box, first the cells are filled with electrolyte, then the lid is welded to the arrester of the electrolyte-filled cell in an automated process. As a last step in the glove box, the cells are sealed. The sealed cells are transported to a clean room and the weight of the cells is measured. Then, they are moved to the transfer room, where they are washed and dried. After that, the washed cells are annealed in an automatic process at another station in different lab. From this process, it is distinguished whether the cells were produced for internal purposes or for external customers. The cells for external customers are sent to delivery and the cells produced for internal research purposes are sheathed with a sleeve in the next process in the same room. The sheathed cells are automatically formatted and manually stored.

For the exemplary IT landscape implementation, cell assembly is selected and considered as a core process in battery manufacturing. In the ZDB battery cell assembly the production of the battery cells comprises 31 different process steps. The battery cells are processed in six different room environments at 13 different cell assembly workstations. To create an implementation design of the IT landscape for such a highly complex environment, the next section introduces the service-oriented IT system landscape of ZDB.

4. Service-Oriented IT System Landscape

This section describes the developed service-oriented IT system landscape for ZDB. First, the development approach of the ZDB IT system landscape is outlined, then the principal functionalities of this IT system landscape are explained with the help of the ISA-95 functionalities in the next concept section. Third, the implementation of the ZDB IT system landscape is described.

4.1. *Approach*

The development approach involved four different tasks (see Fig. 3). The configuration of the ZDB infrastructure, in terms of cloud computing resources, the development of the software services enabling the tracking of each cell, and the development of equipment emulators facilitating integration tests.

In the following each task is further detailed. The development of the ZDB services followed a three step approach (see Fig. 4).

First, the high-level workflows for battery cell assembly in ZDB are created with the help of the process experts. The created workflows describe the relevant process flow in battery cell manufacturing and are based on the standard operation procedures. The workflows consist of actors as well as different employees on the shop floor, the machines and IT systems as well as MES. Table 1 shows the

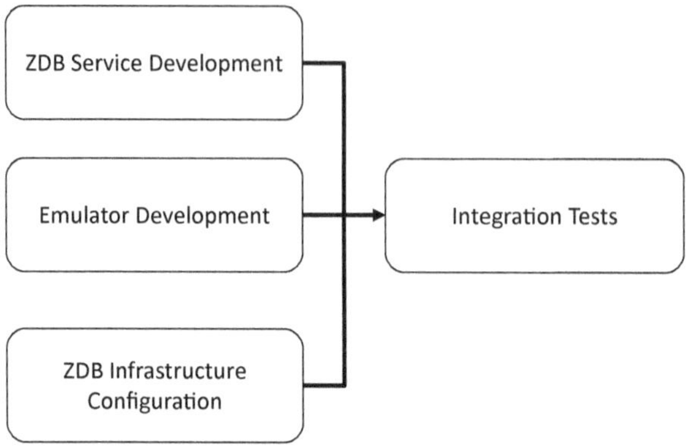

Fig. 3. Approach to develop ZDB service-oriented IT system landscape.

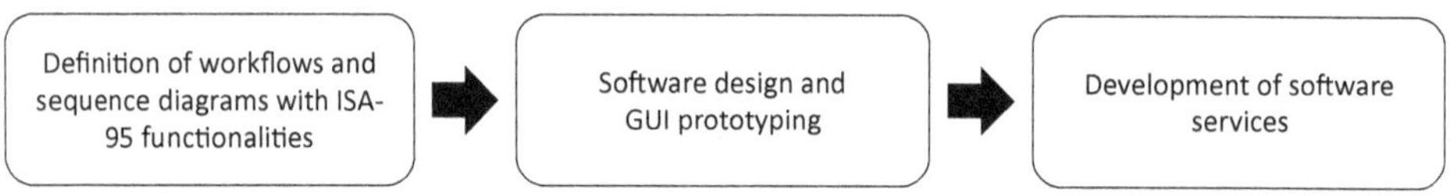

Fig. 4. ZDB service development.

Table 1. Overview of defined workflows.

Workflow Process	Automated	Semi-automated	Manual
Insert coils into the cups and weld coils	x		
Bead the cups (cells)		x	
Weld the arrester to the beaded cell		x	
Vacuum dry the cells		x	
Fill the dried cells with electrolyte		x	
Seal the cells			x
Wash the cells			x
Dry the cells			x
Anneal the cells	x		
Sheathe the cells			x

created workflows for different automated, semi-automated and manual processes. A manual process means that the worker on the shop floor should be a part of the process. The worker works actively and performs the production tasks at the equipment, such as loading the cells into the equipment, setting the recipe parameters at the equipment, starting and stopping the equipment. Only the data acquisition from the equipment proceeds automatically. In the semi-automated processes, the worker has a passive role, but the final decision. For example, he observes whether the equipment has been set with the correct recipe parameters by MES, and can adjust the equipment parameters, before starting the equipment. In these processes, there is harmonized cooperation with the worker and MES. Four processes in the ZDB battery cell assembly are semi-automated. In the automated processes, no interaction with a worker is required.

With the help of these workflows, it was possible to establish greater knowledge of the process itself that helps to determine and analyze which ISA-95 functionalities are required enabling a single cell product tracking with contextual process data.

The ISA-95 is an international standard for the integration of enterprise and control management levels issued by the ISA. It consists of five parts. The first part defines the standard terminology and object models for exchanging information, which can be positioned between level 4 and level 3 of the automation pyramid. The second part presents object model attributes, which are defined by Part 1. In Part 3, four organizational units, production, maintenance, quality and inventory of enterprises, are described. Part 4 defines the object models that determine which information is exchanged between these organizational units. The fifth part standardizes exchanging information structure between the first two higher levels.[3]

In this section, a new IT-landscape concept is designed with consideration of the ISA-95 functionalities of a manufacturing enterprise to implement an exemplary IT-landscape for smart battery cell assembly. Enterprise functions defined by ISA-95 part 1[3] are as follows:

- Order Processing
- Production Scheduling
- Production Control
- Material and Energy Control
- Procurement
- Quality Assurance
- Product Inventory Control
- Product Cost Accounting
- Product Shipping Admin
- Maintenance Management.

According to the created workflows, it is determined that only production control, quality assurance and order processing functions are relevant for the procedures carried out in ZDB. After these workflows are created and the chosen ISA-95 functionalities are determined, five generic sequence diagrams are generated with process experts to find out which software services are required. These sequence diagrams are designed for the following:

- Production order creation
- Manual single-piece production scenario
- Automated single-piece production scenario
- Manual batch production scenario
- Automated batch production scenario.

Then, the software designs for the required software services are drafted and a graphical user interface (GUI) prototype is developed. Based on these findings, the software services are finally developed.

In parallel to the above-described works equipment emulators were developed. Equipment emulators mimic the communication behavior of an equipment. The aim of using emulators is to evaluate performance, robustness and system behavior in advance. The emulators generate rudimentary machine behavior according to state machine, virtual sensors and communication behavior. This behavior is configurable via scripts and mathematical distribution functions. To develop emulators, two modules are required (see Fig. 5). One module is the OPC-UA server that enables the principal communication with other systems by exposing variables that can be read or written. These variables represent input (e.g. dosage amounts) or output (e.g. coil thickness) of the manufacturing process or process parameters that are continuously measured during the manufacturing process (e.g. forces). In order to identify the suitable variable structure (information model) an input/output information analysis is conducted with process experts for all battery cell assembly sub-processes. Based on the results of this analysis a generic schema grouping variables to the same structure (e.g. all input variables are embedded in the container "input") was created in order to construct a consistent interface across all items of equipment. With the help of the OPC-UA

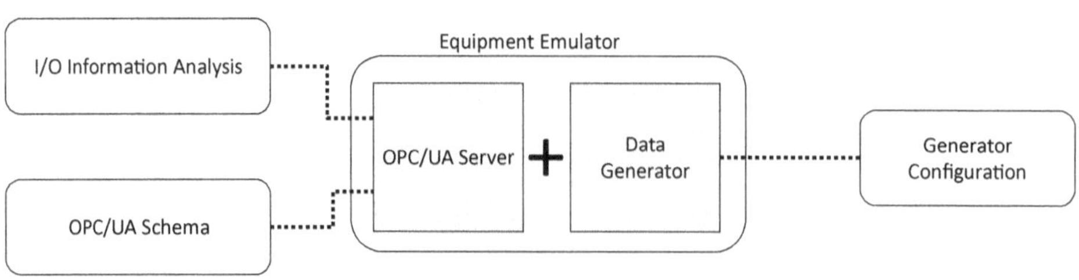

Fig. 5. Equipment emulator development.

schema concept the OPC-UA server of an emulator is configured. The second module of the equipment emulator is a data generator. The data generator is responsible for generating data mimicking the data from a real running equipment. Based on the input from process experts on the expected data distribution of the individual battery cell assembly sub-processes mathematical distributions are configured for each variable of the information model to mimic the behavior of real equipment.

The third task, configuration of the ZDB IT infrastructure, aimed at providing a stable, performance and secure environment for deploying and executing the developed software services and equipment emulators. To accomplish this, the emulators and services should be hosted in a cloud platform. This requirement results from the nature of the ZDB: In the ZDB battery manufacturing processes are not located in same physical room - not even in the same building. Hence, it is necessary that software services supporting the manufacturing processes are accessible from different locations. In addition, experiments carried out on the line should be able to leverage AI technology with highly varying computing demands that cloud technology can more easily cater for than traditional computing.[18] For this purpose Fraunhofer IPAs own cloud platform, the Virtual Fort Knox, was leveraged. This platform allows the flexible provisioning of computing resources in terms of virtual machines and allows the secure segmentation of logical resource segments. After the successful deployment of equipment emulators and software services on the cloud platform, the correct communication is verified through integration tests. For this purpose individual scenarios covering all required user needs were setup and tested.

4.2. *Concept*

ISA-95 provides an established framework for identifying and structuring functions that support manufacturing operations. This framework is leveraged as a foundation for selecting the necessary functionalities for ZDB. As the ZDB is rather laboratory line and its goal is not high-volume production it needed to be made sure that the ISA-95 functions are aligned with the specific ZDB requirements. For this purpose requirements were elicited through interviews with the ZDB process experts. The requirements can be summarized as following:

- track the delivered battery cells individually (in small batches) to enable product transparency and reduce the scrap probability,
- create a product with different battery cell processes (process plan),

- give researchers the opportunity to conduct various experiments with the battery cells,
- support the worker in manual production processes, in which order of the battery cells should be processed,
- support the worker on the shop floor, so that the worker receives the information with which process parameters the equipment should be setup for the selected battery cell,
- collect data in real time from equipment, which support different communication protocols,
- store the context-related data from the production processes in a semantically understandable and analyzable form in real time,
- provide data exchange between different software services and equipment,
- support running and hosting of software services in a secure environment,
- provide the deployment of the software services so that they can be quickly put into operation in the case of code adjustments.

These elicited requirements can be mapped to three ISA-95 enterprise functions: production control, quality assurance and order processing. These ISA-95 functions can be thus used as a guideline for the implementation of the IT system landscape for ZDB. Further guidance could be provided by structuring the roles working in the ZDB and their associated tasks. These provide the frame of specific support needs in the daily work at ZDB. These roles are presented in the following.

Second, workers' roles are described, which are created to divide the production tasks and ZDB functionalities into different working groups. In ZDB, three workers' roles are created:

- Production planner: The production planner associates the jelly-rolls delivered by ZDB suppliers with the workpiece carriers and the experiments created by the engineer so that the battery cell information can be mapped to experiment and workpiece carrier information. This mapping is key to the context-related data collection from the production processes in ZDB.
- Engineer: The engineer creates new production orders and designs new experiments with different variants and process parameters. With the help of this worker's role, the supplied jelly-rolls are digitally represented in ZDB so that the production of the battery cells can start with the jelly-rolls.
- Operator: The operator works on the shop floor and operates the equipment. His task is to set the equipment with the correct process parameters for the selected battery cell.

The next section describes the implementation of the software services, equipment emulators and equipment integration to fulfill the ZDB functional requirements described above.

4.3. *Implementation*

The design of the IT system landscape for ZDB employs the micro-services architecture style. In order to implement the landscape two steps are required. In the first step all software services are implemented. In the second step these services need to linked with other as each service fulfills a particular role.

The first step is divided into two different levels. In the first level, the software systems are introduced, which are involved in production planning and order processing. The focus in the second level lies on the production execution and data collection in battery cell assembly processes. In this level, the quality assurance function of ISA-95 is considered. Additionally, in both levels, one can assign which functionalities of the software services are used by manufacturing workers' roles.

The second step focuses on the IT technology that enables the interaction and data exchange of the micro-services developed in step one and their deployment concept.

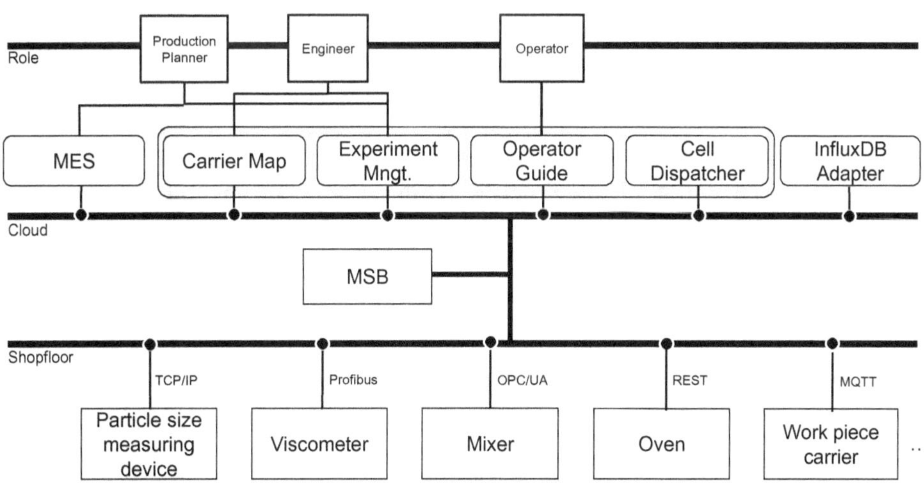

Fig. 6. Implementation of the exemplary IT landscape.

4.3.1. *Micro-services supporting ZDB operations*

Figure 6 illustrates the architecture of the ZDB IT system landscape. It is divided into two levels (cloud and shop floor). On the shop floor level, the equipment is presented, which is connected to the cloud platform via various communication protocols. On the contrary, the four main software services, MES and small-scale software services are depicted at the cloud level.

4.3.1.1. Production planning

MES, two main services such as carrier map and experiment management, and small-scale software services, which are developed for production planning and order processing of ISA 95 functions, are described in this level and it is explained what kind of a role they have in the IT landscape and which workers use these software services.

The MES, which is used in the ZDB for battery cell assembly, supports three ISA-95 functions. These are production control, quality assurance and order processing functions. It is used to create a production plan for battery cells, production orders and tracking the created production orders during battery cell assembly. To create a production plan, the production planner requires access to the MES. The MES also allows to assess the quality of each product after each process step (i.e. scrap or ok). In order to enable the interaction of the MES with other software services, an **MES adapter service** is required. With the help of this service, production orders are first created automatically in the MES. It is also used for cell-specific tracking of battery cells. The service receives process start-and-stop messages from the MES per battery cell and forwards this information to other services. The MES adapter service works as a bridge between the MES and software services. It enables the digital transformation of the automation pyramid to the SOA, so that every MES functionality can be invoked by other software services directly. Thus, not all functionality required by ZDB has to be provided by the MES itself. Tailored software services can access data from the MES and implement additional functionalities.

In order to track the battery cells individually, the battery cells need a unique ID. The delivered jelly-rolls have a serial number, which can be used to track the battery cells via MES. Therefore, new production orders should be created in MES for the delivered jelly-rolls. The engineer sees the delivered jelly-rolls and creates new production orders for them in MES using a **carrier map software service**. It enables one to create new production orders in the MES with a unique ID for each battery cell and communicates with the MES adapter service, so that new production orders for battery cells are created and mapped with the serial numbers of delivered jelly-rolls. Additionally, the carrier map maps the order information and battery cell ID to the workpiece carrier information so that the position of the battery cell on the workpiece carrier is firmly defined. The workpiece carrier consists of 25 slots and is used to deliver battery cells to various working stations in the battery cell assembly process. This mapping enables straightforward tracking and processing of a battery cell during the manufacturing process increasing production efficiency and reducing scrap probability.

The ZDB is built to support development and research activities in battery cell manufacturing. As the goal of the ZDB is to explore new possibilities improving the product quality and/or manufacturing process stability, frequently machine parameters or the battery cell materials are changed. Therefore, **experiment management** is developed to give researchers the opportunity to conduct various experiments with the battery cells. The experiment management software service consists of two parts. The first part is operated by a production planner. Here, the experiments and the variants are created. Then, the process parameters for a single battery cell per production process are defined for the selected variant. The second part is used by an engineer. He assigns experiment variants to single cells on the workpiece carrier so that both information sources are mapped together. It enables the ZDB researcher to evaluate the results after the battery cell manufacturing. To assign the battery cells to experiments in experiment management, data from carrier map and experiment management are required. **Cell assignment** micro-service was developed to link the two different datasets for the cell assignment.

4.3.1.2. Production execution and data collection

In the following the services for addressing the functionalities required for production execution and data collection are explained and their role in the IT landscape is examined.

The **cell dispatcher** handles the communication between the workpiece carrier and the MES. It supports the worker in manual production processes and decides which battery cell should be processed next on the workpiece carrier and sends this information on to the workpiece carrier. The workpiece carrier finds the correct battery cell position and turns on the LED with the correct battery cell position so the operator can pick the correct battery cell. In the next step, the battery cell is picked and the cell dispatcher triggers a process start with the ID of the picked battery cell in the MES. After the battery cell processing at the current station is complete and the battery cell is placed back in its location in the workpiece carrier, an automatic process stop message is sent from the Cell dispatcher to the MES.

The cell dispatcher decides which battery cell should be processed next. After this step, the recipe for this battery cell should be displayed and sent to the equipment. In manual manufacturing processes, the operator has a task to set up equipment with the right recipe. In the ZDB, recipes are often adjusted and overwritten to try out new battery manufacturing process parameters. If there is no digital assistance, it is difficult for the operator to find the right recipe for the current production order. Therefore, the operator requires support on the shop floor. The **operator guide** software service supports the operator on the shop floor level so that the operator receives the information of which process parameters should be used for the equipment setup for the selected battery cell. The process parameters for the selected battery cell and the correct equipment are displayed on a user interface.

After the equipment is set up with the right recipe, it begins to work and generate data in real time. Reading data from the equipment is still a big challenge despite ongoing standardization work in the field of equipment communication. Therefore, **different equipment adapter services** are developed to communicate with the equipment via different communication protocols such as Profibus, MQTT, REST, Websocket and OPC-UA. These services have a major task to read equipment data and send the data to another software service. For equipment which supports the OPC-UA communication protocol, a generic **OPC-UA client** is developed. The OPC-UA client communicates with the OPC-UA servers to other services such as the data collector service.

It is another challenge in manufacturing to enable context-related data storage from the production processes in a semantically understandable and analyzable form. To collect data from the equipment software services in real time with a timestamp and map the data with the experiment and workpiece carrier data, a data collector service was implemented and instantiated for each equipment. The data collector service gathers only process-relevant data from the equipment and maps these data to experiment and carrier data of the processed battery cells. After that, the mapped data are sent to InfluxDB adapter service.

The **InfluxDB adapter** service allows that the equipment data, which are mapped to the experiment data and carrier data, are stored in the Influx database as a time-series. This provides a valuable digital representation (dataset) of battery production processes for the further development of AI-based use cases and evaluation purposes.

4.3.2. *Communication and hosting of the software services*

The previous section describes the functionalities and tasks of various software services and equipment in the ZDB. These services and equipment should be able to communicate together and exchange their data. To enable this integration between the services and equipment, the manufacturing service bus (MSB) is used.

The MSB enables the software systems to be interconnected via a bus system, reducing the number of connections required and ensuring that communication between systems is configurable and transparent. With the MSB, the services can be connected using so-called integration flows. The MSB provides a generic self-description and data transformation. Therefore, no use case-specific data model description is needed for the services.

The whole service integration in the MSB can be performed on a GUI, and with drag and drop technology, facilitating the linking of all services. All developed software services include an MSB connector so that they can communicate with each other. To define the required integration flows for connecting the software services, five different sequence diagrams representing the occuring interactions are modeled. With the help of the described sequence diagrams, it is determined that various integration flows are required to enable the implementation concept described above. Table 2 shows the sequence diagrams with the number of corresponding integration flows.

As described in Fig. 6, the software services are hosted on a cloud platform.

The deployment concept of the ZDB landscape is based on the usage of virtual machines (VMs). The VMs are software programs that virtualize a computer hardware system and behave as a real computer. The advantages of using VMs are described as follows: VMs "are now seen as cost-effective techniques for organizing computer systems resources to provide extraordinary system flexibility and support for certain unique applications".[18] Figure 7 displays the deployment concept of different services in the cloud platform of ZDB. The services are divided

Table 2. Sequence diagrams with corresponding integration flows.

Sequence Diagram	Integration Flow Number
Production order creation	3
Manual single-piece production scenario	29
Automated single-piece production scenario	29
Manual batch production scenario	24
Automated batch production scenario	24

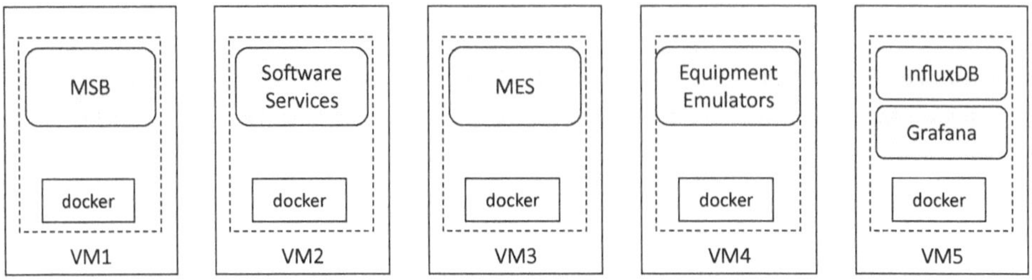

Fig. 7. Implementation concept of different services in the cloud platform.

into five different VMs according to their tasks in the ZDB. The separation of services in different VMs provides flexible and fast integration of services into the proper setup in the ZDB. It also provides the categorization of instances that have different tasks in the ZDB. With the help of this VM separation, IT admins can quickly find and select the right VM to perform their tasks in case of problems and maintenance tasks. VM separation also has the advantage in the ZDB that development and testing of services can continue while, for example, another VM is being configured by an IT admin. Both works can run in parallel as they do not disturb each other.

In order to further reduce deployment complexity and to waste less time on bug fixing container virtualization is also applied for ZDB by leveraging Docker. Docker is an open-source software project and automates the deployment of software services as portable, self-sufficient containers, which can be hosted in a cloud platform or on-premises.[19] Running via Docker container enables the software services to be quickly put into operation in the case of code adjustments and a consistent and isolated environment from other software services so that independent of the location where the software service is deployed, all stays consistent, resulting in tremendous productivity: less time troubleshooting and more time rolling out new features. As for ZDB the presented micro-services are implemented and maintained by different developers, these advantages can be fully exploited. New releases of the micro-services will be even more frequent once AI-based services are explored.

5. Discussion

The IT system landscape was tailored for the battery cell assembly process. However, the ZDB is going to consist of the entire battery cell manufacturing process chain. The question that needs to be discussed is if it could be adapted to the battery electrode manufacturing and battery cell finishing processes. This suitability and the limitations of the presented IT system landscape are discussed in the following.

The developed IT system landscape aims to enable battery cell tracking in small batches in battery cell assembly processes. In electrode manufacturing, it is observed that the characteristic of the processed material along the process chain changes. First, a slurry (liquid) is created, then the films are coated from different batches of slurry to produce a coated roll for the anode and cathode. These product changes increase the complexity level of product tracking in production. Therefore, the developed IT system landscape cannot be transferred one to one for these electrode manufacturing processes. But, some functionalities described in the concept section can still be accomplished for electrode manufacturing using few software services of the developed ZDB IT system landscape. For example, operator support on the shop floor with a visualization tool, planning experiments for research purposes or storing the context-related data from the production processes in a semantically understandable and analyzable form in real time can be applied with minor changes to existing software systems. But, for product tracking, a new IT system landscape concept should be designed and MES features should be investigated with the new concept's requirements. Additionally, new software services for product tracking should be developed. In battery cell, the cells are stored in a warehouse, where they are regularly tested while they are charged and discharged. The battery cell is fully assembled and its shape and handling characteristic does not change any more. Only the quality of the cells is checked during the charging and discharging processes. Therefore, the entire IT system landscape of the ZDB for cell assembly can be transferred and used for these process steps with minor adjustments of the software services.

6. Summary and conclusion

This chapter introduces a service-oriented IT system landscape, which supports battery manufacturing in ZDB. First, the historical evolvement of IT system architectures is described and the Industry 4.0 service-oriented architecture is introduced as a model for an IT system landscape. Then, battery cell assembly — a core process of battery manufacturing — is explained. After that, the center for battery cell manufacturing, ZDB, is introduced. As for the next sections, the approach and the ISA-95 aligned concept of the service-oriented IT system landscape in the ZDB are described. Lastly, the implementation of the IT landscape in ZDB is explained in detail.

In conclusion, it is shown that the described service-oriented IT system landscape accomplishes the ZDB functionalities, which are defined in the concept section, establishes a solid foundation for the further development of the AI-based use cases and other analysis processes in flexible battery production. The presented IT system landscape can be also applied to battery cell finishing. However, to support electrode manufacturing major adaptations are necessary to cope with particularities of continuous processes.

References

1. Turetskyy A, Thiede S, Thomitzek M, Drachenfels N von, Pape T, Herrmann C. Toward data-driven applications in lithium-ion battery cell manufacturing. Energy Technol. 2020;8(2):1900136.
2. Schnell J, Nentwich C, Endres F, Kollenda A, Distel F, Knoche T, *et al.* Data mining in lithium-ion battery cell production. J Power Sources. 2019;413:360–6.
3. ISA. Enterprise-control system integration: part 1: models and terminology. Research Triangle Park, N.C.: ISA; 2000 [cited 2020 Apr 23]. Available from: https://www.isa.org/isa95/
4. Siepmann D. Industrie 4.0–Technologische Komponenten. In: Roth A, editor. Einführung und Umsetzung von Industrie 4.0: Grundlagen, Vorgehensmodell und Use Cases aus der Praxis. Springer Gabler, Berlin Heidelberg; 2016. pp. 47–72.
5. Bildstein A, Seidelmann J. Industrie 4.0-Readiness: Migration zur Industrie 4.0-Fertigung. In: Bauernhansl T, Hompel M, Vogel-Heuser B, editors. Industrie 4.0 in Produktion, Automatisierung und Logistik: Anwendung Technologien Migration. Springer Vieweg, Wiesbaden; 2014. pp. 581–97.
6. Meudt T, Pohl M, Metternich J. Die Automatisierungspyramide–Ein Literaturüberblick; Forschungsbericht; 2017.
7. VDI. VDI 5600 Fertigungsmanagementsysteme (Manufacturing Execution Systems — MES), Blatt 1, Verlag des Vereins Deutscher Ingenieure; 2016.
8. Bundesministerium für Bildung und Forschung. Was ist Industrie 4.0?: Menschen, Maschinen und Produkte sind direkt miteinander vernetzt: die vierte industrielle Revolution hat begonnen.; O.J. [cited 2021 Mar 25]. Available from: https://www.plattform-i40.de/IP/Navigation/DE/Industrie40/WasIndustrie40/was-ist-industrie-40.html
9. Mittal S, Khan MA, Romero D, Wuest T. Smart manufacturing: characteristics, technologies and enabling factors. Proc Inst Mech Eng, Part B J Eng Manuf. 2019;233(5):1342–61.
10. Saldivar AAF, Li Y, Wei-Neng C, Zhan Z-H, Zhang J, Chen LY. Industry 4.0 with cyber-physical integration: A design and manufacture perspective. 2015.
11. Hübner I. Quo vadis Automatisierungspyramide? 2019 [cited 2021 Mar 26]. Available from: https://www.smart-production.de/digital-factory-journal/news-detailansicht/nsctrl/detail/News/quo-vadis-automatisierungspyramide
12. Götz B, Schel D, Bauer D, Henkel C, Einberger P, Bauernhansl T. Challenges of production microservices. Procedia CIRP. 2018;67:16772.
13. Mazzara M, Meyer B. Present and ulterior software engineering. Cham: Springer; 2017. Available from: https://link.springer.com/book/10.1007/978-3-319-67425-4
14. Heimes H, Kampker A, Lienemann C, Locke M, Offermanns C, Michaelis S, *et al.* Lithium-ion battery cell production process; 2019 [cited 2021 Apr 19]. Available from: https://www.researchgate.net/publication/330902286_Lithium-ion_Battery_Cell_Production_Process
15. Kampker A, Burggräf P, Deutskens C, Heimes H, Schmidt M. Process alternatives in the battery production. Electrical Systems for Aircraft, Railway and Ship Propulsion (ESARS). 2012.
16. Birke KP, Weeber M. Zentrum für Digitalisierte Batteriezellenproduktion; 2021 [cited 2021 Mar 24]. Available from: https://s-tec.de/zentren/zentrum-fuer-digitalisierte-batteriezellenproduktion/

17. Vogel-Heuser B, Bauernhansl T, Hompel M ten, editors. Handbuch Industrie 4.0. 2., erweiterte und bearbeitete Auflage. Berlin: Springer Vieweg; 2017. (Springer Reference Technik). Available from: https://link.springer.com/book/10.1007%2F978-3-662-53248-5
18. Goldberg RP. Survey of virtual machine research. Computer. 1974;7(6):34–45.
19. Microsoft. What is Docker? 2018 [cited 2021 Mar 25]. Available from: https://docs.microsoft.com/en-us/dotnet/architecture/microservices/container-docker-introduction/docker-defined

https://doi.org/10.1142/9789811245626_0009

Chapter 8

Digitalization Concepts for the Supply Chain Management in Battery Manufacturing

Peter Grüninger*, Martina Schiffer†, Roman Weiner‡ and Dominik Nemec§
Fraunhofer Institute for Manufacturing, Engineering and Automation IPA
Nobelstr. 12, 70569 Stuttgart, Germany
**peter.grueninger@ipa.fraunhofer.de*
†martina.schiffer@ipa.fraunhofer.de
‡roman.weiner@ipa.fraunhofer.de
§dominik.nemec@ipa.fraunhofer.de

Logistics and supply chain management are subject to a constant process of change and development leading towards a global, network-oriented field of activity. Individual customer wishes and demands for shorter delivery times as well as increasing cost pressure are both current and future challenges for manufacturing companies. Also, the battery industry is affected by this dynamic change. This chapter therefore covers specific challenges along the entire supply chain in the battery industry as well as solution approaches in the form of digitalization potentials for overcoming these. Corresponding concepts and technologies for implementing digital supply chain management are presented along the Plan, Source, Make, Deliver and Return processes of the SCOR model. In particular, a digitalization concept for testing raw material quality in battery cell production is detailed based on the slurry manufacturing process, in which inline and online measurement and testing equipment can be integrated to determine raw material quality characteristics. The approach underlies a cloud-based database for shared storage of characteristic data in a material ID. Supplier information on the condition of the raw material can be digitally registered before delivery and thus viewed in an integrated manner during incoming goods inspections. The concepts presented here offer the potential to optimize logistical targets such as capacity utilization, on-time delivery, inventory and lead times, and thus increase efficiency along the supply chain.

1. Introduction

Logistics as well as supply chain management (SCM) have already undergone significant changes in recent years. Traditional transport, handling and storage processes with a strict functional orientation have been transformed into a global, network-oriented field of activity. Today, supply chain management makes a significant contribution to the success of a company. The future challenges for manufacturing companies lie primarily in individual customer requirements, the demands of shorter delivery time and the increasing cost pressure. Based on these challenges, increasing globalization as well as the required environmental protection, companies today have to cope with more complex supply chain networks. Even battery cell production is not unaffected by these challenges.[1] The development of the battery industry is considered to be very dynamic and the expectations of the players involved in the supply chain are high.[2] In particular, the rising demand for lithium-ion batteries is leading to a shortage of raw materials and thus to an enormous increase in raw material prices.[3] This results in challenges not only for supply chain management but also for internal processes such as quality management and purchasing. Over the last five years, the price of lithium as a raw material has increased by more than 150%. According to a report by the USGS, the American authority for cartography, 56% of the extracted lithium is used for battery production. This trend is rising.[1] A major factor of the high consumption of raw materials in lithium-ion battery production is the variation of the manufacturing process in terms of quality.[4] Cell production cannot guarantee a consistently high quality and this results in a high scrap rate.[5] This is often due to raw material and coating defects, which result in a lower energy density.[4,6] In order to avoid high scrap rates and make lithium-ion battery production more resource-efficient, it is essential to ensure high raw material quality right from the start of the process chain. At present, the quality of raw materials is usually only checked at the supplier's premises. Inspections of incoming goods and monitoring of the storage conditions, such as humidity and temperature, are often not carried out at the electrode manufacturer or are only carried out on a random basis. As a result, quality fluctuations of the raw materials cannot be verified and therefore no statements can be made about possible correlations between the raw material quality and the cell quality. Digital transformation is considered a possible solution to avoid these challenges. Specifically, digital transformation is described as "the change in value creation processes through the further development of existing digital technologies, the implementation of new digital technologies, the adaptation of corporate strategies on the basis of new digitalized business models and the acquisition of the necessary competences or qualifications".[7] The digitalization concepts described below are to be divided into two parts: First of all, a general digitalization concept of the supply chain is described (see Fig. 1). In this respect, the

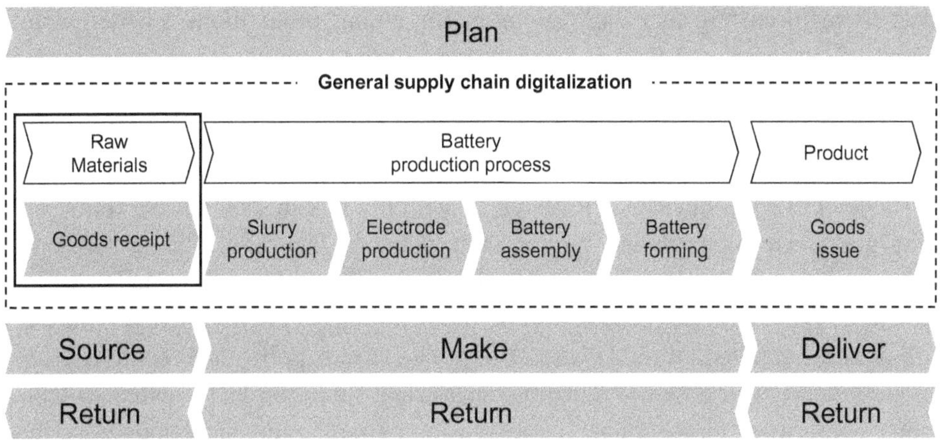

Fig. 1. General supply chain digitalization.

potentials of digital transformation along the supply chain are discussed in particular. The concept of the digitalized supply chain for battery production is based on the SCOR model (Supply Chain Operation Reference Model) with five process elements (planning, procurement, production, delivery and return).[8] In the second part, the quality management of raw materials as an interface between the processes "source" and "make" of the SCOR model is investigated.

2. General Concept for Supply Chain Management Digitalization

According to McKinsey, digitalization will make the supply chain faster, more flexible, more transparent, more accurate and more efficient across all sectors. Sensors are placed on all products, large amounts of data are analyzed and used for forecasts, and robots and drones are used for deliveries. These possible scenarios are just examples of a digitalized and automated supply chain.[9] Digitalization offers the opportunity to control the complexity in SCM better.[10] For an intelligent and comprehensively digitalized supply chain, many modern technology concepts will be available in the future. According to a survey by the Fraunhofer Institute for Manufacturing Engineering and Automation, companies see the greatest opportunities of digitalization with regard to the supply chain in the use of artificial intelligence, sensor technology on/in products and big data applications.[11] The core technologies of artificial intelligence include machine learning, deep learning and predictive analytics. Machine learning algorithms are able to quickly analyze large, diverse datasets in order to improve the accuracy of the demand forecast or to detect anomalies in the production, for example. Machine learning in collaborative supply chain networks also offers benefits like reducing freight costs and improving delivery performance. In order to use artificial intelligence, large amounts of data are required. Big data applications enable, among other things, a better exploitation of synergies and thus an enhancement of the planning information base at the same time. It is also possible to avoid empty journeys in logistics and achieve dynamic route optimization in real time, for example, by taking into account the current traffic situation, the current cost structures dominant on the market or the variable delivery sequences. However, this is only possible by adding third-party data that serve to enrich and improve the planning data base. The sensors on/in products also offer a great opportunity to optimize the supply chain. The identification and localization of goods as well as the recording of environmental influences are possible application fields here. Along the entire supply chain, companies can monitor the flow of goods in real time, measuring, for example, the extent of vibrations, humidity or temperature fluctuations. In particular, the challenges described above must be taken into account and conceptually considered.

The basis of the digitalization concept is established by the supply chain operations reference model (SCOR) (see Fig. 2).[8] The possible improvement potentials of digitalization in respect to battery production are described based on the five process elements of the supply chain.

The process "Plan" describes the activities associated with developing plans for the operation of the supply chain. The planning process involves capturing requirements, collecting information about available resources, balancing requirements and resources to identify planned capabilities and gaps in demand or resources, and identifying actions to correct these limits. The process "Source" describes the ordering and receipt of goods. It includes placing orders or planning deliveries, receiving, validating and storing goods, and accepting the invoice from the supplier. The "Make" processes describe the activities associated with adding

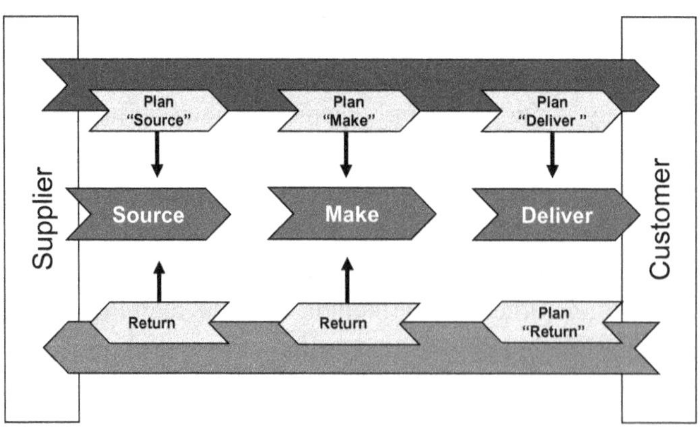

Fig. 2. SCOR model (According to Ref. [8].).

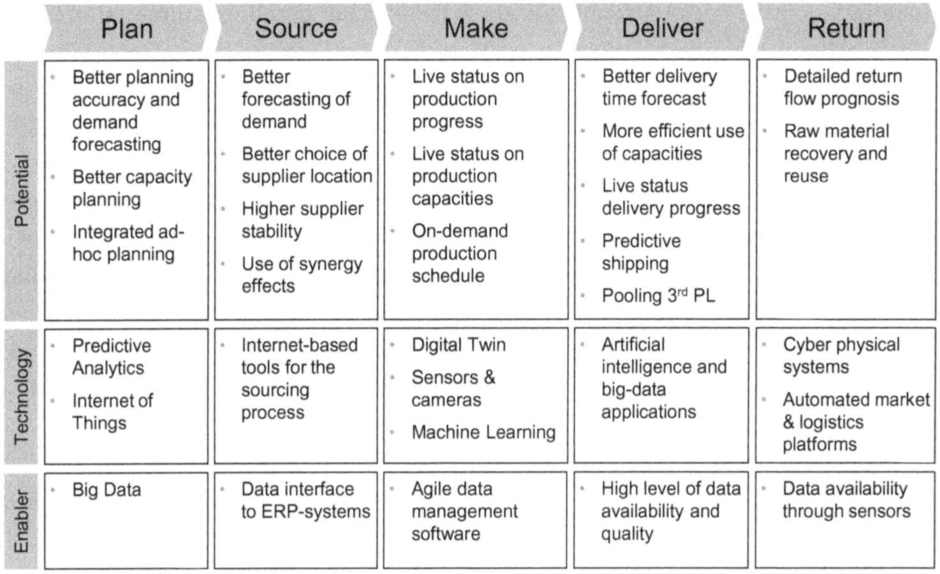

Fig. 3. Digitalization potential of the supply chain.

value to materials or creating content for a service. The transformation of materials is more commonly used as "production" because "make" represents all types of material transformation, e.g. assembly and chemical processing. The "Deliver" processes describe the activities associated with the creation, maintenance and fulfillment of customer orders. The process embodies the receipt, validation and creation of customer orders, scheduling order delivery pick, packaging and shipment, and invoicing the customer. The process "Return" describes the reverse flow of goods. It embodies the identification of the need to return, the disposition decision-making, the scheduling of the return, and the shipment and receipt of the returned goods.[8]

The concept of the digitalized supply chain for battery production has two main focuses — on the one hand, the comprehensive potential that can be achieved by the application of Industry 4.0 technologies, and on the other hand, the potential of recycling battery cells, see Fig. 3.

2.1. *Plan process digitalization*

A large number of raw materials are required for the production of batteries. These include in particular lithium, cobalt, nickel and graphite.[12] The extraction and processing of these raw materials are not only associated with environmental and social consequences but also with a high purchase price, which is reflected in the material costs. This can account for up to 65% of the combined production costs of the battery.[13,14]

By using predictive analytics, better forecasts regarding the demand for raw materials, components and products or services can be more accurately made within supply chains, especially in planning, based on past and ongoing trends.[15] This method captures large datasets (Big Data) and automatically generates predictions based on predictive models that support decision-making through analysis results. These models are generated using machine learning. Algorithms build an optimal statistical model for this purpose in machine learning, which is based on training data.[16] Another advantage is the improved quality of risk management. With the help of predictive analytics, corresponding system models can be used to assess the risk of future disruptive influences within the supply chain.[17]

The use of IoT (Internet of Things) approaches also improves the transparency of the entire supply chain. Not only the localization of the position but also the characteristics of all things within the supply chain can be made available at any desired time.[18] This is ensured by using integrated sensors on, e.g. machines, means of transport, semi-finished products or finished battery cells and connecting them to the Internet. For example, data on position, temperature, humidity or vibrations during transport can be recorded and made available in real time.[19] Other potential benefits in terms of the planning process include reduced costs, better operational performance and the associated reduction of overproduction. In addition, networking through the IoT offers faster responsiveness to changing customer needs or supplier availability.[20]

In order to take advantage of these benefits, an increased availability of data in the form of Big Data from, for example, past demand data for raw materials or machine or operating data in battery cell production is necessary. But, the collection of weather data or traffic data by means of GPS, which play a decisive role for optimal transport planning, must also be taken into account. This is the basis for better transparency of the entire supply chain and enables competitive advantages such as shorter lead times in production or higher on-time delivery through better demand forecast accuracy, more accurate inventory planning, and more efficient planning and design of complex distribution routes.[21]

2.2. *Source process digitalization*

Even today, there are temporary supply bottlenecks along the value chain in battery cell production. These are due to increased demand, for example. In addition, the social and ecological impacts in the extraction of raw materials require measures to ensure environmental, social and safety standards through control and regulatory mechanisms.[22]

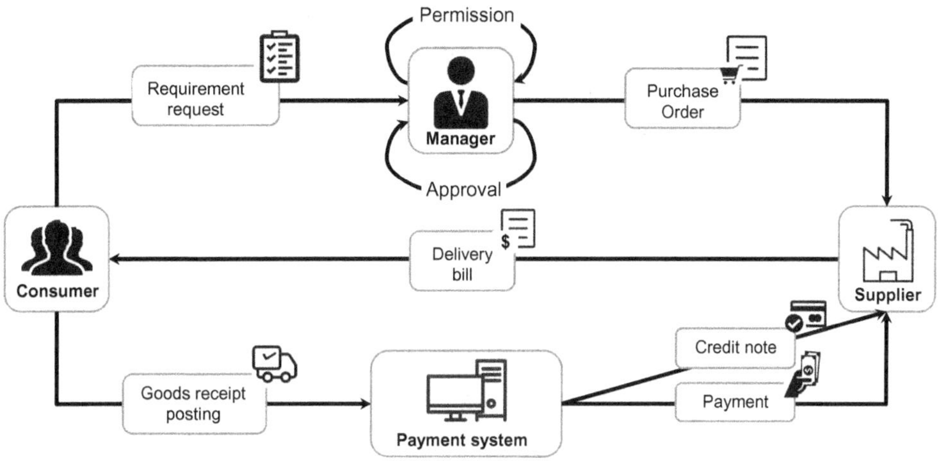

Fig. 4. Idealized, IT-based order-to-pay process according to Ref. [23].

Internet-based tools for the strategic sourcing process provide solutions for ensuring transparency and improved coordination along the value chain. The basic prerequisite for the use of these Internet-based tools is appropriate data interfaces to traditional Enterprise Resource Planning (ERP) systems. They guarantee access to the server-based ERP system and thus support the assurance of the integrity of the data used.[23]

Thus, e-sourcing/e-tendering can be used to adhere to standards in tendering, increase transparency and ensure compliance in the purchasing process. In addition, E-SRM systems offer the possibility of process control and documentation during initial assessment, auditing and prequalification of suppliers as well as during their inclusion in the supplier pool. Factors such as price attractiveness, product quality, delivery reliability, communication behavior and innovative ability are taken into account here. E-SCM systems provide IT support for the flow of information, goods and finances along the value chain and the associated planning and control tasks. Improved coordination between those involved in research and development, value analyses, troubleshooting and quality defect rectification as well as project planning and project execution can be achieved by means of e-collaboration systems.[23]

Another potential lies in the reduction of process costs by shortening the processing time. E-catalog-based procurement systems and purchase-to-pay systems/order-to-pay systems can be used to automate invoice and payment processing in operational procurement,[24] see Fig. 4.

In an order-to-pay system, the consumer orders decentrally via the ordering system and accesses an electronic product catalog. If the items are not in stock, there is no need to check the available inventory. Supplier selection then takes place simultaneously with the filling of the virtual shopping cart. Corresponding cost center data are stored in the system so that approval is given automatically. Rule violations start escalation routines in the respective case. Upon approval, the supplier receives the order digitally and automatically initiates a status for delivery tracking. The ordered items are shipped directly to the requisitioner, who posts the goods receipt and enters any deviations from the order in the system. The goods receipt posting triggers the payment or credit memo process.[24]

2.3. *Make process digitalization*

Global demand for battery capacity is expected to increase to 1,200 GWh in 2030 and a total of 3,500 GWh by 2050.[25] With this order of magnitude, rejects in production due to insufficient product quality are particularly significant. In this case, approximately 5% production scrap would correspond to 18 GWh per year of rejects or 70,000 tons of battery cells.[26] In addition, short-term changes in the battery cell production process due to personalized or customer-specific requirements require rapid software adjustments in the Manufacturing Execution System (MES). Agile data management software for battery production is needed to manage the data. MES with a micro service architecture offer better scalability and agile extensibility. Extensibility is provided by encapsulating the individual micro services. This offers the possibility to dynamically integrate or replace additional functionalities.[27] The advantage here is increased responsiveness by ensuring that each service can be designed, developed and deployed independently. This increases the flexibility of the existing MES.[28] In addition, individual services such as energy management can be integrated. The potential here lies in the collection of data to increase energy efficiency in the battery production process. This service is necessary, for example, in battery cell production to optimize energy consumption for the welding process or for the use of dry or clean rooms. By training optimization models using machine learning, the energy consumption of production equipment can be reduced, resulting in more cost-efficient production.[29]

To increase efficiency in production and thus reduce battery cell waste, tools such as the digital twin and traceability approaches provide opportunities to optimize and improve the production process. Each production step can be recorded and traced in real time by applying built-in sensors and cameras.[26] In addition, mobile technologies such as RFID (Radio Frequency Identification) tags assist in tracking and locating the materials and products within the production process as well as recording their movement data.[30] The corresponding tools receive and store the data. This includes process data of coating, cutting, winding, electrolyte

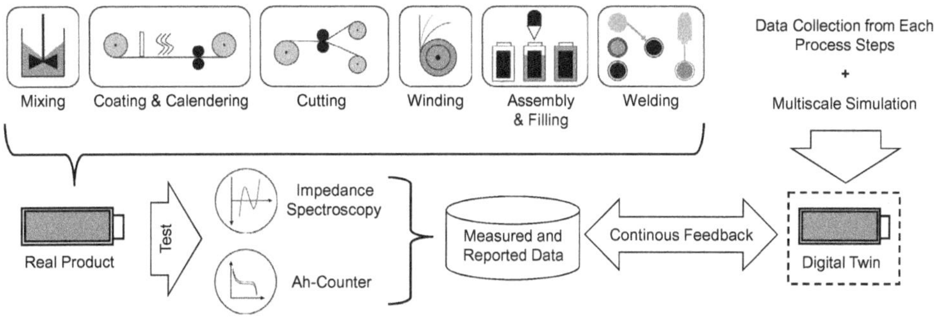

Fig. 5. Digital Twin engineering at cell level according to Ref. [31].

filling and packing, which can later be analyzed and evaluated for performance by engineers.[26,31] With the corresponding process data, a digital twin of the physical battery can be continuously generated, see Fig. 5.

With this, predictions can be made about the aging process of the battery as well as its end-of-life phase if the data quality is appropriate.[26]

2.4. *Deliver process digitalization*

The basis for exploiting this potential is a high level of data availability and quality. Manual data input represents a high error potential in the information flow in many areas, for instance, the high feedback delay and the type errors. This error potential can be eliminated, for example, by applying automated data acquisition using sensors and model calculations. For automated data acquisition, standardized interfaces between systems (e.g. MES) and machines are indispensable.[32] It is therefore essential to ensure that the interfaces in the whole context of the IT concept are compatible with each other. However, for the individual phases of supply chain management, not only do the own internal data along the handling process play an important role but also the data of suppliers/customers as well as third-party data (e.g. weather forecasts, traffic data). In this way, potentials can be exploited, e.g. through better forecasts and capacities in all phases. In general, high data availability creates a transparent supply chain (e.g. real-time status of the production site), which can be utilized by all participants. Also, with the help of central logistics platforms, the capacities of logistics service providers can be better planned and employed.

The digitalization of battery cells also supports the digitalization of the supply chain. Data such as cell voltage, internal resistance, temperature and usage history are considered to be essential parameters for the digitalized battery. These data can be collected and analyzed by the electrical circuit integrated in the product (battery management system). Based on these data, a wide range of battery states such as state of charge (SOC), state of health (SOH) and state of power (SOP) can be determined. By sharing the data with the participants in the supply chain, it will be technically possible to analyze in a global environment, e.g. using artificial intelligence and big data applications. This could be used to draw inferences for sales planning, maintenance and recycling.

2.5. *Return process digitalization*

The resource-efficient recycling of battery products brings challenges to the supply chain management. Not only are new, used and to be recycled batteries involved in the flow of goods but so are the elements disassembled and recovered during the recycling, which must be processed and transported under different conditions. Several different recycling levels are feasible for the battery production:

- The overall product recycling process restores the product.
- Component recycling only reuses parts of the product.
- Material recycling aims at reusing the basic materials.

For the production of lithium-ion batteries, the metal raw materials lithium, cobalt and nickel are essential.[25] Information on the availability and quality of recyclates is therefore necessary for recycling. Currently, there is a lack of information on the purity, type or quantity of admixtures that are relevant for reuse. Similarly, there is a lack of transparency about the available quantity of recyclates on the market, so that economies of scale cannot be exploited. This increases the price of recyclates.[33]

Cyber-physical systems can carry information about the battery cell, such as the composition of the slurry, the electrolyte filling or the auxiliary materials used, through the entire production process. Moreover, for an efficient circular economy, this information must be carried through the entire life cycle of the battery cells and include not only information on the material composition of the battery cells but also environmentally relevant information such as "footprints". The bases for the cyber-physical systems are sensors that record information not only about the product but also about waste. For example, the location and amount of waste generated during battery cell production, as well as its exact material composition, can be recorded in real time. This information can be made available to recycling companies for more efficient procurement and production planning.[34] Currently, information deficits lead to increased transaction and search costs. The effort required to find high-quality secondary material increases if information about the corresponding quality is unclear. This also has an impact on pricing and is associated with increased costs for concluding contracts and guarantees.[33] Internet-based solutions in the form of automated market and logistics platforms (as an "Uber for battery cell waste", so to speak) can therefore reduce search and transaction costs. In addition, they create clarity about existing material quantities and are thus the basis for achieving economies of scale.[34]

3. Digitalization of Raw Material Quality Control

The focus of the digitalization concept for raw material quality control aims at an active feedback of the quality to the supplier and the linking of quality control with subsequent processes (see Fig. 6). This is intended to reduce fluctuations in raw material quality and increase the resource efficiency of Li-ion battery production by a reduction of scrap. By digitalizing the raw material quality control, error sources such as measuring errors (from the supplier and battery cell producer) or errors caused by delivery and storage-related quality changes as well as fluctuations within a batch or between batches can be identified. To illustrate the concept, the slurry production process is envisioned as a dosing system with feed pipes, in which inline and online measuring and testing devices can be integrated to determine the quality characteristics of the raw materials. The focus in this concept is especially on the raw materials of the electrode slurry (active material, binder, additives and organic solvents/water). However, the concept can be extended to other raw materials for cell production.

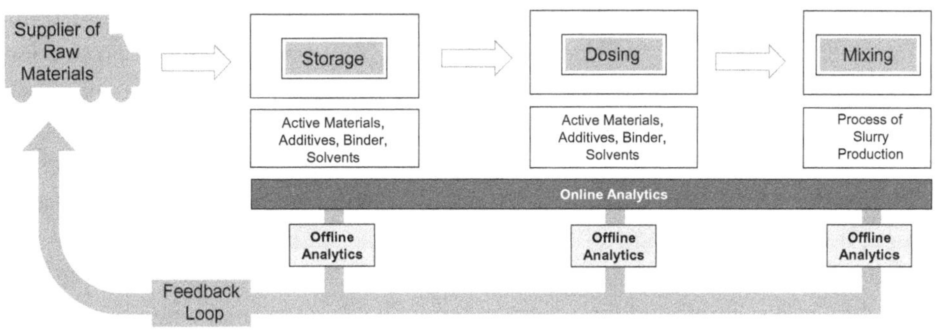

Fig. 6. Quality management of raw materials.

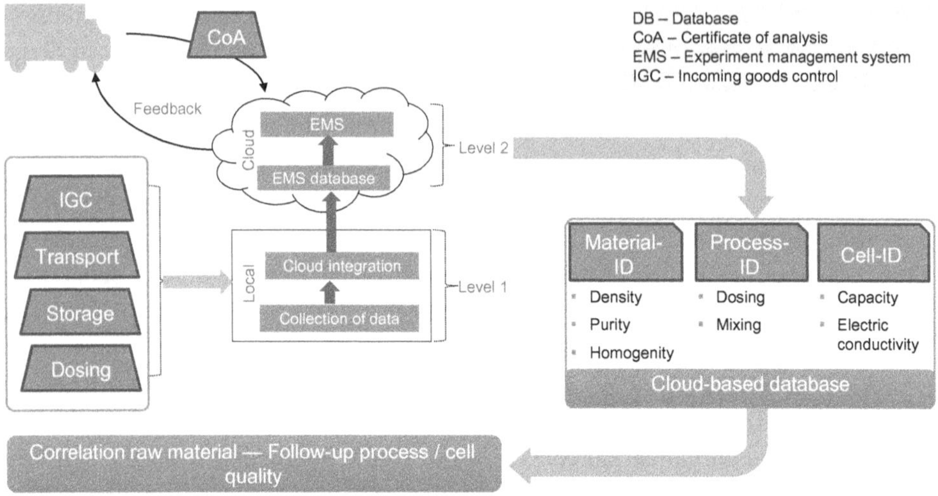

Fig. 7. Digitalization concept of raw material criteria.

3.1. *Conceptual framework*

The overall framework is based on a shared cloud-based information system between raw material supplier and electrode manufacturer (see Fig. 7). Selected quality features of the raw material are recorded both at the supplier and at the electrode manufacturer before the slurry is produced. The data of the incoming goods control (IGC), as well as parameters during transport, storage and dosing, are collected in a central database which serves as a basis for analysis. Using the Experiment Management System (EMS), the electrode manufacturing process can be planned, controlled and documented digitally.

A unique raw material identifier is generated for each raw material in a shared database, in which the quality features are stored in respect to batch and time. This material identifier is carried along the entire process chain of cell production and enables the traceability of specific parameters starting from the incoming goods control of the raw materials. Therefore, each battery cell produced has to be assigned a specific material identifier with the quality features of the raw material. This enables the analysis of the influence of raw material quality (e.g. density, purity, homogeneity and specific surface area of the compounds, see Fig. 8) on the subsequent processes and also the quality of the final cell. As an example, the particle size distribution of the active material can be correlated with electric conductivity of the battery cell.

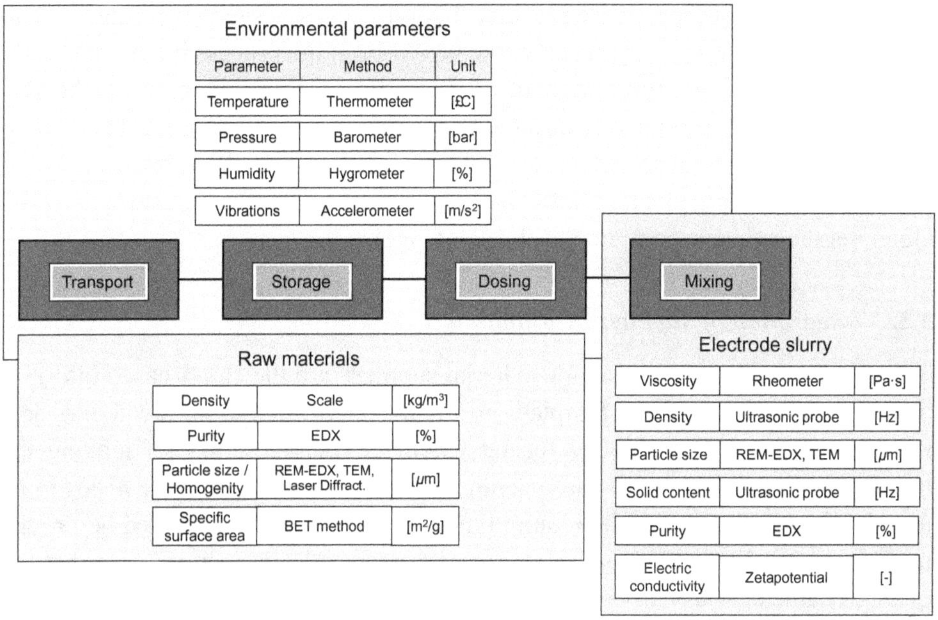

Fig. 8. Quality parameters and analytical methods for electrode slurry manufacturing.

3.2. *Raw material quality features*

To characterize the quality of the raw material, special parameters have to be considered. The particle size distribution, moisture content, bulk density, chemical composition (purity, type and proportion of impurities), specific surface area and specific resistance are important quality characteristics of the active materials. These parameters significantly determine the raw material quality and influence the quality of the battery cell. For this very reason, raw material suppliers also analyze these parameters and usually provide a certificate of analysis. Besides the quality characteristics, further raw material parameters, such as the category of the raw material (e.g. active material, binder), the name, the supplier, the batch number, the CAS number for identification and the storage conditions (air temperature and humidity), are associated with the raw material identifier.

The real-time measurement of a quality feature is outlined below by taking an example of measuring the particle size distribution. It is one of the most essential parameters for describing the quality of the raw material and has an influence not only on the viscosity of the slurry but also on its segregation behavior. By using real-time measurement of the particle size distribution before the slurry production process in the raw state, batch fluctuations can be detected. Raw materials that lie outside the limits of the particle size distribution or the acceptable particle size can thus be identified at an early stage and removed from the process before the slurry is produced. Real-time measurement of the particle size distribution can be performed with a particle size measuring device. The measuring principle is based on laser diffraction. The data output and analysis is done digitally in real time using the manufacturer's dedicated software. The data can also be uploaded into the cloud via the analysis computer at the same time.

3.3. *Integration of supplier information*

To integrate the data from the raw material supplier into the cloud-based information system, a standardized template of a data sheet needs to be created. A data format such as XML or JSON format would be ideally suited for making the transferred data both human and machine interpretable. The template needs to be used by the suppliers with the required quality features and transferred before the raw material delivery. This template is uniform for all raw materials, so that the quality characteristics can be automatically interpreted. Suppliers need to upload this information to the cloud where it is provided to the EMS for assignment in experiments. In this way, the condition of the raw material is recorded and digitally registered before the delivery.

After comparing the quality features provided by the supplier with the raw material data obtained from the electrode manufacturer and analyzing possible correlations between raw material and cell quality, the supplier receives feedback on his raw material quality and can make any adjustments if necessary. Therefore, the supplier can use a comparative table between the quality features of the incoming goods and the feedback during the manufacturing process, for example.

3.4. *Incoming goods inspection and process monitoring*

The recording of quality features in the electrode production process takes place in two different ways (offline and inline/online). Environmental parameters and raw material features during the transport, storing and dosing processes as well as quality characteristics of the electrode slurry during the mixing process are traced (see Fig. 5). In order to be able to assess the influence of the different sources of error on the quality of the raw material separately, the measurements of the quality features are additionally designed to have variable frequencies. The quality features of the raw material (e.g. density, purity, particle size and homogeneity) are determined on the receipt of the goods (offline) and will be measured before, during and after storage via a tracking device as well as before electrode manufacture (offline and inline/online). In order to be able to examine delivery and storage-related quality fluctuations separately, the offline characterization of the quality features should be carried out on the receipt of the goods and before the start of the slurry production process, and the data should be saved in the cloud. In addition, the environmental conditions such as air temperature and humidity during storage should be recorded and assigned the corresponding material identifier.

Random checks of the quality features by sampling are performed on delivery of the goods. The data are then saved in the cloud-based information system in relation to material identifiers and are compared with the manufacturer data. This makes it possible to check the information provided by the manufacturer. Furthermore, the possible quality gaps in the delivery process can be identified. The supplier will then receive a feedback after the analysis of the data. In order to uncover quality losses of the raw materials due to storage conditions, the quality features of the raw material are checked all the time during the storage, especially right before the dosing the raw materials for electrode slurry production, and are saved and analyzed in the cloud. Fluctuations within a batch can be detected by the real-time characterization (inline or online) of the quality features. All the six determined quality features are checked upon receipt of the goods. The chemical composition, specific surface area and specific resistance are also determined offline before the dosing of raw materials for slurry production. The quality features such as particle size distribution, moisture content and bulk density can be measured inline or online directly during the dosing process. Storage conditions (temperature, humidity, pressure, storage time, etc.) are also monitored in real time and can be correlated to the raw material quality parameters.

3.5. *Implementation of offline and inline concepts*

The offline and the inline concepts are described below on the basis of certain quality features. Impurities in the raw materials have a great influence on the cell quality. In order to determine the purity and the type and quantity of impurities in the raw material, the quantitative material analysis of the powders can be performed by using energy dispersive X-ray fluorescence spectroscopy.

For the other quality features, there are also measuring and testing devices which have a digital data output and a capability to save the data in the cloud. As examples of offline measurement, a powder measuring system for measuring the specific powder resistance under a controlled pressure is available as well as a device for measuring the specific surface area using the Brunauer–Emmett–Teller (BET) method. For the combined real-time measurement of moisture content and bulk density by using microwaves, online measuring devices can be used.

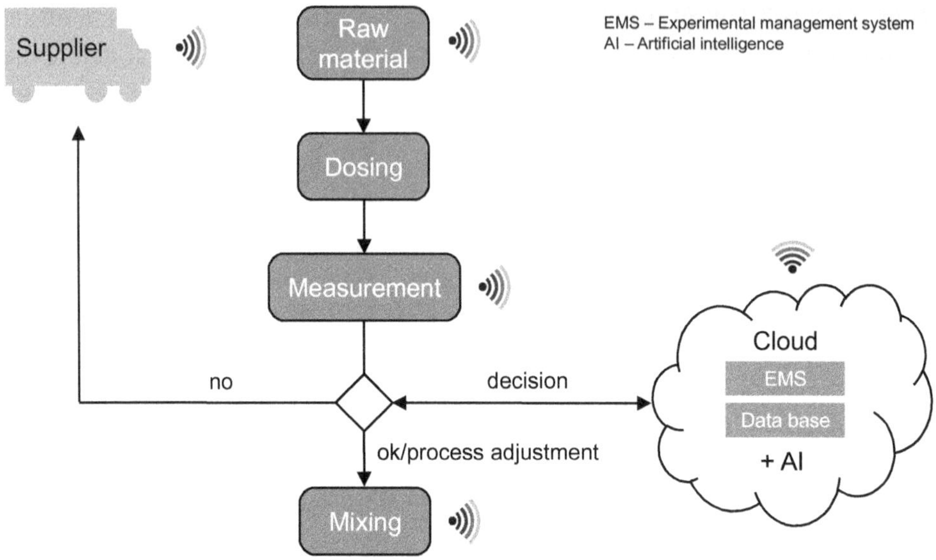

Fig. 9. Implementation of the digitalization concept.

This enables algorithms or other systems to react on the specific characteristics of raw materials in subsequent processes (see Fig. 9). Process parameters and conditions can be adjusted in order to achieve a slurry of high quality integrating the information available on the quality features of the raw material.

4. Conclusion

Both the digitalization concepts for supply chain management and control of raw material quality could identify key factors for resource-efficient battery cell production.

Based on the SCOR model and by applying modern technologies such as artificial intelligence, Big Data and sensor technology, it is possible to reveal strategies to handle the complexity in supply chain management in terms of battery cell production. Based on these introduced potentials, a rough concept with detailed conceptual design for the realization and implementation of the respective potentials is first developed in the next step, depending on the respective logistical target values (capacity utilization, adherence to schedules, stock and lead time).

Furthermore, a digitalization concept for raw material and process controls for the slurry production of electrode materials was introduced, which should reduce raw material consumption. An automated and cloud-based data collection of all process parameters was presented, which uses an experiment management system. The digitalization concept enables the tracking of several raw material quality features during the transport, storage and dosing process giving an active feedback to the supplier of the raw materials. These quality features of the electrode slurry can be correlated to the properties of the raw materials. As a consequence, the manufacturing process of the electrode slurry can be digitally controlled and optimized with a high potential for better quality of the battery cell and resource-efficient production processes.

References

1. USGS. Mineral commodity summaries 2020: U.S. Geological Survey. Reston, VA; 2020. Mineral Commodity Summaries [cited 2020 Sep 5]. Available from: http://pubs.er.usgs.gov/publication/mcs2020
2. Tosarkani BM, Amin SH. A possibilistic solution to configure a battery closed-loop supply chain. Expert Syst Appl. 2018;92(C):12–26.
3. Hohmann M. Globaler Bedarf an Lithium, Kobalt und Nickel zur Produktion von Lithium-Ionen-Batterien im Jahr 2017 und Prognosen für die Jahre 2023 und 2028; 2020. Available from: https://de.statista.com/statistik/daten/studie/982144/umfrage/bedarf-an-wichtigen-metallen-zur-produktion-von-lithium-ionen-batterien-weltweit/
4. Smekens J, Gopalakrishnan R, Steen N, Omar N, Hegazy O, Hubin A, *et al.* Influence of electrode density on the performance of li-ion batteries: Experimental and simulation results. Energies. 2016;9(2):104.
5. Dinger A, Martin R, Mosquet X, Rabl M, Rizoulis D, Russo M; *et al.* Batteries for electric cars: Challenges, opportunities, and the outlook to 2020; The Boston Consulting Group, 2020; p. 1–14.
6. Schnell J, Reinhart G. Quality management for battery production: A quality gate concept. Procedia CIRP. 2016;57:568–73.
7. Seiter M, Grünert L, Berlin S. Betriebswirtschaftliche Aspekte von Industrie 4.0. Wiesbaden: Springer Fachmedien Wiesbaden; 2017.
8. APICS. Supply Chain Operations Reference (SCOR) model; 2020. Available from: https://www.ascm.org/membership-community/corporate-membership/scormark-benchmarking/
9. Alicke K, Rachor J, Seyfert A. Supply Chain 4.0 — The next-generation digital supply chain; 2016. Available from: https://www.mckinsey.com/business-functions/operations/our-insights/supply-chain-40--the-next-generation-digital-supply-chain
10. Fend L, Hofmann J, editors. Digitalisierung in Industrie-, Handels-und Dienstleistungsunternehmen: Konzepte–Lösungen–Beispiele. Wiesbaden: Springer Gabler; 2018.
11. Schiffer M, Wiendahl H-H, Saretz B, Lickefett M, Pietrzak G, Forstmann B. SCM2040 — Wie verändert sich die Supply Chain in der Zukunft? Available from: https://www.ipa.fraunhofer.de/de/Publikationen/studien/supply_chain_management_2040.html
12. Diekmann J, Hanisch C, Froböse L, Schälicke G, Loellhoeffel T, Fölster A-S, *et al.* Ecological recycling of lithium-ion batteries from electric vehicles with focus on mechanical processes. J Electrochem Soc. 2017;164(1):A6184–91.
13. Kwade A, Diekmann J. Recycling of lithium-ion batteries: The lithorec way. Cham: Springer; 2017. (Sustainable Production, Life Cycle Engineering and Management Ser).

14. Berckmans GJ, Messagie M, Smekens J, Omar N, Vanhaverbeke L, van Mierlo J. Cost projection of state of the art lithium-ion batteries for electric vehicles up to 2030. Energies. 2017;10(9):1314.
15. Sarangi S. Business intelligence systems: A necessity for agile supply chains. Parikalpana. 2016;12(2):52.
16. Reitmaier T. Aktives Lernen für Klassifikationsprobleme unter der Nutzung von Strukturinformationen. In: Ausgezeichnete Informatikdissertationen 2015. Bonn: Gesellschaft für Informatik; 2015. p. 239–48.
17. Ghadge A, Dani S, Chester M, Kalawsky R. A systems approach for modelling supply chain risks. Suppl Chain Manage. 2013;18(5):523–38.
18. Geerts GL, O'Leary DE. A supply chain of things: The EAGLET ontology for highly visible supply chains. Decis Support Syst. 2014;63:3–22. Available from: https://www.sciencedirect.com/science/article/pii/S0167923613002352
19. Brett D. Smart move for containers; 2015. Available from: https://www.lloydsloadinglist.com/freight-directory/adviceandinsight/Smart-move-for-containers/62491.htm#.X8y1GshKiUk
20. Cerasis. The future of supply chain, logistics & manufacturing: how technology is transforming industries; 2015. Available from: https://ftp.idu.ac.id/wp-content/uploads/ebook/ip/LOGISTIK/Technology_Manufacturing_SupplyChain_Logistics_eBook.pdf
21. Haddud A, Khare A. The impact of digitizing supply chains on lean operations. In: Khare A, Kessler D, Wirsam J, editors. Marktorientiertes Produkt- und Produktionsmanagement in digitalen Umwelten. Wiesbaden: Springer Fachmedien Wiesbaden; 2018. p. 27–46.
22. Thielmann A, Wietschel M, Funke S, Grimm A, Hettesheimer T, Langkau S, *et al.* Batterien für Elektroautos: Faktencheck und Handlungsbedarf: Sind Batterien für Elektroautos der Schlüssel für eine nachhaltige Mobilität der Zukunft? 2020. Available from: https://www.isi.fraunhofer.de/content/dam/isi/dokumente/cct/2020/Faktencheck-Batterien-fuer-E-Autos.pdf
23. Bogaschewsky R. Digitalisierung in Einkauf und Supply Chain Management. In: Obermaier R, editor. Handbuch Industrie 4.0 und Digitale Transformation: Betriebswirtschaftliche, technische und rechtliche Herausforderungen. Wiesbaden: Springer Fachmedien Wiesbaden; 2019. p. 139–64.
24. Bogaschewsky R, Müller H. Industrie 4.0: Wie verändern sich die IT-Systeme in Einkauf und SCM? Universität Würzburg: Lehrstuhl für Industriebetriebslehre; HTWK Leipzig: Fakultät Wirtschaftswissenschaften; 2016. Available from: http://downloads.cfsm.de/Studien/I40/Studie_Industrie_4_0_2016.pdf
25. Tsiropoulos I, Tarvydas D. Li-ion batteries for mobility and stationary storage applications scenarios for costs and market growth; 2018.
26. Kaus D, Singh S, Wanner J, Weeber M, Birke KP. Digitalization in battery cell manufacturing; 2020.
27. Nadareishvili I, Mitra R, McLarty M, Amundsen M. Microservice architecture: Aligning principles, practices, and culture. 1st ed. Beijing, Boston, Farnham, Sebastopol, Tokyo: O'Reilly; 2016.
28. Zhou T, Li D. Application of microservice architecture in battery monitoring system; 2019.
29. Thiede S, Turetskyy A, Loellhoeffel T, Kwade A, Kara S, Herrmann C. Machine learning approach for systematic analysis of energy efficiency potentials in manufacturing processes: a case of battery production. CIRP Ann. 2020;69(1):21–4.
30. Barata J, Cunha PR. Mobile supply chain management: moving where? 2016.
31. Wessel J, Turetskyy A, Wojahn O, Herrmann C, Thiede S. Tracking and tracing for data mining application in the lithium-ion battery production. Procedia CIRP. 2020;93:162–7.
32. Werner H. Supply chain management: Grundlagen, Strategien, Instrumente und Controlling. 5., überarb. und erw. Aufl. Wiesbaden: Springer-Gabler; 2013. (Lehrbuch).

33. Organisation for Economic Co-operation and Development. Improving recycling markets. Paris: OECD Publishing; 2006. Available from: https://www.oecd.org/env/waste/38093900.pdf
34. Wilts CH, Berg H. Digitale Kreislaufwirtschaft : die digitale Transformation als Wegbereiter ressourcenschonender Stoffkreisläufe; Working Paper. Wuppertal: Wuppertal Institut für Klima, Umwelt, Energie; 2017. In brief. Available from: http://nbn-resolving.de/urn:nbn:de:bsz:wup4-opus-69775

https://doi.org/10.1142/9789811245626_0010

Chapter 9

Evaluation of Visual Assistance System Technologies for Supporting the Information Need of Production Workers

Michael Oberle*, Ozan Yesilyurt†, Andreas Schlereth‡, Sascha Gärtner§ and Florian Maier¶

Fraunhofer IPA, Nobelstraße 12, Stuttgart, Germany

**michael.oberle@ipa.fraunhofer.de*

†ozan.yesilyurt@ipa.fraunhofer.de

‡andreas.schlereth@ipa.fraunhofer.de

§sascha.gaertner@ipa.fraunhofer.de

¶florian.maier@ipa.fraunhofer.de

This chapter investigates which digital visual information media are suitable to support production workers in battery production. First use cases representing typical tasks in battery manufacturing with an information need and task-specific requirements are defined. Corresponding requirements are defined based on various use cases. The suitability of the technologies is finally determined by a weighted calculation of the ratings assigned by experts. This procedure is described as an independent method. In summary, the statements can be made that tablets and smart watches enable generic use to support workers with visual information. Stationary screens are a very good alternative, but are only suitable for specific use cases. AR technologies, in sum, are not suitable to support workers in a production environment. Difficult operation, high effort for setup and little compatibility with the user are the limiting factors here.

1. Introduction

The number of electric cars will rise to 54.3 million by 2030 worldwide. In total, nearly forty-four million vehicles with a rechargeable energy storage will be sold in the next ten years.[1] A key factor for the success and usage of electric cars is a high-quality battery cell. The rising demand for electric car leads to a high demand of these battery cells. In consequence, battery cell production becomes a crucial production process for electric vehicle manufacturers.

Production is a transformation process in which raw material and resources are used to manufacture finished products. Production workers with tools and machines in factories carry out this value-adding process.[2] Therefore, production workers play a major role in production and should be supported to achieve this goal with high efficiency and effectiveness. Digitalization is seen as a revolutionary approach to support production workers and increase the speed and accuracy of production. Under the term "Industry 4.0", Digitalization in the production industry is described as the fourth industrial revolution after the invention of the steam engine, industrialization and the beginning of the computer age.[3]

This chapter analyzes the most suitable digital visual information tools for use cases in battery cell production. Section 2 defines the terms and underlying concepts for this chapter and presents the state of the art of digital visual information media. Section 3 introduces typical use cases for production workers in battery cell manufacturing and identifies requirements for which visual information tools. Section 4 evaluates the considered tools for different solution approaches in different battery manufacturing processes. After an evaluation of the visual information tools, the last section concludes with a summary and outlook.

2. State of the Art

The term production worker is first defined in the state of the art. Then, the battery manufacturing processes are described, which are considered in this chapter. Finally, existing solutions are reviewed and analyzed. It was identified that there is a lack of knowledge in how to support the production workers visually in battery cell manufacturing.

2.1. *Production workers in manufacturing*

Production workers (blue-collar employees) are commonly defined as employees involved in manufacturing, assembly and associated operations such as material handling, logistics, machine repair and maintenance, and other jobs that are closely related to those described above.[4]

Modern manufacturing systems consist of machines and information technology systems of a high technological level.[5] However, a human factor is still playing an important role for the operation of most manufacturing systems. To react quickly in production to adapt production capacity and change product features, production workers are considered as a cost-effective option to expensive automated digital production solutions.[6] If production employees are actively involved in manufacturing process and there are instabilities in the process, the production employees must be considered as a possible factor when analyzing the causes.

Given the improvement in manufacturing techniques and the overall increase in quality, human error has naturally been the subject of even greater focus of interest, as it affects the efficiency of production systems to a greater extent than ever before.[7] Thus, it is important to support production workers in solving problems that occur while executing their assigned tasks. In this chapter, supporting the information need of production workers in battery manufacturing is addressed. For a better understanding of the following subsections and use cases, the general steps for battery cell manufacturing are introduced in the next section.

2.2. *The battery manufacturing process*

Battery manufacturing consists of three main processes.[8] From a high-level perspective, battery manufacturing consists of electrode manufacturing, cell assembly, and formation and quality control. Table 1 represents the main battery manufacturing processes with their sub-processes.

In the first process step, the electrodes are manufactured. This step consists of the following sub-processes: mixing of the active material binder and conductive agent for preparing a slurry, and coating and drying of the slurry on a carrier material. In the following, the electrodes are compressed before they are slit (or punched) into the desired shape.

The cells are assembled in the second process step. This includes the winding or stacking of the electrode sheets together with the separator. The coil or stack is

Table 1. Production processes of battery cell manufacturing.

Step	Sub-step
Electrode manufacturing	• Mixing • Coating & Calendaring • Slitting
Battery cell assembly	• Winding/Stacking • Packaging • Electrolyte filling
Battery cell finishing	• Formation • Degassing and sealing • Aging

then placed into packaging and sealed before the cell is activated by injecting electrolyte.

In the last process step, the cells are repeatedly charged and discharged (formation) for forming the solid electrolyte interface. Gases building up during formation are pressed out of the cell and the cell is finally sealed. The final step in the battery cell manufacturing is aging. The battery cell open circuit voltage is monitored while the cells are stored 1 to 3 weeks. This allows the identification of short circuits in quality control.[8]

2.3. *Review on the support of production workers*

The tasks that production workers fulfill in manufacturing lines require specific skills related to the task. However, they also require the correct information for applying these skills as the complexity due to product varieties in manufacturing would be difficult to handle. For example, machine operators need to be capable of executing the procedure for starting a machine. Setting up the machine for the current workpiece requires specific information of how the machine should process that particular workpiece. Information required by production workers can be provided in a variety of ways. Visual assistance systems display information to production workers on a screen. Acoustic assistance systems convey information via audio signals that production workers listen to. Haptic assistance systems enable the transmission of information that can be felt by the production worker (e.g. through vibration). According to a study investigating the potential of these different systems for different areas of a production organization (commissioning, service, packaging, support, manufacturing, assembly and training) the highest potential is perceived for visual assistance systems. Over 80% of study participants see the utility of visual assistance systems in all areas. The potential of acoustic assistance systems is perceived to be significantly lower (below 60% in any area). Haptic assistance systems are perceived as even less useful.[9] The lower perceived potential of acoustic systems can be explained by noisy environments. This can significantly impair the ability to understand the information provided by the acoustic assistance systems. The low potential of haptic assistance systems can be explained by the amount of information that can be conveyed. Due to the limited capability of conveying information, only very coarse-grained information, e.g. general feedback if a task was executed correctly or not, can be conveyed. The results of the study indicate that visual assistance systems have the highest potential for supporting production workers in battery manufacturing.

Visual Assistance Systems are a quite broad category. Stationary displays, projectors, augmented reality headsets, mobile devices such as smartphones and tablets, and smart watches fall in this category. The characteristics of these devices, such as the form factor, interaction type or portability, strongly vary among these devices. As a consequence, not all devices are suited equally for all production worker tasks. In order to identify potential patterns for the utility of specific devices, a systematic literature review for identifying use cases has been conducted. Due to the practical nature of the topic, the analysis of existing approaches not only focuses on academic literature but also includes reports on industrial projects. To survey relevant use cases of industrial projects in this context, databases of public institutions are searched. Industry 4.0 use cases from industrial organizations are presented publicly by the German Federal Ministry for Economic Affairs and Energy[8] and the "Allianz Industrie 4.0", a company network initiated and funded by the federal state of Baden-Wuerttemberg.[10] Combining both resources, a total of 534 projects on Industry 4.0 use cases are presented ([8]: 157,[10]: 377). From these, 23 of the projects dealt with support for the production workers on the shop floor.[11–33]

Academic research on visual assistance systems supporting production workers was analyzed by searching the scientific literature database Scopus. To identify the literature, a search matrix of keywords and synonyms or closely related terms was created to narrow the search to the field of worker assistance systems in manufacturing (see Table 2).

These terms are merged vertically with "OR" and horizontally with "AND" resulting in a meta search string as presented in Table 3.

Table 2. Research matrix.

Terms		
Production	Worker	Assistance system
Manufacturing	Employee	Support system
Assembly	Human centered	
Operations		

Table 3. Meta database search string.

Database	Search String
Scopus	(Production OR Manufacturing OR Assembly OR Operations) AND (Worker OR Employee OR Human centered) AND (Assistance System OR Support System)

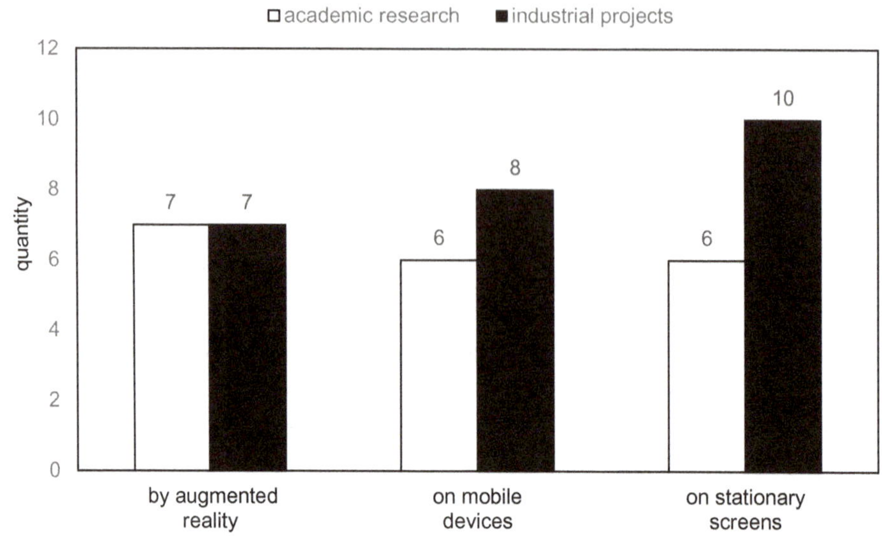

Fig. 1. Count of different visual information tools to support employees on the shop floor.

Applying the search strings, 73 papers could be retrieved. 15 of these papers deal with the topic in question and are relevant from timely perspective (define as published between 2012 and 2020).[34–47] The implementation of the assistance systems can be categorized into three information types of conveying information[48]:

- Information on mobile devices
- Information on stationary screens
- Information by augmented reality (AR).

Figure 1 illustrates the number of occurrences of device categories in the retrieved industrial projects and scientific articles. According to these findings, stationary screens are more often used and chosen than other technologies to visualize information for production workers on the shop floor. A reason for this might be the prevalence of these devices, which makes them an easy implementation choice. The low use of augmented reality to display information may be related to its cumbersome use especially in terms of operation and interaction.

However, this distribution only shows the general use of the various technologies to support the worker with visual information. Based on the evaluation, however, it is not possible to say whether the technology is used in a specific context or task. For this reason, the task of this chapter is to estimate which technology is suitable for specific use cases. The following section introduces the methodology for evaluating the fit of the solution approaches to specific tasks in battery manufacturing.

3. Approach for Evaluating Visual Information Methods

The approach used for the evaluation of visual information methods for addressing the information needs of production workers is based on the theoretical foundation of the human-centered design. In the following, first, this theoretical foundation is introduced before the instantiation of key elements of the human-centered design, production worker use cases in battery manufacturing, use case specific requirements and solution approaches are explained.

3.1. *Human-centered design process*

ISO 9241-210 is an international standard that describes processes for the design of serviceable systems. Part 210 of the standard addresses in detail the user-oriented (human-centered design) approach in development projects and covers usability and user experience. Human-centered design is an approach to the development of interactive systems that aims to make systems usable and useful. Such a design process should concentrate on the users, their needs and requirements and apply knowledge and techniques of ergonomics in the field of usability.[49] To make production workers as effective as possible, task-specific information must be presented appropriately so that the focus is on meeting the objective. The operation of the device or technology itself must not become the focus. In other words, the solution providing the information must be usable and useful for the specific task. Hence, the human-centered design approach is pursued as a foundation for the solution approach evaluation. This approach is a multi-stage iterative process (see Fig. 2).

The human-centered design approach starts with the planning of the design process. In the case presented in this chapter, technologies that can be used to

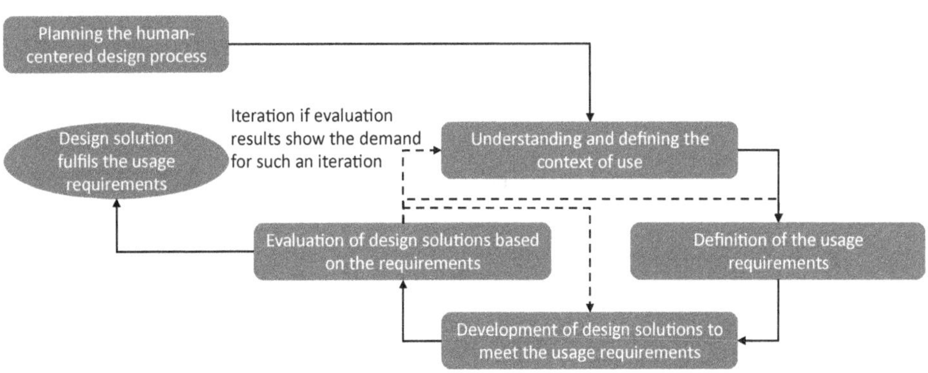

Fig. 2. Human-centered design approach as a multi-stage iterative process according to Ref. [49].

solve known problems from battery manufacturing are identified. Then, the context of use, here, the support production workers in their daily tasks, is defined and analyzed for thorough understanding. Based on this understanding, the usage requirements are defined for all processes. Alternatives solutions are designed to meet the usage requirements. These solutions are then evaluated by utility value analysis based on the requirements. In the human-centered design approach, multiple iterations are possible for the evaluation of results that demand further design. However, as only a general technology evaluation is the goal of this study, multiple iterations are not required. This would be relevant for designing specific solution instances in which there are more options (such as specific placement of user interface elements) in the design space than the underlying technology selection. Hence, the highest rank solution approach can be seen to support the usage requirements.

3.2. *Use cases in battery manufacturing*

The definition and understanding of the context of use are crucial for the design of a useful and usable solution. The scope for selecting appropriate solutions for supporting production workers is focused on different use cases involving human production workers in battery manufacturing. A total of four use cases are defined and shown in Table 4 together with the occurring frequency and the worker type. The detailed description of the use cases is given below.

3.2.1. *Use Case 1 — Material use*

Processes in battery manufacturing, in particular slurry mixing, cell packaging and electrolyte filling, consume materials which need to be replaced from time to time by production workers. Hence, production workers need to know the

Table 4. Use cases in battery manufacturing including occurring frequency and worker type.

Use Case	Occurring Frequency	Worker Type
Material use (UC1)	Once per charge	Trained worker
Parameter setup (UC2)	Once per batch	Setup man
System maintenance (UC3)	Once/twice a day (scheduled)	Technician
Process monitoring (UC4)	Continuously	Trained worker

kind, dosage and procedure for application of materials to the specific machine. These instructions need be quickly and efficiently accessible for production workers.

3.2.2. *Use Case 2 — Parameter setup*

To date, not all Legacy battery manufacturing equipment features a machine-to-machine interface enabling the automatic machine process parameter setup in the case of product changes or other necessary adjustments of process parameters. In this case, production workers need to be provided with the correct information for configuring the equipment with correct parameters for the current product to be produced.

3.2.3. *Use Case 3 — System maintenance*

Equipment in battery manufacturing, as in other production lines, requires regular maintenance. This includes inspection, replacement of parts and precision checks in order to ensure the optimum operation conditions for the equipment.[50] Production workers, typically maintenance specialists, need to be provided with information about which equipment to maintain next and the maintenance reason. This can be because of either a regular maintenance schedule or an abnormal system behavior due to a failure.

3.2.4. *Use Case 4 — Process monitoring*

The finished product of manufacturing processes underlies process variations caused by machine wear or change in process inputs such as material. Depending on the reason and magnitude of the process variations, corrective actions or scrapping of the material needs to be initiated. For example, minor variations in the layer thickness of the electrode process could result in necessary adjustments of the slot nozzle and foil speed, whereas major variations would require the blocking of the foil for further use in the manufacturing line. In order to address this task, production workers need information from sensors of the manufacturing process and information about how to deal with specific variations.

Each use case describes a specific task for production workers. Within these use cases, there are specific requirements for the usage of solutions for supporting the production worker effectively. The definition of these requirements is described in the following section.

3.3. *Usage requirements*

Usage requirements play a key role for the evaluation of suitable solutions for supporting operators. In order to determine appropriate requirements, a total of three battery experts with a long-standing experience of operating battery manufacturing lines were consulted for verification. All requirements can be evaluated on a discrete scale. The following list (Table 5) gives a brief overview of the requirements. A short description simplifies the interpretation of the requirements. This description will be supplemented in further course in order to avoid different interpretations in the independent rating of the experts.

3.4. *Technological solutions*

The solution space for supporting production workers is defined to include only visual assistance systems. Visual assistance systems have been identified as offering the most potential for providing effective support to production workers resulting in less errors.[9] The evaluation of alternative solutions is further limited by market availability. Solutions that have not been field tested (technological readiness level ≥ 5) are not considered for potential solution candidates. Visual assistance systems can be categorized by the method used for conveying the production task-relevant information to production workers.[48] These categories (see Table 6) serve as alternative solutions in the human-centered design process:

Table 5. Resulting requirements of expert survey and their short description.

	Requirement	Description
R1	Hands-free operation	Interaction with the technology without occupying both hands
R2	Mobility	Portability of the technology enabling a radius of action
R3	Expected user compatibility	Prevalence of the technology leading to compatibility with the general population
R4	Interaction richness	Amount of information that can be conveyed by the user to the technology
R5	Customization effort	Effort required to customize/update the technology for the use case
R6	Interaction robustness	Error-proneness of input (e.g. inaccurate recognition) and output e.g. influence of light)
R7	Routine setup	Effort required for routine operational readiness of the technology
R8	Collaboration capability	Possibility of simultaneously interaction on the same information display
R9	Directed distribution of information	Limitation of scope of potential information recipients
R10	Mean time to awareness	Time needed for recognizing a change in information by the user
R11	Required footprint	Space occupied by technology in the use case environment

Table 6. Considered technological solutions to possibly fulfill previously defined requirements.

Technological Solution			Description
Mobile Device Screens T1	**T1.1**	**Tablet/Phablet**	Mobile devices are used to present the information in a suitable way and to offer interaction possibilities. Tablets and smartphones are grouped under mobile devices, while smart watches are evaluated independently, as they allow free hands, for example.
	T1.2	**Smart Watch**	
Stationary T2	**T2.1**	**Touch Interface**	The usage of stationary screens at the workplace is divided into screens with touch-sensitive display and without touch display.
	T2.2	**Mouse & Keyboard**	
Augmented Reality T3	**T3.1**	**Smart Glasses**	Task-relevant information is overlaid on the workers' environment (by a projector or smart glass) or a visual representation of the environment (by mobile device).
	T3.2	**AR Tablet**	
	T3.3	**Projector**	

Each of the presented categories can be, as explained, further differentiated by the device being used. Hence, there are in total seven different solutions that are evaluated.

3.5. *Method for evaluating technological solutions*

The selection of the best fitting solution from the available alternatives requires an informed decision. The goal of the evaluation is a general recommendation for solution alternatives in battery manufacturing use cases. As we conduct the evaluation based on general use cases and not specific use cases for a particular battery manufacturer, a qualitative evaluation method, the utility analysis is chosen. Utility analysis assigns a value to each alternative depending on its characteristics. The total value of each alternative is calculated by the weighted sum of each characteristic. The alternative with the highest value (score) is regarded as the best decision among the available alternatives.[51] The utility analysis was performed as follows.

3.5.1. *Selection of experts*

Qualitative scoring of alternatives is prone to subjectivity. In order to arrive at rational evaluation, three experts are selected to conduct the utility analysis independently.

3.5.2. *Scoring of weights*

The previously introduced requirements are scored with respect to their importance for each use case. Each expert scores individually on a Likert scale from 0 to 2,[51] whereby 0 represents no relevant impact and 2 high relevance of the requirement for the use case. In addition, the scale values from 0 to 2 are more precisely defined for each requirement in the form of continuous texts. Without a detailed description, there is a possibility that the experts will misinterpret the given scale values. Detailed descriptions for every use case ensure that there are no misinterpretations (see Table 7).

3.5.3. *Scoring of technological alternatives*

All technological alternatives are scored by each expert individually in terms of fulfilling the requirement on a three-point Likert scale whereby 0 indicates that the technology does not satisfy the requirement at all. A score of 1 means that the criteria for the requirement are only partially met by the technology under consideration. Accordingly, the score 2 is assigned as soon as a technology has all the criteria to meet the requirement. The descriptions for the rating of a requirement are listed below. This again serves the purpose of avoiding misunderstandings and creating a uniform perspective and is done for all requirements (see Table 8).

Table 7. Scale of the weights describing the importance of the requirement R1 — Hands-free operation.

Value	Importance	Description
0	No relevant impact	Hands-free operation plays no role in the use case. Workers do not need both hands free for fulfilling the tasks required for the use case.
1	Noticeable impact	Workers do not strictly require both hands free for most of the time of the use case, but holding a device makes the task more cumbersome.
2	Decisive impact	Workers require both of their hands for fulfilling the task (e.g. for holding a tool) for most of the time of the use case.

Table 8. Rating score of technologies describing the fulfillment of requirement R1 — hands-free operation.

Score	Rating	Description
0	Not met	Device or technology does not satisfy requirements for hands-free operation.
1	Acceptable	There are certain limits or various options of how device/technology can be operated.
2	Fully addressing	Device or technology can be used without hands.

3.5.4. *Conducting expert ratings*

Three terms have been defined for the evaluation so far. The use cases describe which activity is performed by which type of employee and how often. The requirements describe the required attributes which can be more or less relevant for the use case and which can be fulfilled by the technological alternatives.

Each expert is presented with the following scheme for evaluation: Tables 9 and 10. These two tables have to be filled out by each expert. In Table 9, each column corresponds to one of the four defined use cases. The rows represent the eleven requirements defined earlier. The expert must now assign a weight (0, 1 or 2) to each requirement for all use cases.

Table 10 is completed according to a similar scheme. The columns in this case represent the technological alternatives. Column 1 corresponds to technology T1.1, which according to the assignment corresponds to the tablet. As defined before, there is a scale from 0 to 2 used as well. The expert assigns a value depending on the extent to which the requirement is met by the technological alternative.

3.5.5. *Summarizing the expert ratings*

Interviewing several independent persons inevitably results in different conclusions. This objectivity is desired, but a consensus must be found for further evaluation. For this reason, the following basic rules are introduced in Table 11.

In the present case, a total of three experts are interviewed. This means that three variants of Tables 9 and 10 each must be evaluated according to the above scheme. In total, two overall tables are created. One summarizes the weights from Table 9. The second one (Table 10–technology) summarizes the technology ratings.

Table 9. Weights of all requirements showing their importance per use case.

	Use Cases			
Requirements	Material Use (UC1)	Parameter Support (UC2)	System Maintenance (UC3)	Process Monitoring (UC4)
Hands-free operation (R1)	(0,1,2)	(0,1,2)	...	...
Mobility (R2)	(0,1,2)	...	...	...
...	...	...	...	...
Required footprint (R11)	...	...	...	...

Table 10. Technology ratings per requirement.

	Mobile Device Screens		Stationary Screens		Augmented Reality		
Requirements	Tablet/ Phablet (T1.1)	Smart Watch (T1.2)	Touch Interface (T2.1)	Mouse & Keyboard (T2.2)	Smart Glasses (T3.1)	AR Tablet (T3.2)	Projector (T3.3)
R1	(0,1,2)	(0,1,2)	(0,1,2)	…	…	…	…
R2	(0,1,2)	(0,1,2)	…	…	…	…	…
…	…	…	…	…	…	…	…
R11	…	…	…	…	…	…	…

Table 11. Defined rules to find consent between different experts' voting (only valid under consideration of three experts).

		Example				
Rule	Description	Rating Expert 1	Rating Expert 2	Rating Expert 3		Result/ Consent
No. 1	Consent — All experts assigned the same value	2	2	2	→	2
No. 2	Majority vote — Two experts assigned the same value	1	0	1	→	1
No. 3	Every expert assigned a different value	2	1	0	→	1

3.5.6. *Calculation of weighted scores*

Now, given the two overall tables, it is possible to calculate the weighted scores for each technology depending on the use case. Considering Use Case 1 "Material Use", this is done by taking the first column of the weight from Table 9 and multiplying each cell value with cell values of each column of technology from Table 10. If this process is repeated for each use case, there will be four weighted technology tables in total.

3.5.7. *Calculation of total scores and rank*

To calculate the total score, only the sum of each column (technology) must be formed. This is done by adding up all the elements of the column. This procedure is carried out within a use case for each technology and enables the suitability to be compared because the higher the total score, the better the technology was assessed by the experts together with the weighting. This results in seven overall ratings for each use case, one for each technology. Within the use case, the technologies can be evaluated according to their rank.

4. Results

In the previous section, the use cases, requirements and technologies were presented in detail. Together with the evaluation method, the technologies can now be evaluated in relation to the use case. For this purpose, the calculation of the weighted scores is shown in the following using an example.

4.1. *Execution of expert voting*

All three interviewed experts filled out Tables 9 and 10 according to the scheme presented above. However, due to the limited size of the document, these tables are not shown here. Instead, the summarized tables (Tables 12 and 13) are shown according to the rules (Table 11) introduced above.

To illustrate the application of the calculation scheme, the process is carried out for Use Case 1 "Material Use" as an example. The first column of the weight in Table 12 is set off against the columns of the technology in Table 13 accordingly. This results in Table 14. After performing the column multiplication, the total score of each column can be calculated by adding

Table 12. Importance rating of requirements regarding different use cases (after concatenating all expert ratings).

	Use Cases			
Requirements	Material Use (UC1)	Parameter Support (UC2)	System Maintenance (UC3)	Process Monitoring (UC4)
Hands-free operation (R1)	2	1	2	0
Mobility (R2)	1	0	2	0
Expec. user compatibility (R3)	1	1	0	2
Interaction richness (R4)	1	1	2	0
Customization effort (R5)	0	1	2	1
Interaction robustness (R6)	1	2	2	0
Routine setup (R7)	2	2	1	2
Collaboration capability (R8)	0	0	2	2
Dir. distribution of info. (R9)	1	1	2	0
Mean time to awareness (R10)	1	1	2	2
Required footprint (R11)	1	1	2	0

Table 13. Technology ratings per requirement (after concatenating all expert ratings).

	Mobile Device Screens		Stationary Screens		Augmented Reality		
Requirements	Tablet/ Phablet (T1.1)	Smart Watch (T1.2)	Touch Interface (T2.1)	Mouse & Keyboard (T2.2)	Smart Glasses (T3.1)	AR Tablet (T3.2)	Projector (T3.3)
R1	0	1	0	0	2	0	2
R2	2	2	0	0	2	1	0
R3	2	1	2	2	0	1	0
R4	1	0	1	2	1	1	2
R5	1	1	2	2	0	0	1
R6	1	1	1	2	1	0	0
R7	1	1	2	2	0	0	2
R8	2	1	2	2	0	1	2
R9	2	2	1	1	2	2	0
R10	1	2	0	0	2	1	1
R11	1	2	0	0	2	1	0

Table 14. Evaluation matrix for technology alternatives considering Use Case 1 "Material Use".

	Mobile Device Screens		Stationary Screens		Augmented Reality		
Requirements	Tablet/ Phablet (T1.1)	Smart Watch (T1.2)	Touch Interface (T2.1)	Mouse & Keyboard (T2.2)	Smart Glasses (T3.1)	AR Tablet (T3.2)	Projector (T3.3)
R1	2·0 = 0	2·1 = 2	2·0 = 0	2·0 = 0	2·2 = 4	2·0 = 0	2·2 = 4
R2	1·2 = 2	1·2 = 2	1·0 = 0	1·0 = 0	1·2 = 2	1·1 = 1	1·0 = 0
R3	1·2 = 2	1·1 = 1	1·2 = 2	1·2 = 2	1·0 = 0	1·1 = 1	1·0 = 0
R4	1·1 = 1	1·0 = 0	1·1 = 1	1·2 = 2	1·1 = 1	1·1 = 1	1·2 = 2
R5	0·1 = 0	0·1 = 0	0·2 = 0	0·2 = 0	0·0 = 0	0·0 = 0	0·1 = 0
R6	1·1 = 1	1·1 = 1	1·1 = 1	1·2 = 2	1·1 = 1	1·0 = 0	1·0 = 0
R7	2·1 = 2	2·1 = 2	2·2 = 4	2·2 = 4	2·0 = 0	2·0 = 0	2·2 = 4
R8	0·2 = 0	0·1 = 0	0·2 = 0	0·2 = 0	0·0 = 0	0·1 = 0	0·2 = 0
R9	1·2 = 2	1·2 = 2	1·1 = 1	1·1 = 1	1·2 = 2	1·2 = 2	1·0 = 0
R10	1·1 = 1	1·2 = 2	1·0 = 0	1·0 = 0	1·2 = 2	1·1 = 1	1·1 = 1
R11	1·1 = 1	1·2 = 2	1·0 = 0	1·0 = 0	1·2 = 2	1·1 = 1	1·0 = 0
Total	12	14	9	11	14	7	11
Rank	3	1	6	4	1	7	4

up the values of each element in one column. Based on these values, the ranking can finally be performed in the last row. The evaluation matrix (Table 14) exists four times in total (one table per use case). The only difference is the use of the column from the weight (Table 12) with which the calculation is carried out.

4.2. *Interpretation of total scores*

For interpretation purposes, the scores (penultimate row of Table 14) are presented as a bar chart. However, looking at the absolute values makes it difficult to compare the different use cases. This is because different maximum values can be achieved as a result of the different weightings per use case.

This is easily seen when each value of a column (use case) of Table 12 is multiplied by the maximum value 2 and then all values are added up. For Use Case 1 and Use Case 2, the maximum score is 22, for Use Case 3 it is 38 and for Use Case 4 it is 18.

The maximum values can be used to standardize the scores achieved for each technology. This makes it easy to compare the different use cases, since a uniform scale is used on the Y-axis. Likewise, it becomes clear how large the difference is to reach the maximum value. This analysis is initially carried out individually for each use case.

4.2.1. *Use Case 1 — Material use*

Figure 3 shows the distribution of scores across the different technologies. According to the illustration, the smart watch and smart glasses are the most suitable technologies for this use case. This is initially plausible since it was assumed in this use case that information must be provided quickly and precisely during the mixing process. Ideally, the operation is performed without the use of the worker's hands. The tablet/phablet, the stationary screen with mouse and keyboard, and the projector achieve somewhat lower ratings because hands-free use is not possible here.

The touch interface and the AR tablet are unsuitable for the present use case with their number of points. Although the touch interface and the tablet are controlled by touch gestures, the tablet offers the advantage of mobility. In many cases, this is easier than working with a stationary screen. The AR tablet takes last place due to poor handling and control. The presentation of the required information via AR is not effective in this case.

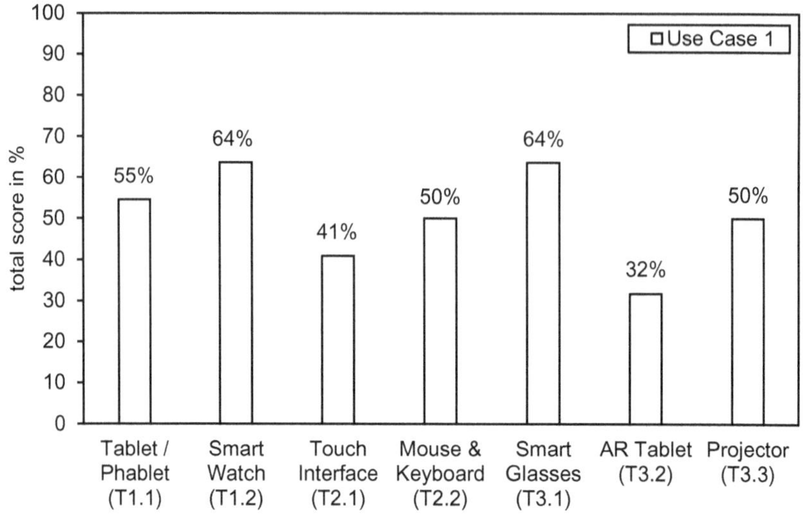

Fig. 3. Total score of alternative technologies considering Use Case 1 "Material Use" and a maximum total score of 22.

4.2.2. *Use Case 2 — Parameter support*

In this application, it is necessary to clearly display the process parameters to be set to the user without errors. This is necessary, for example, when setting up the machine or when correcting settings during the process. Figure 4 shows the total scores achieved by the individual technologies. The highest score was achieved by the stationary screen with mouse and keyboard. Here, 68 percent of the total score was achieved. The technology is very easy to use. In addition, information can be accessed quickly and precise inputs and outputs are possible. The tablet, together with the smart watch, the touch interface and the smart glasses, is at a similar level. These technologies have in common the fact that interaction is intuitive but not necessarily error-free. In this use case, however, fast and at the same time secure interaction is required. This can be seen in the rating of the importance of requirements of interaction robustness (R6) and routine setup (R7) — compare Table 12.

The AR technologies tablet and projector are in the last two ranks. A sufficiently accurate interaction with the technologies and reproduction of the parameter is not given here.

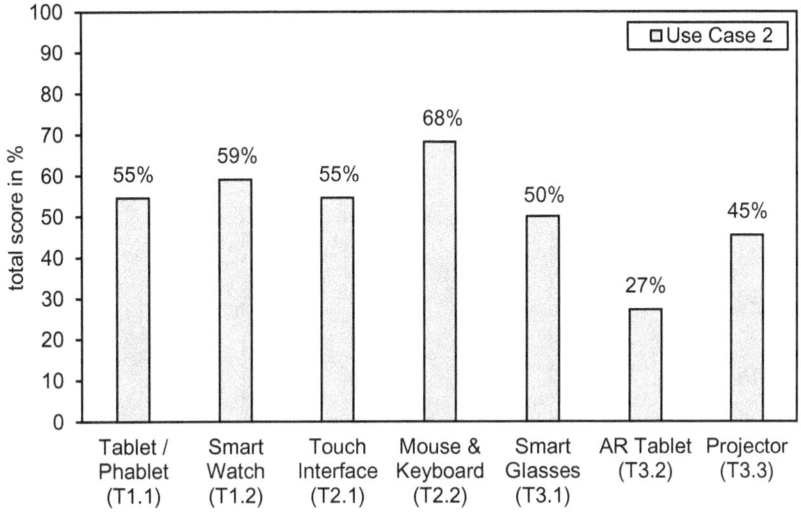

Fig. 4. Total score of alternative technologies considering Use Case 2 "Parameter Support" and a maximum total score of 22.

4.2.3. *Use Case 3 — System maintenance*

The weightings assigned to the third use case are very interesting. The experts almost exclusively assigned the highest priorities to most of the requirements. This is due in particular to the fact that a lot of interaction with the device takes place during maintenance work. Both inputs and outputs are made. Often, content must also be shared with colleagues. Since this use case is carried out exclusively by specialist personnel, the requirement of expected user compatibility (R3) is correspondingly low. The duration and effort of the routine setup (R7) are also of lower relevance, since this use is executed less frequently and not several times per shift.

The smart watch achieves the two best ratings together with the smart glasses, closely followed by the tablet/phablet. The smart watch has many advantages due to its small footprint and the ability to be worn on the wrist. This provides maximum mobility, and information can be provided immediately to the user and make them aware of it. Smart glasses have similar advantages. The biggest advantage is that they can be used without the use of hands. The control can therefore take place without contact, which brings great advantages in maintenance. The stationary screens score worse due to their limited mobility and

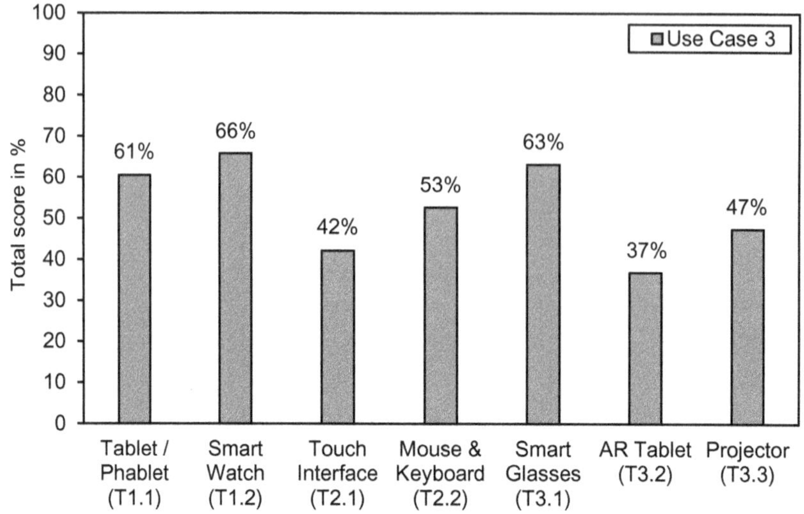

Fig. 5. Total score of alternative technologies considering Use Case 3 "Process Maintenance" and a maximum total score of 38.

comparatively cumbersome way of interacting. The AR tablet and projector are also not good alternatives. Here, interaction and the display of information are also a problem.

4.2.4. *Use Case 4 — Process monitoring*

For this use case, the experts require a high level of user compatibility in particular. Any failures must be detected as quickly as possible so that the process can be adjusted immediately. The prioritization of the requirement of mean time to awareness (R10) is correspondingly high. In addition, the information should be made visible to many people. The maximum score of 18 is comparatively low because requirements such as hands-free operation (R1), mobility (R2) and interaction richness (R4) as well as interaction robustness (R6) are not considered important. In summary, it can be said that it is not so much the operation of the technology that is relevant, but rather the presentation and the fastest possible visibility of the information.

Based on these weightings, stationary screens achieve the best results with 78 percent in each case. Even across the board, this is relatively the best suitability of a technology with regard to a use case. Stationary screens offer the advantage that they display information permanently and in a very simple way. They can communicate visually with the user very easily. The tablet also meets important

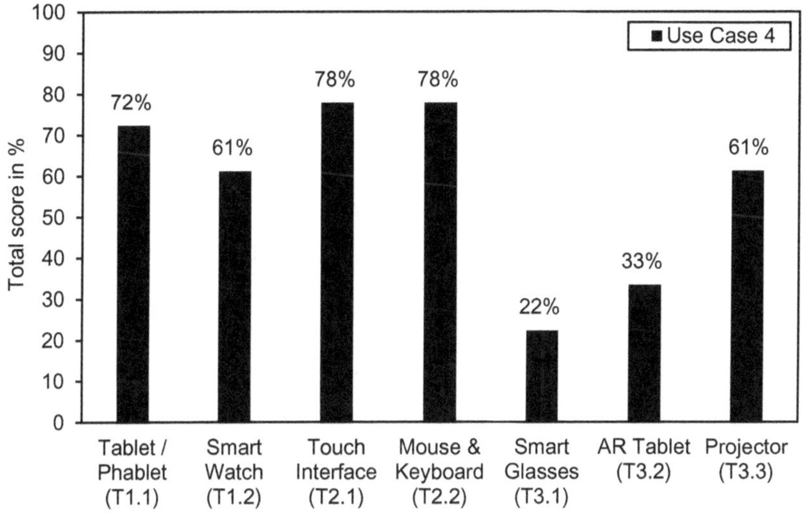

Fig. 6. Total score of alternative technologies considering Use Case 4 "Process Monitoring" and a maximum total score of 18.

requirements such as high user compatibility (R3) and collaboration capability (R8). It is not as good at quickly alerting the worker to information.

The smart watch fulfills many requirements only in the average, i.e. just sufficiently. In total, however, this is enough to achieve a score of 61 percent. The projector achieves the same score. However, user compatibility (R3) is rated very low by the experts, although it is considered very important. The smart glasses and the AR tablet achieve very poor results due to the distribution of the weightings. When one looks at Tables 12 and 13, it is noticeable that the strengths of the two technologies are almost inverted in relation to the importance of the requirements.

4.2.5. *Comprehensive view*

The illustration of the plots in Fig. 7 makes it possible to assess the general suitability for several use cases. Here, for example, it becomes clear that the tablet and the smart watch are suitable for a wide range of applications. This also applies to the AR tablet and the projector, although it should be noted here that their suitability is in a lower range. Stationary screens, just like the smart glasses, are suitable for a more specific application in the respective use case. This can be clearly seen in the fluctuating height of the bars in the diagram.

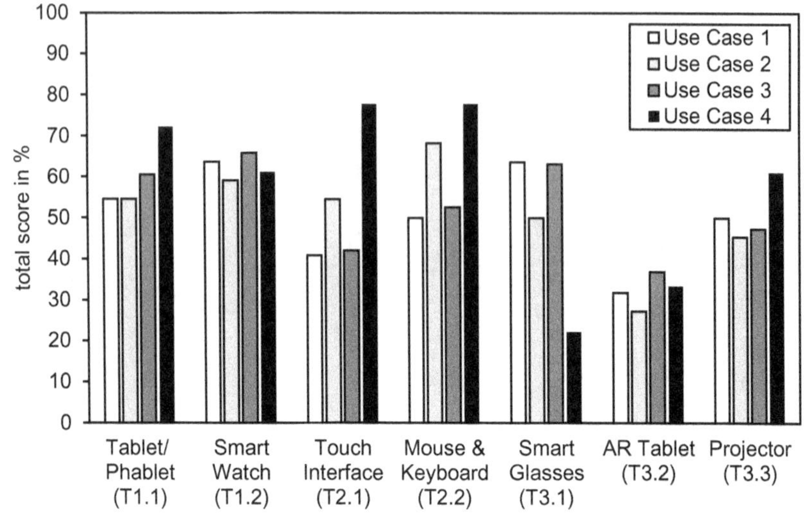

Fig. 7. Total score of alternative technologies considering all four Use Cases.

5. Conclusion and Outlook

This section examines the use cases for visual support of production workers in battery manufacturing and how these use cases can be effectively supported by different digitalization concepts. Furthermore, an evaluation method is presented to determine which visual information tools are best suited for production workers.

First, the use cases that production workers experience during battery manufacturing in the workshop are introduced. Then, the requirements for these use cases are determined. Afterwards, solution approaches for these use cases are developed, in consideration of the requirements. Finally, the visual information tools are evaluated via the proposed method. The results of the evaluation are clearly shown in Fig. 7 of Section 4. The results state that the tablet/phablet and smart watch technologies have a tendency toward generic usability. Stationary screens, however, also achieve very good results, but these are more dependent on the use case. The AR technology of smart glasses also shows great suitability in specific use cases. However, the AR tablet and the projector are not considered to be suitable overall.

For the final selection of a technology in a practical case, an additional cost analysis is required in addition to this technical analysis. The costs incurred are strongly dependent on the progress of the technology and the quantity required. Another factor that should not be underestimated is the human factor. If training is required for operation, this must also be considered.

In addition to the cost or economic considerations, it should be noted that the present analysis is a snapshot. In particular, the assessment of technologies for the requirement for expected user compatibility may change over time if new technologies become established in people's everyday lives in the near future. For this reason, it is possible that the overall ratios will shift.

Furthermore, it makes sense to expand and diversify the group of experts. Having a high number of experts ensures that the opinion of the general public is reflected by the sample. It is also important to interview experts on the store floor as well as process experts. This shows whether and to what extent the views of the two parties differ from each other.

In summary, the proposed method offers a good starting point for evaluating technologies for one specific use case. After the plausibility check of the results, it is conceivable, for example, to set up a practical test scenario in which the two best technologies are compared with each other.

References

1. https://de.statista.com/statistik/daten/studie/973273/umfrage/prognostizierte-anzahl-der-neuzulassungen-von-elektroautos-weltweit/ bis 2030; 2019.
2. Westkämper E, Löffler C. Strategien der Produktion: Technologien, Konzepte und Wege in die Praxis. Berlin, Heidelberg: Springer; 2016. Available from: http://gbv.eblib.com/patron/FullRecord.aspx?p=4530194
3. Bauer W, Hämmerle M, Bauernhansl T, Zimmermann T. Future work lab. Springer Vieweg, Berlin, Online ISBN: 978-3-662-55890-4. pp. 171–87.
4. International Labour Office. Key indicators of the labour market, 2001–2002. 1st ed. Geneva: Psychology Press; 2002. ISBN: 9780415939522.
5. Kłosowski G, Gola A, Świć A. Application of fuzzy logic controller for machine load balancing in discrete manufacturing system. In: Jackowski K, Burduk R, Walkowiak K, Woźniak M, Yin H, editors. Intelligent Data Engineering and Automated Learning–IDEAL 2015: 16th International Conference, Wroclaw, Poland; 2015 Oct 14–16. Proceedings. Cham, Heidelberg, New York, Dordrecht, London: Springer; 2015. pp. 256–63 (Lecture Notes in Computer Science; vol 9375).
6. Mossa G, Boenzi F, Digiesi S, Mummolo G, Romano VA. Productivity and ergonomic risk in human based production systems: a job-rotation scheduling model. Int J Prod Econ. 2016;171:471–7.
7. Nowakowski T, Mlynczak M, Jodejko-Pietruczuk A, Werbinska-Wojciechowska S, editors. Safety and reliability: methodology and applications. CRC Pr I Llc; 2014.
8. Allianz Industrie 4.0 Baden-Württemberg. 100 Orte für Industrie 4.0 in Baden-Württemberg; 2020 [cited 2020 Mar 2]. Available from: https://www.i40-bw.de/de/100-orte/
9. Kasselmann S, Willeke S. Interaktive Assistenzsysteme: International Performance Research Institute; Institut für Integrierte Produktion Hannover; 2017 [cited 2020 Feb 19]. Available from: https://www.iph-hannover.de/_media/files/downloads/Projekt_40-Ready_Technologie-Kompendium.pdf
10. Bundesministerium für Wirtschaft und Energie. Landkarte Anwendungsbeispiele Industrie 4.0: Bundesministerium für Wirtschaft und Energie; 2020 [cited 2020 Mar 2].
11. Adtance GmbH & Co. KG. Adtance support [cited 2020 Jan 10]. Available from: https://www.plattform-40.de/PI40/Redaktion/DE/Anwendungsbeispiele/341-adtance-gmbh/beitrag-adtance-gmbh-support.html

12. ASYS Group. ASYS PULSE Softwarelösung: Intelligentes Assistenzsystem für die Produktionsumgebung; 2016 [cited 2020 Jan 10]. Available from: https://www.i40-bw.de/de/100orte/asys-group/
13. aucobo GmbH. Mobile Maschinenbedienung mit Smartwatches; 2017 [cited 2020 Jan 10]. Available from: https://www.i40-bw.de/de/100orte/aucobo-gmbh/
14. Balluff GmbH. Transparenz durch Mold ID-System; 2016 [cited 2020 Jan 9]. Available from: https://www.i40-bw.de/de/100orte/balluff-gmbh/
15. BASF SE Ludwigshafen. Mobile Bereitstellung von Informationen in Produktion und Instandhaltung [cited 2020 Jan 10]. Available from: https://www.plattform-i40.de/IP/Redaktion/DE/Anwendungsbeispiele/042-augmented-reality/beitrag-augmented-reality-mobile-solutions-basf.html
16. Bosch Rexroth AG. Hydraulikventil-Montage Homburg: Multi-Produktmontagelinie–Industrie 4.0 in der Kleinserienfertigung [cited 2020 Jan 13]. Available from: https://www.plattform-i40.de/IP/Redaktion/DE/Anwendungsbeispiele/235-hydraulikventil-montage-homburg-bosch-rexroth/beitrag-hydraulikventil-montage-homburg-bosch-rexroth.html
17. Cioplenu GmbH. Cioplenu–Die Software für digitale Arbeitsanweisungen, Checklisten und Protokolle: Variantenvielfalt beherrschen und Mitarbeiter entlasten: Mit einer Software für die Erstellung und Nutzung von digitalen Arbeitsanweisungen und Checklisten [cited 2020 Jan 13]. Available from: https://www.plattform-i40.de/PI40/Redaktion/DE/Anwendungsbeispiele/508-cioplenu/beitrag-cioplenu.html
18. E. Zoller GmbH & Co. KG. Werkzeugüberwachung im Fertigungsprozess; 2016 [cited 2020 Jan 10]. Available from: https://www.i40-bw.de/de/100orte/e-zoller-gmbh-und-co-kg/
19. Era-contact GmbH. "Papierlose Fertigung" — Softwarelösung zur Optimierung der Fertigungsprozesse bei der era-contact; 2017 [cited 2020 Jan 10]. Available from: https://www.i40-bw.de/de/100orte/era-contact-gmbh/
20. Essert GmbH. Augmented Automation: Allianz Industrie 4.0 Baden-Württemberg; 2016 [cited 2020 Jan 10]. Available from: https://www.i40-bw.de/de/100orte/essert-gmbh/
21. Fischerwerke GmbH & Co. KG. Prozessleitsystem für die Mischerei; 2018 [cited 2020 Jan 10]. Available from: https://www.i40-bw.de/de/100orte/fischerwerke-gmbh-und-co-kg/
22. Gambro Dialysatoren GmbH. Intelligente Vernetzung von Fertigungsdaten steigert die Effizienz in der Produktion: Intelligente Produktionssteuerung [cited 2020 Jan 13]. Available from: https://www.plattform-40.de/PI40/Redaktion/DE/Anwendungsbeispiele/302-gambro/beitrag-gambo.html
23. GPIO Solutions GmbH. Assemble-by-Light; 2017 [cited 2020 Jan 10]. Available from: https://www.i40-bw.de/de/100orte/gpio-solutions-gmbh/
24. Grundig Business Systems GmbH. Grundig Business Systems GmbH: Traceability: Grundig Business Systems setzt auf Traceability und Prozessüberwachung [cited 2020 Jan 13]. Available from: https://www.plattform-i40.de/PI40/Redaktion/DE/Anwendungsbeispiele/080-grundig-business-systems-setzt-auf-traceability-und-prozessueberwachung/beitrag-grundig-business-systems-setzt-auf-traceability-und-prozessueberwachung.html
25. HARTING KGaA. Flexibles Montagekonzept durch autonome Fertigungskomponenten: Modulare Produktionsplattform ermöglicht Losgröße 1 Fertigung für die individualisierte Herstellung von Steckverbinderkomponenten [cited 2020 Jan 13]. Available from: https://www.plattform-i40.de/IP/Redaktion/DE/Anwendungsbeispiele/227-flexibles-montagekonzept-durch-autonome-fertigungskomponenten/beitrag-flexibles-montagekonzept-durch-autonome-fertigungskomponenten.html
26. Kinemic GmbH. Gestensteuerung mittels Wearables und Smartwatches; 2017 [cited 2020 Jan 10]. Available from: https://www.i40-bw.de/de/100orte/kinemic-gmbh/
27. memex GmbH. Qualifizieren und Instruieren auf dem Shopfloor: mit Utility Film; 2017 [cited 2020 Jan 10]. Available from: https://www.i40-bw.de/de/100orte/memex/

28. Optimum datamanagement solutions GmbH — Kognitive Assistenzsysteme; 2017 [cited 2020 Jan 10]. Available from: https://www.optimum-gmbh.de/industrie-40/kognitive-assistenzsysteme
29. Picavi GmbH. Picavi Pick-by-Vision beim Kosmetikhersteller Dr. Babor im Echtbetrieb: Picavi ist das erste Unternehmen weltweit, das Pick-by-Vision — die Kommissionierung mit Datenbrillen — von einer Vision zur marktreifen Lösung für die Intralogistik entwickelt hat; [cited 2020 Jan 13]. Available from: https://www.plattform-i40.de/PI40/Redaktion/DE/Anwendungsbeispiele/339-picavi-gmbh/beitrag-picavi-gmbh.html
30. Schnaithmann Maschinenbau GmbH. Navigationssystem für den Arbeitsplatz; 2016 [cited 2020 Jan 10]. Available from: https://www.i40-bw.de/de/100orte/schnaithmann-maschinenbau-gmbh/
31. TRUMPF Werkzeugmaschinen GmbH + Co. KG. Die digital vernetzte Blechfertigung; 2016 [cited 2020 Jan 9]. Available from: https://www.i40-bw.de/de/100orte/trumpf-werkzeugmaschinen-gmbh-und-co-kg/
32. Volkswagen AG. Einsatz der Pick-by-Vision-Lösung in den Volkswagen-Werken Wolfsburg und Emden; [cited 2020 Jan 13]. Available from: https://www.plattform-i40.de/IP/Redaktion/DE/Anwendungsbeispiele/261a-xpick-ubimax-volkswagen/einsatz-der-pick-by-vision-loesung.html
33. WS System GmbH. Optimierung der Produktion und Qualitätssicherung mit Werkerführungslösung xMake und Smart Glasses bei WS System GmbH; [cited 2020 Jan 13]. Available from: https://www.plattform-i40.de/IP/Redaktion/DE/Anwendungsbeispiele/261b-werkerfuehrung-in-der-produktion-ws-system-ubimax-gmbh/einsatz-von-datenbrillen-gestuetzter-werkerfuehrung.html
34. Aksu V, Jenderny S, Kroll B, Röcker C. A digital assistance system providing step-by-step support for people with disabilities in production tasks. In: Ahram TZ, Falcão C, editors. Advances in usability, user experience and assistive technology: proceedings of the AHFE 2018 International Conferences on Usability &User Experience and Human Factors and Assistive Technology, held on 2018 Jul 21–25, in Loews Sapphire Falls Resort at Universal Studios, Orlando, Florida, USA. Cham: Springer; 2019. pp. 775–785 (Advances in Intelligent Systems and Computing; vol 794).
35. Hahn J, Ludwig B, Wolff C. Mixed reality-based process control of automatic printed circuit board assembly lines. In: Extended Abstracts of the 2018 CHI Conference on Human Factors in Computing Systems. Erscheinungsort nicht ermittelbar: Verlag nicht ermittelbar; 2018. pp. 1–6.
36. Haslgrubler M, Gollan B, Ferscha A. A cognitive assistance framework for supporting human workers in industrial tasks. IT Prof. 2018;20(5):48–56.
37. Langer M, Soffker D. Situation-based process guiding and supervision assistance system for semi-automated manufacturing processes. In: 2012 IEEE International Conference on Systems, Man, and Cybernetics. [Place of publication not identified]: IEEE; 2012. pp. 820–5.
38. Muller R, Horauf L, Bashir A. Cognitive assistance systems for dynamic environments. In: Proceedings, 2019 24th IEEE International Conference on Emerging Technologies and Factory Automation (ETFA): Paraninfo Building, University of Zaragoza, Zaragoza, Spain, 2019 Sep 10–13. Piscataway, NJ: IEEE; 2019. pp. 649–56.
39. Müller R, Vette-Steinkamp M, Hörauf L, Speicher C, Bashir A. Worker centered cognitive assistance for dynamically created repairing jobs in rework area. Procedia CIRP. 2018;72:141–6.
40. Nelles J, Kuz S, Mertens A, Schlick CM. Human-centered design of assistance systems for production planning and control: the role of the human in Industry 4.0. In: 2016 IEEE International Conference on Industrial Technology (ICIT): Proceedings: The Howard Plaza Hotel Taipei, Taipei, Taiwan, 2016 Marc 14–17. Piscataway, NJ: IEEE; 2016. pp. 2099–104.
41. Paelke V, Röcker C. User interfaces for cyber-physical systems: challenges and possible approaches. In: Marcus A, editor. Design, User Experience, and Usability: 4th International Conference, DUXU 2015, Held as a Part of HCI International 2015, Los Angeles, CA, USA,

2015 Aug 2–7; proceedings. Cham: Springer; 2015. pp. 75–85 (Lecture Notes in Computer Science; vol 9186).
42. Saggiomo M, Loehrer M, Kerpen D, Lemm J, Gloy Y-S. Human-and task-centered assistance systems in production processes of the textile industry: determination of operator-critical weaving machine components for AR-prototype development. In: 2016 49th Hawaii International Conference on System Sciences (HICSS). [S.l.]: IEEE; 2016–2016. pp. 560–8.
43. Rauh S, Zsebedits D, Tamplon E, Bolch S, Meixner G. Using Google Glass for mobile maintenance and calibration tasks in the AUDI A8 production line. In: 2015 IEEE 20th Conference on Emerging Technologies & Factory Automation (ETFA): 2015 Sep 8–11, City of Luxembourg, Luxembourg. Piscataway, NJ: IEEE; 2015. pp. 1–4.
44. Schafler M, Lacueva-Pérez FJ, Hannola L, Damalas S, Nierhoff J, Herrmann T. Insights into the introduction of digital interventions at the shop floor. In: PETRA 2018: The 11th ACM International Conference on PErvasive Technologies Related to Assistive Environments: 2018 Jun 26–29, Corfu, Greece: Conference Proceedings. New York, NY, USA: ACM; 2018. pp. 331–8 (ICPS).
45. Söffker D, Fu X, Hasselberg A, Langer M. Modeling of complex human-process interaction as framework for assistance and supervisory control of technical processes. Int J Inf Technol Web Eng. 2012;7(1):46–66.
46. Tao W, Lai Z-H, Leu MC, Yin Z, Qin R. A self-aware and active-guiding training & assistant system for worker-centered intelligent manufacturing. Manuf Lett. 2019;21:45–9.
47. Waechter M, Loeffler T, Bullinger AC. Towards a mobile assistance system to raise productivity in maintenance. In: Nunes IL, editor. Advances in Human Factors and Systems Interaction: proceedings of the AHFE. [S.l.]: Springer; 2019. pp. 56–64 (Advances in Intelligent Systems and Computing).
48. Hinrichsen S, Riediger D, Unrau A. Assistance Systems in Manual Assembly: 6th International Conference Production Engineering and Management; 2016 [cited 2020 Feb 19]. Available from: https://www.researchgate.net/publication/311535944_Assistance_Systems_in_Manual_Assembly
49. DIN EN ISO 9241-210:2011-01, Ergonomie der Mensch-System-Interaktion_- Teil_210: Prozess zur Gestaltung gebrauchstauglicher interaktiver Systeme (ISO_9241-210:2010); Deutsche Fassung EN_ISO_9241-210:2010. Berlin: Beuth Verlag GmbH.
50. Kosicka E, Gola A, Pawlak J, editors. Application-based support of machine maintenance. Elsevier; 2019. (IFAC PapersOnLine52-10).
51. Likert R. A technique for the measurement of attitudes, 1932.

https://doi.org/10.1142/9789811245626_0011

Chapter 10

Managing Environmental Sustainability of Battery Cell Manufacturing Lines

Martin Reisinger[*,†,‡] and Michael Oberle[*,¶]
[*]*Fraunhofer-Institute for Manufacturing Engineering and Automation IPA, Nobelstr. 12, 70569 Stuttgart, Germany*
[†]*Institute for Energy Efficiency in Production EEP, Nobelstraße 12, 70569 Stuttgart, Germany*
[‡]*martin.reisinger@ipa.fraunhofer.de*
[¶]*michael.oberle@ipa.fraunhofer.de*

Sustainability assessment of battery cell manufacturing is difficult due to the lack of data. The rise of digitalization provides an opportunity to establish a foundation for facilitation, collection and processing of sustainability data, which in turn enable carbon accounting along the value chain of battery manufacturing systems. We present a conceptual framework for enabling data-based Life Cycle Assessment (LCA) for battery cells or even battery packs based on ISO 20140. The framework aims at automated tracking energy consumption directly from the shop floor and the associated carbon dioxide (CO_2) emissions in the LCA by employing standardized process models fueled by a data warehouse and simulation. This builds the foundation for enabling production system and production strategy benchmarks and analysis and could also enable the vision for a Life Cycle Energy Label for battery cells and vehicles.

1. Introduction

Responsible manufacturing and the reduction of greenhouse gas emissions are two goals for sustainable development[2] and should therefore be increasingly focused on by science and industry. In general, the awareness for the potential of digitalization for minimizing energy consumption and CO_2 emissions in Industry is growing. In its current main report,[3] the German Advisory Council on Global Change points out that the increased availability of data by Industry 4.0 could also lead to further development potentials for resource efficiency. The German Industry 4.0

Platform even promotes sustainability, more specifically, climate change mitigation and the circular economy, as a part of the vision for Global Digital Ecosystems.[4]

According to a study by the Boston Consulting Group,[5] "digitally enhanced structures and processes" in battery manufacturing could reduce the overall costs of battery manufacturing in Germany in comparison to the current manufacturing in China by up to 20%. This impressive number refers to a landed cost comparison of battery manufacturing for electric vehicle suppliers measured in cell costs (excluding materials) per kilowatt hour of capacity (kWh). A closer look at the cost items shows that in a digitally enhanced battery manufacturing scenario in Germany, the site-specific energy cost is higher even with smart production (+1.9 $/kWh in Germany) compared to China. The biggest lever for the named 20% cost reduction seems to be based on the reduction of logistic cost (–7.5 $/kWh in Germany) in comparison to manufacturing in China. However, the study does not quantify the improvement potential of smart manufacturing in energy efficiency or environment sustainability in general as its focus is on location-specific energy cost structures.

The fact that this is not further detailed can be explained by a lack of available data and by data quality for sustainability assessments of battery manufacturing systems. From an industry and policy perspective, there is a need for a location- and manufacturing system-specific sustainability assessment in cell manufacturing. The comprehensive study of the European Commission's science and knowledge service about the Circular Economy Perspectives for the Management of Batteries used in Electric Vehicle concludes that currently "it is difficult [...] to ensure that the most sustainable approach in cell manufacturing has been adopted" and underlines the importance of using international standards for such an assessment.[6]

2. Definitions and Problem Statement

Environmental Sustainability is a broad field in the context of battery manufacturing.[6] This publication focuses on two elements of environmental sustainability — the energy consumption and the related CO_2 emissions in the battery manufacturing value chain.

The value chain of battery manufacturing consists of the following elements[6]:

- Raw material extraction (mining/recycling) and processing.
- Cell component manufacturing.

- Cell manufacturing.
- Battery pack manufacturing.

The term digital manufacturing refers to the uptake of Cyber-Physical Systems (CPSs) in manufacturing. CPSs in general "reflect a vision of a technology that deeply connects our physical world with our information world".[7] For battery manufacturing the application of this paradigm enables improving the yield through traceability of all battery cell components by inline quality monitoring and adaptive manufacturing process control.

These data could also be used for tools to measure and optimize environmental sustainability, presenting an opportunity for tackling the basic challenges for sustainability measurement tools: the lack of data or the high effort to acquire the data.

In the following, examples are presented underlining this foundational challenge on different levels:

- Sustainability optimization on process level: Value stream-oriented tools[8] to optimize resource efficiency on a process level by naming the "time consuming data collection" as a weakness.
- Assessment of sustainability on manufacturing system level: The general ultra-efficiency Frameworks for the benchmark of sustainability on a branch-specific manufacturing system level names "high effort for data acquisition" as a challenge.[9] This is also a challenge for a certification based on the comparable UL 880 Framework.
- Assessment of sustainability of products on a value chain level: A significant effort related to data acquisition and data quality assurance is required for analysing the energy consumption and energy cost structure in a value chain of one product group (e.g. insulation material). An approach addressing this issue could be a dynamic data basis in form of a Life Cycle Inventory (LCI). This could reduce the effort for data acquisition and increase data quality.[10] An LCI can be described as the structured database for conducting Life Cycle Assessment (LCA).

Despite the limited access to real data, the last few decades of research produced a broad set of available tools. The review of toolsets in[11,12] and the assessment framework for Mass Customization and Sustainability[13] provide an overview. Given the number and variety of tools, the focus of applied research should now be rather on evolving existing tools and the applying these tools in CPS as smart services. The automated Life cycle Cost Calculation (LCC) in CPS for single energy consuming manufacturing equipment components as described in[14] provides a good example. As LCC and LCA can be integrated to support each other,[15] there is potential for automating LCA in the context of CPS. This in turn can improve the data quality and allow its integration with the management system.[16]

A similar idea is outlined in the German Battery Production Equipment Roadmap 2030.[17] The Roadmap names a "live LCA" as a solution to improve data availability and quality for more precise LCA in a general technology assessment. However, it identifies the use of LCA as a tool to optimize sustainability on a "shop floor" level with integrated software as the R&D field. This article builds on this idea and outlines a concept for a primary data-based Life Cycle Assessment (LCA) based on the increased data availability in digital battery manufacturing. The development of this concept is guided by the following research questions:

i. What are the current challenges for Carbon Accounting in battery manufacturing?
ii. What is the primary data-based LCA and what is the state of research in this field?
iii. Could an already available solution be combined for a framework for data-based LCA in battery manufacturing?

Each question is addressed in a specific subsection in the following.

3. Situation Analysis in Carbon Accounting in Battery Manufacturing

To outline the challenges for LCA in Battery manufacturing, we use the first subchapter to present a situation analysis for the relevance of measuring and comparing CO_2 Emission of battery manufacturing. In the second subchapter, we focus on the current challenge of LCAs in battery manufacturing: the availability of sufficient data.

3.1. *The relevance of measuring and comparing CO_2 of battery manufacturing systems*

Actual LCA or Product Carbon Footprint (PCF) analyses show that in transferring the manufacturing of vehicles from the combustion engine to e-mobility, a significant portion of the CO_2 emissions is shifted from the operating phase of the vehicles to the manufacturing of the vehicles. However, the potential for reducing CO_2 emissions in cell manufacturing is significant in comparison to the manufacturing of other vehicle components like the vehicle body.[18] This can probably be explained by the fact that vehicle body manufacturing has had decades for its optimization, whereas battery manufacturing on a large scale is a relatively new development.

For battery manufacturing, the associated CO_2 emissions are of high relevance,[19] as in the current transport policy, electric mobility is the focused solution for the decarbonization of the transport system. However, the current regulatory focus in transport is on the CO_2 emission in the use phase of a vehicle.[20] An integrated regulatory approach also considering the emission from manufacturing or the embodied energy in the product is not yet in place. Additionally, the Batteries Directive[21] currently does not tackle a full life cycle approach. The current EU pilot project on the Product Environment Footprint (PEF) only addresses consumer goods, including mobile phone batteries.[22] The BDI criticizes the interim results of the pilot project as "the current status of the PEF methodology and the quality of the available input data are not mature enough for a broad, routine application".[23] In addition PCE has been criticized for limited comparability.[24]

This means that there is little regulatory framework in place to guide the battery industry toward the sustainable development goals mentioned in the introduction. However, the general transformation of the linear to a circular economy with a stronger focus on products with long-lasting design, maintenance, repair, reuse, remanufacturing, refurbishing and recycling[25] will lead to cycle-oriented assessment of CO_2. This assessment approach will allow the quantification of the benefits of circular approaches for electric vehicle batteries in future mobility systems.

Currently, there seems to be a growing relevance for carbon accounting from an industry perspective with voluntary announcements stating decarbonization plans.[26] However, today the economical relevance of carbon neutralization through CO_2 compensation is relatively low. The current price level for CO_2 on the European emissions trading system is €21 per ton of CO_2 in average in Q1 2019[27] and prices for offset projects on a voluntary basis are in the range of €14–€21 per ton of CO_2.[28] This means that the production of an e-car (10 tons of CO_2) would cost about 200 € compensation.

3.2. *Insufficient data availability and unsure data quality for LCA*

The development of the Cumulative Energy Demand (KEA) in the early 80s[29] started the discussion about embodied energy in products. KEA is used for technology assessment and is described in detail in a guideline from the German engineering association, VDI.[30] The assessment is mainly aimed at comparing different technical planning scenarios in regard to the embodied energy through the manufacturing and end of life of the technical systems at the energy and building system level.[31] It does not support the comparison of products, like batteries, produced by different manufacturing systems, although the assessment procedure has elements comparable to the LCA method used in carbon accounting.

LCA is mainly used for technology assessment and its results are often a basis for political or industrial strategy developments or decision on technological alternatives.

Key factors influencing the CO_2 emissions in battery manufacturing include electrical and thermal power consumption[32] and the respective CO_2 emission factors used for the assessment. In the context of battery manufacturing, about 50% of CO_2 emissions are directly allocated to cell manufacturing and the remaining 50% to the upstream value chain of battery production.[33] However, there are varying numbers, for example, in[34] cell manufacturing itself only accounts for 20% of the CO_2 emission.

In general, the data availability of the material-specific energy consumption and the resulting CO_2 emissions of the current Li-ion battery generation have been low for several years.[35,36] This is also presented as a fundamental problem in the current study on the decarbonization of the transport sector in Germany.[19]

There is little information about the manufacturing yield in cell manufacturing. In the ramp-up phase of the cell production line only 15% yield is achieved.[35] Established production lines have a higher yield but are far from the 90% yield of long-running optimized lines of the semiconductor industry.[17] In addition, there is no information or guideline in LCA studies about how this yield and its potential share of lower-quality cells is allocated to the CO_2 emissions of the cell manufacturing systems. Furthermore, there is no standard for how energy consumption of the cell manufacturing system is allocated to the cells, when the system is not in the production status. This includes maintenance, missing material or maybe low demand for the products, which can be at least 10% of the time.[17]

As stated in the beginning of this chapter, the biggest challenges for the carbon neutrality of battery manufacturing lie in the upstream value chain and not only in cell manufacturing itself. The largest share in the current value chain is the energy input for the production of cathode materials.[19] This means that the current raw material extraction, transport and fossil fuel-powered energy-intensive processing of cathode powder like NMC111 can have a share of nearly 40% of the cumulated CO_2 Emission in battery manufacturing.[34] The switch to lithium hydroxide (LiOH•H2O) is expected to further increase the CO_2 emissions required due to additional material processings steps.[38] Despite this high impact of raw material on CO_2 Emission, there is only very limited publicly available life cycle inventory (LCI) data. From over 79 LCAs on Li Batteries, only nine LCAs have published specific life cycle data and most of these are based on the same data source.[39]

Furthermore, in basic research on materials and processes of the next-generation battery, there is scant data, as mentioned, for example, in.[40] In consequence, it is difficult to quantify the potential of new developments, as, for example, new bio-based electrode materials[41] or solid-state designs with improved energy densities,[42] to reduce CO_2 emissions.

In summary, the scientific and publicly accessible database for electricity demand and CO_2 emissions from battery manufacturing is poor.

4. Primary Data-Based Carbon Accounting and State of the Art

In order to address the scarcity of available data, the effort for data acquisition required for LCI as the basis for a product-related LCA needs to be reduced significantly. Ideally, as much information as possible should be extracted from the cell manufacturing system itself and the associated value chain as this would also increase the quality of the database.

One the one hand, the improved data quality and resolution in an LCI should empower LCA as a management tool for the analysis of changes on the sub-processes level of the cell manufacturing system. Individual optimization results due to implemented changes in process operation parameters, process adjustments, reengineering or even the changes in the associated industrial energy dry room and energy supply system should trigger the automatic update of the LCI. Simulated optimization scenarios from digital twins should generate LCI Scenarios to evaluate the impact of possible adjustments.

On the other hand, the improved data baseline should enable a comparison of product-related LCA results on the level of different battery manufacturing plants and organizations. Primary data from the production process steps in the value chain should be included without disclosing production cost structures and production strategies (value streams etc.) to other market participants. This would improve data quality significantly, as these data are currently based on generic LCI databases. Furthermore, this would also enable primary data-based carbon accounting at a company level for tracking the status in regard to decarbonization goals and sustainability reporting. The aim is to reduce cost and improve comparability in sustainability reporting up to the level of real-time CO_2 accounting in the company for the management.

In the following, a literature review of the state of the art in the context of primary data-based carbon account is presented. The first subsection addresses the question of how primary data from the production process steps in the supply chain could be included. The second subsection addresses the question of how the granularity of an LCA could be scaled to the process level.

4.1. *Integrating primary data from the supply chain*

The Greenhouse Gas Protocol (GHG) standard is often used for sustainability reporting at a company level. The emissions of the suppliers are assigned here to the SCOPE 3 emissions.[43]

Daimler, one of the world's premium car manufacturers, also uses this standard. In 2018, Daimler calculated a value of 2.9 million tons for its internal emissions of CO_2 and for the passenger car supply chain 17.4 million tons of CO_2. Converted to passenger car sales (Daimler AG, 2019b), this amounts to 7.5 tons of CO_2 per passenger car.[44–46] It can be assumed that this value will increase significantly by switching the product portfolio from combustion engines to e-mobility, as lightweight construction and battery production are much more energy-intensive. Therefore, increasingly, emissions in upstream supply chains need to be integrated into the climate strategy of automotive OEMs. Daimler plans to "make CO_2 targets an important criterion for supplier decisions and contracts" and to agree upon "concrete CO_2 measures with suppliers".[47] At the current time, the way in which this criterion will be implemented is unclear. The following scenarios are conceivable:

i. Publication of an annual sustainability report with CO_2 targets based on existing Carbon Accounting Standards.
ii. A supply-specific indication of the production system-specific energy consumption and CO_2 emissions per produced unit as well as reporting of the measured values on delivery, if applicable.
iii. The supplier is also responsible for the emissions from the supply chain upstream of the supplier, including the commissioned logistics companies. This means that the supplier also passes the requirements on to his suppliers or pays for their emission compensation projects with regard to these emissions.

For a comparison of the climate neutrality of suppliers, however, a high data quality would certainly be required. The requirements for this could extend to detailed energy consumption data,[43] which might have to be provided by suppliers in the context of one of the SCOPE 3 reports. However, detailed energy consumption data are often considered a trade secret by suppliers, as they provide insight into cost structures.[48]

It is also unclear how the data will be exchanged.[49] For example, existing structures in product compliance could be extended. In the automotive industry, there is a de facto industry standard for material data systems — the International Material Data System (IMDS). Here, suppliers document the material composition of individual components at the component level, for example, right down to information on the proportion of recycled material. IT systems are in place at the supplier level to systematically maintain the data.[50] The IMDS could therefore be further developed into a production system-specific LCI database.

According to the research below, these considerations already exist in the R&D area:

- In the project "Live LCA", an extension of the IMDS is not addressed, but the IMDS is so far only named as a data source.
- Within the framework of a DFG project, the concept of "Environmental Product Lifecycle Management" was presented. By integrating an LCA with the mySAP Business Suite and by automating the inclusion of energy and material flows in the LCI as much as possible, complexity is reduced. In this way, the LCA can be integrated into existing workflows of administrative personnel. The IMDS is regarded as an information supplier for the automated creation of a product and production system-specific LCI, but not explicitly for the renewed data transfer of energy and material flows.[51]
- The dissertation "Automotive Product Life Cycle Assessment" describes how IMDS is used as a basis for LCIs at Volkswagen.[51] Furthermore, there is a concept for partially automating the generation of an LCI.[52]
- A case study at Volvo shows a 90% time saving when using IMDS data for LCI generation.[53]

In the following, further concepts regarding data exchange are named:

- A study by "Agora Verkehrswende" proposes a Clearing House for anonymous data aggregation in the supply chain. The European Platform on Life Cycle Assessment could provide foundational approaches for this idea.[19]
- Within the framework of Industry 4.0 standardization, sustainability data could be passed on in the so-called administration shell.[54] In the context of energy, ISO 20140-5 is proposed as a characteristic for the administrative shell (in the figure on page 19).[55]
- New innovative systems with open, distributed and secure communication protocols based on distributed ledger technology (DLT) could make it possible to aggregate sustainability data in supply chains in a traceable and tamper-proof manner.[56,11]
- Circularize is a start-up that has developed a protocol for tracking materials along the supply chain.[57,58] Smart questioning enables stakeholders to ask

questions about a product and receive trustworthy answers. The data are stored decentrally at the individual companies and the response only includes the data aggregated along the value chain. This could be a solution for the disclosure of cost structures or details about supply chains which are relevant to competition. The focus of the current application is traceability of raw materials, like plastic in the automotive supply chain.[58]

- Under the term Supply Chain Carbon Management,[59] there are already approaches in the scientific field to select suppliers based on their carbon footprint. However, this only addresses specific evaluation criteria in the electronics industry.[60]
- The use of graphs[61] to link data could also be a foundation for a technical solution for carbon accounting.

The analysis of the state of the art shows that there are some early approaches for using IMDS as the basis for LCA in the automotive industry. In addition, there are basic ideas of linking and exchanging data, but these have not yet been applied in the context of LCA. In summary, there is a general lack of experience and solution concepts that are available to focus on the battery manufacturing supply chain and especially for the implementation of the balancing of SCOPE 3 emissions.

4.2. *Integrating data at the product level*

A key challenge for a product-related labeling, also called Product Carbon Footprint (PCF) analysis, is the lack of global and standardized systems.[62] The high costs of implementation also represent an obstacle.[48] The effort required for data collection in internal company processes depends heavily on the available data in the energy management system. Waste or scrap quantities are also of great importance. These are recorded in the Inventory Management System. A further challenge for PCF is the apparent accuracy of the determined values. This is caused by methodical variation, data uncertainty and different system limits. Product comparisons of different manufacturers without a specific standard for the accounting rule for a product category are not meaningful.[63] Automated generation of PCF through data integration with ERP Systems[64,65] is the first step for basing a PCF on more reliable and comparable data for a product category. However, ERP systems typically provide data only on a coarse process level. For a finer process level, a process structure in Manufacturing Execution Systems (MES) would be better suited. However, if an MES is capable of providing data for sub process level depends on its underlying process information model. Process models in ERP and MES are typically proprietary and thus limit comparability if used as a basis for LCA. Furthermore, the process model must enable mapping of the measured or simulated data to a process structure to serve as a database. In the following, approaches are presented which enable the automation of carbon accounting at the product level and thus enable a resolution of LCAs on a sub-process level.

The standard ASTM E3012–16 (Standard Guide for Characterizing 2 Environmental Aspects of Manufacturing) Processes is an information model to describe the structure and procedures of production processes in relation to their environmental impact. The aim of the standard is to enable the collaboration of manufacturing teams by facilitating communication and exchange and providing compatibility with analytical services. The standard allows the decomposition of the production process into individual process steps to a so-called "Unit Manufacturing Process" (UMP). For these UMPs, inputs are defined in terms of energy (electrical, thermal, compressed air), material (workpiece/batch, operating materials) and resources (equipment incl. process parameters, tools, software) as well as outputs (products, by-products and scrap). In comparison to the value stream, the approach allows the assignment of sustainability data to processes with variable throughput times. The E3012 standard describes this information largely in a textual nature that is encoded an in an XML schema.[66] Although physical transformations that occur between different manufacturing operations and KPIs can be captured using mathematical expressions modeled in MathML, E3012 models cannot be used directly for computational purposes due to their textual nature.[67] In order to enable the use of the captured information as a basis for analytical, data-driven simulation models for the optimization of production processes, the information needs to be transformed to other modeling formats aimed at such analysis. This principle has been demonstrated by the formal mapping of E3012 models and the MOca (MOdel Composition and Analysis) domain-specific language. In this way, operational models for analytical purposes such as Design of Experiments, Discrete Event Simulations (DES) and manufacturing optimization studies can be constructed.[67] This transformation from static models to dynamic models enables linking of UMP models to traditional LCA workflows. Generation of LCI data by leveraging UMP models allows for more accurate production models in LCA. For this purpose, E3012 models need be mapped to the ecoSpold2 format which conforms to ISO 14048 and is aimed at curating LCI datasets. This is necessary as the formats are incompatible due to their different purposes. Although the workflow for mapping between the two format requires significant human input, this shows the potential of increasing the accuracy of the LCA down to individual production steps. Furthermore, this would enable manufacturers to conduct automated design-space exploration allowing one to better understand changes in the production process, for example, the evaluation of a change in the number of teeth in a gearbox design, and the associated change in the production process and the resulting environmental impact.[68]

In the following, projects and frameworks are presented which are relevant for sustainability data management in the context of battery manufacturing:

- The cooperation project “Live LCA” started in 2017 and aims at the integration of different existing software solutions,[69] which enable companies to combine available material and energy consumption data for live calculation of LCA and material flow cost accounting (MFCA) and thereby reduce, e.g. the costs for Environmental Product Declarations (EPDs) and LCAs by 90%. Methodologically, LCA and MFCA are linked with primary data from production, for example, from energy management.
- From 2003 to 2008, the “Network Life Cycle Data” project has provided the scientific basis for data provision as an essential basis for any life cycle analysis. The challenge in using SAP data to generate company-specific LCI is the “semantic gap” between these systems. This means that an employee is required to integrate the relevant DAP data into the LCI. Furthermore, the feasibility depends strongly on the production data acquisition.[70]
- In a vision on the use of Big Data in LCA, it is pointed out that local or production system-specific characterization factors could be mapped.[71]
- An architectural concept already exists for providing resource-event data for DES modeling to reduce the effort required for decision support.[72]
- In the context of the transformation of the automotive industry and the resulting changes in production systems, a holistic assessment of environmental impacts is required.[73] A framework for holistically assessing the environmental impacts of different production system variants has been proposed.[73] However, it lacks the specification of the data management and corresponding system architecture. Although the ISO 20140 standard is not referred to in this context, it can be viewed as implementation variant of such an assessment framework. A specification of data management and architecture cannot be excluded from the framework. ISO 20140 is also not named in this context, but the standard basically addresses an implementation variant of these frameworks.
- The integration of energy data from battery production processes in a data warehouse should enable the data-based optimization of the environmental impact (energy consumption, emissions) of the production system. So far, only a first use case in relation to economic efficiency (product quality &

costs) has been published.[74] This data-based approach is conceptually transferred by [75] to the entire life cycle of a battery cell.

In summary, integration of data at the product level is limited by the availability of the data at the process operation level and by the transformation process that is required for LCA. The integration of data directly from the shop floor is envisioned by leveraging CPS as a data source for LCA. However, there is a lack of solution concept for the workflow from leveraging machine data to LCA.

5. Concept for a Sustainability Data Management System as an Enabler for Primary Data-Based LCA

The third Version of the German Industry 4.0 Standardization Roadmap[54] describes data and sustainability management as a strategic development. The term sustainability data management has not yet been further defined in this context, even not in the fourth version of the Standardization Roadmap.[76] In our view, a sustainability data management system should enable the aggregation and evaluation of manufacturing data in order to create energy efficiency and sustainability profiles in different views from manufacturing systems. Views can be process related (single energy consuming manufacturing equipment components, subprocess, process, manufacturing system) or product related. An example for such a product-related profile is the proposed primary data-based LCA (see Section 4). In this context, the ISO 20140 series is an important standardization activity. In essence, it is a method for standardized data management to capture, aggregate and analyze energy and resource consumption data in manufacturing systems.[1] The ISO 20140-1 series of standards names several benefits, which can be summarized in three points:

- Benchmarking against a generic production system or different production systems.
- The analysis of scenarios for optimization.
- Setting targets for improvements and monitoring shop floor operations.

The five-part series of standards is currently being developed in an international standardization working group. Figure 1 shows an overview of ISO 20140 and represents the current state of development. The application of the ISO 20140 series could enable a manufacturing system site-specific certification for valid sustainability quantification on a cell and battery pack system level. By validity, we mean that it should be possible to compare manufacturing system-specific

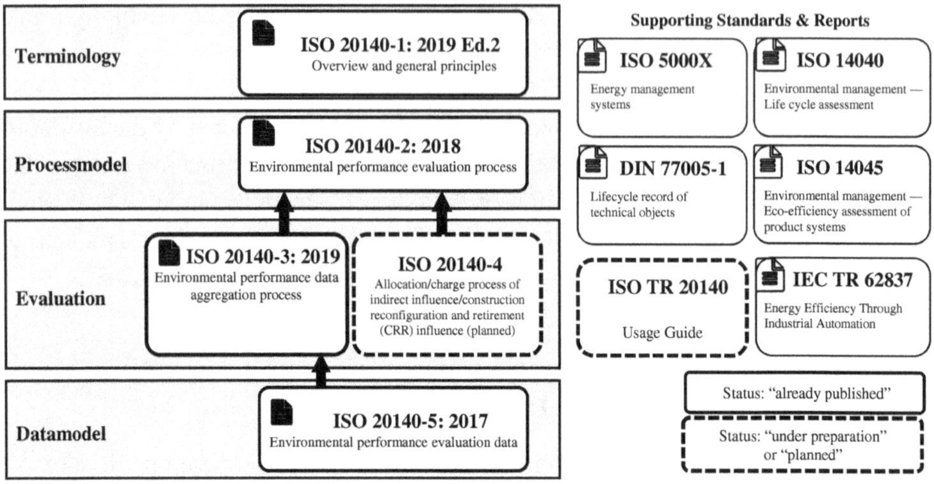

Fig. 1. Overview in reference to Ref. [1]: ISO 20140 Automation systems and integration — Evaluation of energy efficiency and other factors affecting the environment in production systems.[1]

product-related profiles. The term manufacturing system-specific refers to the entire site-specific life cycle of a production system, including the energy and supply system. In the long term, the environmental impacts of manufacturing the production equipment, energy supply system and of the construction of the building, or their redesign and the decommissioning, could also be included. This entire production system-specific life cycle consideration can in turn be integrated into the life cycle from a product design perspective. Transferring the data to the upstream product design processes enables the analysis of the effects of design variants in the context of the production system (see Fig. 2).

A sustainability data management system (see Fig. 3 for conceptual view) should be aggregated automatically and compliant with ISO 20140 to enable its mapping to the standardized process model and to production runs. This enables the creation of life cycle inventory data specifically by production batch number and its associated scrap as a basis for LCAs. In this way, the actual manufacturing process is analyzed in contrast to estimated numbers. On the basis of the standardized process model, simulation results can also be integrated into the analysis process in order to evaluate different production strategies. Our assumption is that production system-specific Life Cycle Inventory data could be structured in the same way as the basic data of event-oriented simulations, in which energy and resource consumption must be stored for each production step.

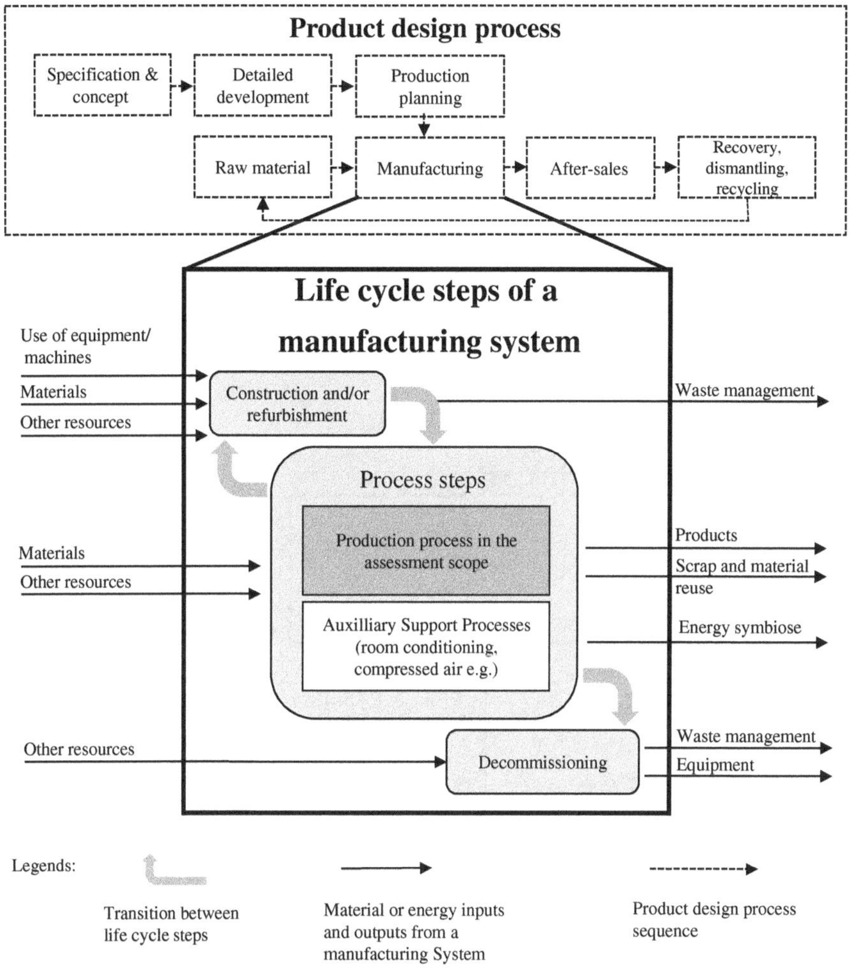

Fig. 2. Integration of life cycle perspectives of the manufacturing system and product design in Ref. 72.

Within the value chain, companies are not willing to provide their primary data-based LCIs to central databases. By extending the circularize protocol[58,57] for Life Cycle Inventory data, it could be possible to retrieve these data in aggregated form from several companies without being able to draw direct conclusions about cost structures, since these data are only available to the automotive OEM in aggregated form. This could enable the integration of primary data from the whole value chain in the LCA.

In order to implement the concept, further steps would be required.

In the following, key next steps are highlighted:

- Gap analysis for the process-specific data acquisition of sustainability data based on actual input from quality management, maintenance and energy management.
- Data mapping to production batch number or data linkage within the traceability concept in order to be able to assign each sub-component of a produced battery cell to its creation parameters and sustainability data.

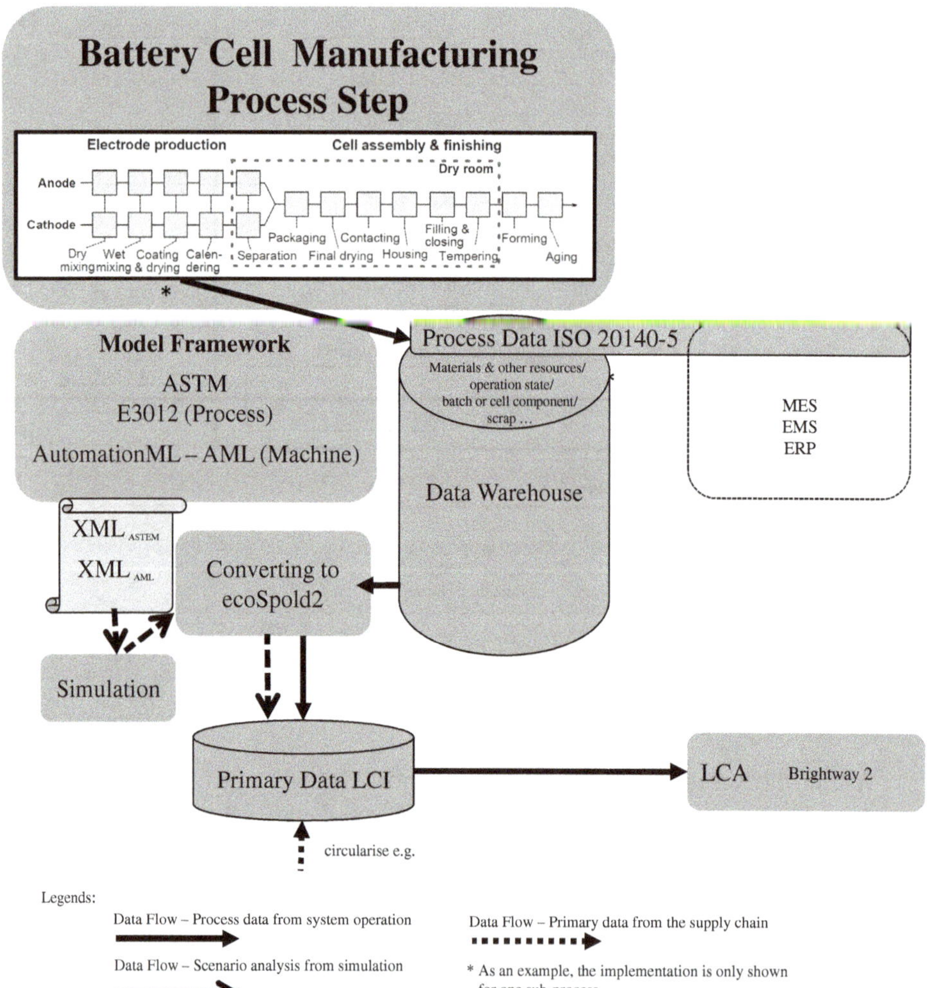

Fig. 3. Sustainability data management system concept for a primary data-based LCA in Refs. [67, 77, 78].

- Mapping procedure for sustainability data to the usable units per batch in order to be able to adequately assign the scrap (e.g. to a batch).
- Application of the Unit Manufacturing Process (UMP) model according to ASTM E3012–16 and the already existing tools to the battery production.
- Analysis of whether hybrid modeling from ASTM E3012 and AutomationML or the Asset Administration Shell simplifies data collection by integration of machine and process level.
- Tool development for the generation of production system-specific LCI datasets based on the ASTM models of ISO 20140 data management.
- Coordination of data formats and concept for a circularize protocol specification for data transmission in the supply chain for the presented use case (cumulated energy consumption kWhel, kWhth, CO_2 emission factors, cumulated CO_2 production, CO_2 life cycle production).

6. Life Cycle Energy Label for Battery Cells and Vehicles

Based on the presented concept, the creation and certification of a CO_2 or energy consumption label for battery cells and vehicles could be possible. Figure 4 shows an example of the energy consumption to be covered by the label.

The extension of CO_2 emission standards from vehicle operation to the production of e-vehicles could be part of a future climate protection policy. And therefore, such a label would be presented when selling vehicles. Furthermore, a production system-specific CO_2 emission or energy label for battery cells could enable the regulation of the import of batteries on the basis of sustainability criteria in the event of protectionist tendencies in the global economy or in the event of an internationally divergent climate protection policy. The introduction of a CO_2 import duty as a protective duty for products from economies without emissions trading based on the LCA approach has also been discussed for years. So far, this has been classified as impracticable.[77] Currently, however, this debate is being reevaluated by possible new climate protection policy requirements. In the context of the Green Deal, for example, the EU could introduce a border-crossing compensation for imports, in which the price of imports reflects the CO_2 footprint of the product. In its latest study on e-mobility, the German Energy Transition Think Tank "Agora Verkehrswende" proposes to introduce a CO_2 limit for the production for the European market.[19] Furthermore, such a label could be an assessment component for the allocation of EU/state/federal state subsidies for the establishment of highly efficient battery cell production or for the promotion of electric cars. A first-time reference implementation of the presented concept could pave the way for such regulatory visions.

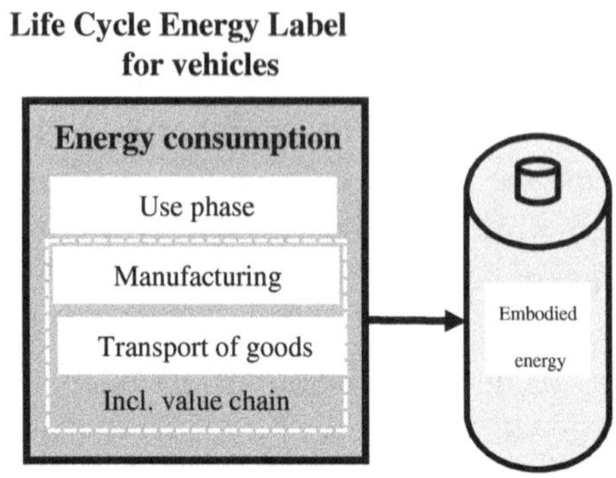

Fig. 4. Concept of a life cycle energy label (*in the further development, recycling could also be included).

7. Conclusion

In our opinion, the ongoing development of the European battery manufacturing capacity with a new generation of digital battery cell manufacturing plants offers the opportunity to design, implement and test management systems and tools to measure, evaluate and optimize the environmental sustainability of the whole value chain in battery manufacturing. A sustainability assessment approach, as described in our concept, linking the manufacturing site and manufacturing system-specific impact to its products, which is applied on the basis of international standards, could enable site-specific certification of the results. This would enable a valid sustainability quantification and comparison at the cell and battery pack system level. An international rollout could be driven through industry, based on sustainable procurement from manufacturers who integrate the battery systems in their products. Another driver could be stronger sustainability requirements by environmental policies, e.g. for imported goods from international value chains, products themselves or for the further development of sector-specific policies for the decarbonization.

References

1. Reisinger M, Dierolf C, Schneider C, Sauer A, Hörcher G. Intelligente Druckluftsysteme und Datenmodellierung. wt-online 6-2019. 2019;109:508–12.
2. United Nations. Transforming our world: the 2030 Agenda for Sustainable Development: A/RES/70/1 2015 [cited 2020 Feb 3]. Available from: https://sustainabledevelopment.un.org/content/documents/21252030%20Agenda%20for%20Sustainable%20Development%20web.pdf
3. Messner D, Schlacke S, Fromhold-Elisabith M, Grote U, Matthies E, Pittel K, *et al.* Unsere gemeinsame digitale Zukunft: Empfehlungen. Berlin: Wissenschaftlicher Beirat der Bundesregierung-Globale Umweltveränderungen (WBGU); 2019.

4. Plattform Industrie 4.0. 2030 Vision for Industrie 4.0: Shaping digital ecosystems globally; 2019 [cited 2020 Jul 27]. Available from: https://www.plattform-i40.de/PI40/Redaktion/EN/Downloads/Publikation/Positionspapier%20Leitbild%20(EN).pdf?__blob=publicationFile&v=6
5. Küpper D, Kuhlmann K, Wolf S, Pieper C, Xu G, Ahmed J. The future of battery production for electric vehicles; 2018 [cited 2019 Aug 26]. Available from: https://www.bcg.com/de-de/publications/2018/future-battery-production-electric-vehicles.aspx
6. Hill N, Clarke D, Blair L, Menadue H. Circular economy perspectives for the management of batteries used in electric vehicles. Luxembourg: Publications Office of the European Union; 2019.
7. Lee EA, Seshia SA. Introduction to embedded systems: a cyber-physical systems approach. 2nd ed. Cambridge, Massachusetts, London, England: MIT Press; 2017.
8. Sheehan E. A simulation-based method for improving material efficiency within the constraints of existing production systems [Dissertation]; 2017.
9. Waltersmann L, Kiemel S, Bogdanov I, Lettgen J, Miehe R, Sauer A, *et al.* Benchmarking holistic optimization potentials in the manufacturing industry — A concept to derive specific sustainability recommendations for companies. Procedia Manuf. 2019;39:685–94.
10. Albrecht S. Systematischer Ansatz zur Analyse der Energiekostensensitivität von Produktionssystemen unter Berücksichtigung der Wertschöpfungskette im Rahmen der Ganzheitlichen Bilanzierung [Zugl.: Stuttgart, Univ., Diss., 2014]. Stuttgart: Fraunhofer Verl.; 2014. (Forschungsergebnisse aus der Bauphysik; vol 17). Available from: http://www.bookshop.fraunhofer.de/buch/241390
11. Brundage MP, Bernstein WZ, Hoffenson S, Chang Q, Nishi H, Kliks T, *et al.* Analyzing environmental sustainability methods for use earlier in the product lifecycle. J Cleaner Prod. 2018 [cited 2019 Aug 15];187.
12. Wulf C, Werker J, Ball C, Zapp P, Kuckshinrichs W. Review of sustainability assessment approaches based on life cycles. Sustainability. 2019 [cited 2020 Jan 7];11(20):5717.
13. Boër CR. Mass customization and sustainability: an assessment framework and industrial implementation. London, New York: Springer; 2013.
14. Schneider C, Reisinger M, Adolf T, Heßberger N, Sauer A. A novel 'automated hardware upgrade service' for manufacturing systems. SCAP — Stuttgart Conference on the Future of Automobile Production 2020 In Press 2020.
15. Shi J, Wang Y, Fan S, Ma Q, Jin H. An integrated environment and cost assessment method based on LCA and LCC for mechanical product manufacturing. Int J Life Cycle Assess. 2019 [cited 2020 Aug 4];24(1):64–77.
16. Bicalho T, Sauer I, Rambaud A, Altukhova Y. LCA data quality: a management science perspective. J Cleaner Prod. 2017 [cited 2019 Dec 19];156:888–98.
17. VDMA Batterieproduktion. Roadmap battery production equipment 2030. Frankfurt am Main; 2018 Update 2018 [cited 2019 Aug 19].
18. Reisinger M. Life cycle energy labels for products. Berlin; 2016. (Industrial Efficiency 2016; Panel 2. Sustainable production design and supply chain initiatives) [cited 2019 Aug 26]. Available from: https://www.eceee.org/library/conference_proceedings/eceee_Industrial_Summer_Study/2016/2-sustainable-production-design-and-supply-chaininitiatives/life-cycle-energy-labels-for-products/2016/2-162-16_Reisinger_presentation.pdf/
19. Helms H, Kämper C, Biemann K, Lambrecht U, Jöhrens J, Meyer K. Agora Verkehrswende–Klimabilanz von Elektroautos: Einflussfaktoren und Verbesserungspotential; 2019 [cited 2019 Aug 14]. Available from: https://www.agora-verkehrswende.de/fileadmin/Projekte/2018/Klimabilanz_von_Elektroautos/Agora-Verkehrswende_22_Klimabilanz-von-Elektroautos_WEB.pdf
20. The European Parliament and of the Council of the European Union. REGULATION (EU) 2019/ 631 of 17 April 2019 — Setting CO2 emission performance standards for new passenger

cars and for new light commercial vehicles, and repealing Regulations (EC) No 443/2009 and (EU) No 510/2011 [cited 2020 Jul 27].

21. The European Parliament and of the Council of the European Union. DIRECTIVE 2006/66/EC of 6 September 2006 on batteries and accumulators and waste batteries and accumulators and repealing Directive 91/157/EEC [cited 2020 Jul 28].
22. European Commission. The environmental footprint transition phase; 2019 [cited 2019 Aug 16]. Available from: https://ec.europa.eu/environment/eussd/smgp/ef_transition.html
23. BDI. Product Environmental Footprint (PEF): Industrie am weiteren Prozess angemessen beteiligen! 2019 [cited 2019 Aug 16]. Available from: https://bdi.eu/artikel/news/product-environmental-footprint-pef-industrie-am-weiteren-prozess-angemessen-beteiligen/
24. Bach V, Lehmann A, Görmer M, Finkbeiner M. Product Environmental Footprint (PEF) pilot phase — comparability over flexibility? Sustainability. 2018;10(8):2898.
25. Geissdoerfer M, Savaget P, Bocken NM, Hultink EJ. The circular economy — A new sustainability paradigm? J Cleaner Prod. 2017 [cited 2020 Jul 28];143:757–68.
26. Schneider C, Büttner S, Sauer A. Klimaneutralität in der produzierenden Industrie. wt-online (in Press). 2020.
27. Bundesministerium der Finanzen. Bericht des Bundesministeriums der Finanzen über die Tätigkeit des Energie- und Klimafonds (EKF; Kap. 6092) im Jahr 2018 und über die im Jahr 2019 zu erwartende Einnahmen- und Ausgabenentwicklung; 2018.
28. Teucher T. CO2-Emissionen: Kompensationsanbieter im Vergleich: Hintergrundinfos & Anbieter-Vergleich. Inspiring sustainable business; 2017 [cited 2019 Aug 26]. Available from: https://www.tinateucher.com/wp-content/uploads/2017/12/co2-emissionen-kompensation sanbieter-vergleich.pdf
29. Seebacher F. Energieproduktivität als Indikator für Nachhaltigkeit im Innovationsprozess; 2006. (Berichte aus dem Maschinenbau).
30. VDI 4600. Cumulative energy demand (KEA) — Terms, definitions, methods of calculation. VDI-Gesellschaft Energie und Umwelt; 2012.
31. Mahler B, Idler S, Nusser T, Gantner J. Energieaufwand für Gebäudekonzepte im gesamten Lebenszyklus [cited 2020 Aug 2].
32. Regett A. Klima- und Ressourcenwirkung von Elektrofahrzeugbatterien: Begleitdokument zum Artikel "Klimabilanz von Elektrofahrzeugen — Ein Plädoyer für mehr Sachlichkeit" Begleitdokument. München: Forschungsstelle für Energiewirtschaft e.V. FfE; 2019.
33. Philippot M, Alvarez G, Ayerbe E, van Mierlo J, Messagie M. Eco-efficiency of a lithium-ion battery for electric vehicles: influence of manufacturing country and commodity prices on GHG emissions and costs. Batteries. 2019;5(1):23.
34. Dai Q, Kelly JC, Gaines L, Wang M. Life cycle analysis of lithium-ion batteries for automotive applications. Batteries. 2019 [cited 2020 Aug 2];5(2):48.
35. Gaines L, Sullivan J, Burnham A, Belharouak I. Life-cycle analysis of production and recycling of lithium ion batteries. Trans Res Rec. 2011;2252(1):57–65.
36. Gaines L, Sullivan, John, Burnham, Andrew, Belharouak I. Life-cycle analysis for lithium-ion battery production and recycling; 2010 [cited 2020 Jan 31].
37. Jordan F. How predictive quality analytics increased battery cell manufacturing yield; 2019 [cited 2020 Aug 2]. Available from: https://elisaindustriq.com/how-predictive-quality-analytics-increased-battery-cell-manufacturing-yield-by-16-percent/
38. Grant A. The CO2 impact of the 2020s battery quality lithium hydroxide supply chain. Jade Cove. 2020 [cited 2020 Aug 2].
39. Cusenza MA, Bobba S, Ardente F, Cellura M, Di Persio F. Energy and environmental assessment of a traction lithium-ion battery pack for plug-in hybrid electric vehicles. J Cleaner Prod. 2019;215:634–49.

40. Deng Y, Li J, Li T, Gao X, Yuan C. Life cycle assessment of lithium sulfur battery for electric vehicles. J Power Sources. 2017;343:284–95.
41. Lee YJ, Yi H, Kim W-J, Kang K, Yun DS, Strano MS, *et al.* Fabricating genetically engineered high-power lithium-ion batteries using multiple virus genes. Science. 2009;324(5930):1051–5.
42. Lee Y-G, Fujiki S, Jung C, Suzuki N, Yashiro N, Omoda R, *et al.* High-energy long-cycling all-solid-state lithium metal batteries enabled by silver–carbon composite anodes. Nat Energy. 2020;5(4):299–308.
43. Bhatia P, Cummis C, Brown A, Rich D, Draucker L, Lahd H. Corporate value chain (scope 3) accounting and reporting standard: supplement to the GHG protocol corporate accounting and reporting standard. Greenhouse Gas Protocol; 2011 [cited 2019 Aug 26]. Available from: https://ghgprotocol.org/sites/default/files/standards/Corporate-Value-Chain-Accounting-Reporing-Standard_041613_2.pdf
44. Daimler AG. Nachhaltigkeitsbericht 2018; 2019. Available from: https://www.daimler.com/dokumente/nachhaltigkeit/sonstiges/daimler-nachhaltigkeitsbericht-2018-de.pdf
45. Daimler AG. Mercedes-Benz. Nummer 1 im Premiumsegment; 2019 [cited 2019 Aug 16]. Available from: https://media.daimler.com/marsMediaSite/de/instance/ko/Mercedes-Benz-schafft-achtes-Rekordjahr-in-Folge-und-bleibt-die-Nummer-1-im-Premiumsegment.xhtml?oid=42197316
46. Daimler AG. Scope 3 Emissionen; 2018 [cited 2019 Aug 16]. Available from: https://www.daimler.com/dokumente/nachhaltigkeit/sonstiges/2018-scope-3.pdf
47. Källenius O. Ambition2039: Unser Weg zu nachhaltiger Mobilität; 2019 [cited 2019 Aug 16]. Available from: https://media.daimler.com/marsMediaSite/de/instance/ko/Ambition2039-Unser-Weg-zu-nachhaltiger-Mobilitaet.xhtml?oid=43348842
48. The Economist. Following the footprints; 2011 [cited 2019 Aug 16]. Available from: https://www.economist.com/technology-quarterly/2011/06/04/following-the-footprints
49. BCG. The factory of the future [cited 2020 Jul 30]. Available from: https://www.bcg.com/en-us/publications/2016/leaning-manufacturing-operations-factory-of-future
50. Betz M. Product compliance und material reporting–aber richtig! thinkstep; 2018 [cited 2019 Aug 16]. Available from: https://www.software-journal.de/author/firma_thinkstep/
51. Abele E, Feickert S, Kuhrke B, Clesle F-D, editors. Environmental product lifecycle management-customizing the enterprise specific manufacturing processes; 2006. (Proceedings of LCE 2006).
52. Koffler, C. Methoden der Erstellung und Auswertung von automobil-spezifischen LCAs (Ökobilanzwerkstatt 2005). Bad Urach, 15-16.06.2005. [cited 2019 Aug16] URL: https://www.oekobilanzwerkstatt.tu-darmstadt.de/media/oeko/2005/praesentationen_abstracts_der_arbeitsgruppe_c_oebw_1/C5_Koffler.pdf
53. Dahllöf L, editor. How the chemical and material management in the automotive industry became a quality improver for LCA studies in the Volvo Group; 2013.
54. DIN e. V. and DKE Deutsche Kommission Elektrotechnik. German Standardization Roadmap Industrie 4.0–Version 3. Berlin: DIN/DKE; 2018 Version 3 [cited 2019 Aug 26]. Available from: https://www.din.de/blob/95954/97b71e1907b0176494b67d8d6d392c54/aktualisierte-roadmap-i40-data.pdf
55. BMWi. Struktur der Verwaltungsschal–Fortentwicklung des Referenzmodells für die Industrie 4.0-Komponente. Berlin: Plattform Industrie 4.0; 2016 [cited 2019 Aug 16].
56. CLI. Climate Ledger Initiative [cited 2019 Aug 16]. Available from: https://climateledger.org/en/
57. Licht J, Jong Td, Oudshoorn K, Pasotti P. Circularise: whitepaper [cited 2020 Aug 3].
58. Tramutola L. Data sharing for a circular economy [cited 2020 Aug 3].
59. Goodarzi S, Fahimnia B, Sarkis J. Supply chain carbon management. Sydney: University of Sydney; 2019. Available from: https://ses.library.usyd.edu.au/bitstream/2123/20261/1/ITLS-WP-19-07.pdf

60. Hsu C-W, Kuo T-C, Chen S-H, Hu AH. Using DEMATEL to develop a carbon management model of supplier selection in green supply chain management. J Cleaner Prod. 2013;56:164–72. Available from: http://www.sciencedirect.com/science/article/pii/S095965261100343X
61. Hedberg TD, Bajaj M, Camelio JA. Using graphs to link data across the product lifecycle for enabling smart manufacturing digital threads. J Comput Inf Sci Eng. 2020 [cited 2020 Jan 31];20(1):21001.
62. Banerjee A. Re-engineering the carbon supply chain with blockchain technology. Infosys; 2018.
63. Hottenroth H, Joa B, Schmidt M. Carbon Footprints für Produkte: Handbuch für die betriebliche Praxis kleiner und mittlerer Unternehmen. Institute for Industrial Ecology Pforzheim; 2013.
64. Horng-Huei W, Chih-Hung T, Chia-Chu L, Neng-Hao K, Yi-Chan C. The carbon footprint calculation model of the integrated ERP framework — Green production. Int J Oper Logist Manage. 2015 [cited 2020 Aug 4].
65. Funk B, Möller A, Niemeyer P. The prospects of product carbon footprints in ERP systems: he prospects of product carbon footprints in ERP systems environmental protection proceedings of the 22nd International Conference Environmental. Aachen: Shaker; 2008 [cited 2020 Aug 4].
66. Rebouillat L, Barletta I, Mani M, *et al.* Understanding sustainability data through unit manufacturing process representations: a case study on stone production. Procedia CIRP. 2017;57:686–91.
67. Kulkarni A, Bernstein WZ, Lechevalier D, Balasubramanian D, Denno P, Karsai G. Towards operational use of unit manufacturing process models. In: Proceedings, 2019 IEEE International Conference on Industrial Technology (ICIT): Melbourne Convention and Exhibition Centre, Melbourne, Australia, 2019 Feb 13–15. [Piscataway, New Jersey]: IEEE; 2019. pp. 818–23 [cited 2019 Aug 15].
68. Bernstein WZ, Tamayo CD, Lechevalier D, Brundage MP. Incorporating unit manufacturing process models into life cycle assessment workflows. Procedia CIRP. 2019 [cited 2019 Aug 15];80:364–9.
69. Genest A, Klein M, Schiffleitner A, Hedemann J. Live LCA: how Industry 4.0 changes the way we perform life cycle assessment [cited 2019 Aug 26]. Available from: https://www.google.com/url?sa=t&rct=j&q=&esrc=s&source=web&cd=1&cad=rja&uact=8&ved=2ahUKEwiKjdvogKDkAhWBL1AKHciZASgQFjAAegQIBRAC&url=http%3A%2F%2Favnir.org%2Fdocumentation%2Fcongres_avnir%2F2017%2FPPT%2FUsine_du_futur_IFU_Congres_avniR_2017.pdf&usg=AOvVaw32MGGumoEBuHbMjk3Op_y4
70. Möller A, wohlgemuth V, Lang-Koetz C. Gewinnung und Ntuzung von Lebenszyklusdaten im betrieblichen Konztext: Projektbericht; 2007 [cited 2019 Aug 21].
71. Cooper J, Noon M, Jones C, Kahn E, Arbuckle P. Big data in life cycle assessment. J Ind Ecol. 2013;17(6):796–9.
72. Boulonne A, Johansson B, Skoogh A, Aufenanger M. Simulation data architecture for sustainable development. In: Simulation Conference (WSC), Proceedings of the 2010 Winter: Date, 2010 Dec 5–8. Piscataway, N.J.: IEEE; 2010. pp. 3435–46.
73. Gebler M, Cerdas F, Kaluza A, Meininghaus R, Herrmann C. Integrating life-cycle assessment into automotive manufacturing — A review-based framework to measure the ecological performance of production technologies. In: Schebek L, Herrmann C, Cerdas F, editors. Progress in life cycle assessment. Cham: Springer International Publishing; 2019. pp. 45–55 (Sustainable Production, Life Cycle Engineering and Management) [cited 2019 Aug 14].
74. Turetskyy A, Thiede S, Thomitzek M, Drachenfels N von, Pape T, Herrmann C. Toward data-driven applications in lithium-ion battery cell manufacturing. Energy Technol. 2019 [cited 2020 Jan 31];12:1900136.

75. Singh S, Wanner J, Glanz C, Riexinger G, Birke KP. Development and utilization of a framework for data-driven life cycle management of battery cells. 2019 [cited 2020 Jan 31].
76. DIN e. V. and DKE Deutsche Kommission Elektrotechnik. German Standardization Roadmap Indutrie 4.0 — Version 4; 2020 [cited 2020 Jul 27].
77. iPoint. LCA of lithium-ion batteries — There's more to it than you think. Reutlingen; 2018 [cited 2019 Aug 16]. Available from: https://www.ipoint-systems.com/blog/lca-of-lithium-ion-batteries-theres-more-to-it-than-you-think/

Part IV

Optimization of Battery Cell Manufacturing Lines with Digital Tools

https://doi.org/10.1142/9789811245626_0012

Chapter 11

Productivity and Quality Propagation Modeling in Battery Manufacturing

Friedrich-Wilhelm Speckmann
Fraunhofer-Institute for Manufacturing Engineering and Automation
Nobelstr. 12, 70569 Stuttgart, Germany
friedrich-wilhelm.speckmann@ipa.fraunhofer.de

Finite fossil fuels and the need for reduced carbon dioxide emissions have led to an increase in battery electric vehicle sales. To meet the demand of this steadily growing industry segment, large amounts of battery cells are required. Therefore, large-scale battery cell manufacturing has become a major part in the pursuit of a sustainable transportation sector. However, battery manufacturing is a complex procedure and there are multiple challenges that have to be solved in order to provide a high productivity and quality. The implementation of digital tools in the battery manufacturing lines can support these aims. Additionally, this chapter proposes a quality flow model and derived mathematical equations for the optimization of a serial battery manufacturing line with inspection and repairing stations.

1. Online Data Acquisition — Sensor Integration for Quality Control and Increased Productivity

The demand for batteries for vehicles and mobile applications is ever increasing. To meet the demand, large-scale battery manufacturing has become a crucial part of development. On the one hand, it is necessary to produce a high volume of battery cells, and on the other hand, the product quality is of utmost importance. Therefore, the key concerns of any manufacturing enterprise are productivity and quality. However, these two aspects are opposing each other. Thus, the aim is to enhance productivity without reducing battery quality.

Battery manufacturing is a complex multi-stage process in which productivity and quality are the most critical aspects. There has been extensive research on these two features but, in most cases, independently. Many scientific studies focus on the analysis of the throughput in manufacturing lines.[1–3] Here, the production layout is strongly simplified. Other works regard the more complex layouts of assembly/disassembly lines.[4,5] Jacobs *et al.* additionally integrate rework loops for correction of defective parts.[5] Nevertheless, the detailed layer is relatively low because no split and merge processes are explicitly incorporated in the manufacturing line. In later years, Ju *et al.* introduced these aspects in the quality flow model in automotive paint shops.[6] There are various other, more specific aspects such as multiple products[7] or multiple failure models[8] that have been studied in detail. Most of these studies are based on two analytical models, aggregation and decomposition.

Besides the research on productivity, there have been studies on multiple quality aspects in recent years. Variation propagations, such as dimension errors and sheet stretch forming processes, have gained much attention.[9,10] These studies give insight into the field of error propagation in a multi-stage production line. Furthermore, the influence of errors in various production stages on the final product has been investigated. However, battery manufacturing consists of different stages with a variety of quality control features that are deviating from each other. Here, not every stage requires an identical form of quality control. Joining, dimensions and resistance, for example, cannot be determined with the help of the physical quality model alone. Data-driven approaches are an alternative way of quality calculations, although many of them are only applicable to particular scenarios or do not include in-between repairs.

On the contrary, recent studies' attention is on the integration of product quality in regard to production system design. There is evidence that production system design can influence product quality, and both can be affected by quality control.[1,11] Furthermore, layout design issues and their imposition on the product quality have been analyzed.[1,12] A significant shortcoming of these approaches in battery manufacturing is that the interaction of various components with the system is not implemented. Consequently, the models cannot factor in machine breakdowns or finite buffers.

Inspections in the form of quality checks during the manufacturing process are another vital aspect that has to be regarded in modeling. To analyze such a system, it is required to consider quality as well as productivity variations and their interactions. A basic model of an asymptotically reliable two-machine, one-buffer model has been investigated by Jacobs *et al.*[13] The model has since increased in complexity as it has been extended to longer manufacturing lines,[14] and analytical evaluation has been formulated.[15] However, these approaches assume a

Bernoulli-Markovian-type error behavior, and thus, there is no correlation between consecutive parts. But, the Markovian-type failure is commonly present in production systems, e.g. the quality of a piece depends on the previous one. The concept of quality-quantity coupling (QQC) has been introduced by Meerkov and Zhang.[16,17] It is commonly applied to lines with multiple machines and inspection points, where the quality ratio can decrease with regard to machine efficiency. QQC can be used as a starting point for positioning and quality bottleneck identification, as done by Arinez *et al.* for an industrial paint shop line.[18] Nevertheless, these studies do not incorporate the repairs into the modeling process. Repair loops and bottlenecks in production lines have been investigated by Biller *et al.*[19] but without regard to quality. Quality bottlenecks have been the focus of multiple studies[17,20,21] and use bottleneck indicators based on raw factory floor data, which makes them advantageous in actual industrial applications.

All in all, multiple studies are focusing on different manufacturing challenges but without an analytical model of integrated productivity and quality in a multi-stage serial production line. However, these are mainly employed for battery manufacturing and have not been studied explicitly. Therefore, this chapter will focus on the integrated analysis of productivity and quality in a multi-stage, serial battery manufacturing line with multiple buffers, inspection and repair points.

2. Serial Line Battery Production

This section proposes a commonly employed battery assembly system consisting of a serial manufacturing line, multiple inspections and repair stations. The system structure will be used as a basis for the detailed investigation of productivity combined with quality. Here, the correlation of the throughput and the quality failure rate is the central component of an analytical model. However, the complex battery manufacturing process's multiple manufacturing lines are decomposed into separate serial lines, and one of these is studied in detail.

In this battery manufacturing system, each component is inspected at each station. If the component shows no unwanted behavior during the quality inspection, it continues in the main assembling line. Otherwise, a non-conforming part is routed to the respective repair station. Following this, the restored components are returned to downstream operations and can be processed in the next production step. Figure 1 shows the layout of the discussed serial manufacturing line with quality inspection and repair stations.

To understand and work with this figure, all notations and explanations are as follows:

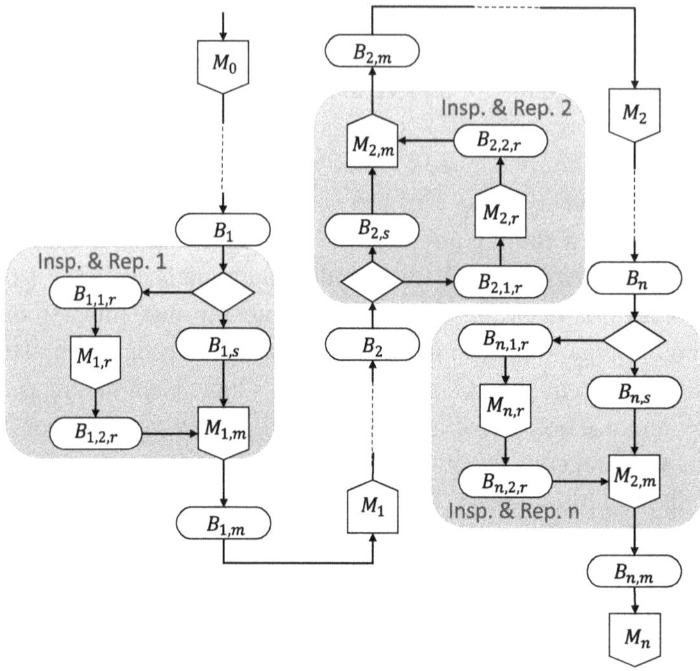

Fig. 1. Schematic layout of a serial battery cell manufacturing line with integrated quality inspection points, buffers and repairing stations.

The main serial production line consists of M machines (M_0, M_1, M_2, . . . ,M_n). All machines have two possible states, up and running or down with no production at all. If running, each machine can produce 1 part per time step.

- There is a buffer before the split inspection, inside the mainline during the inspection and after the merging process (B_1, $B_{1,s}$ and $B_{1,m}$). Additionally, there are buffers before and after the repairing process ($B_{1,1,r}$ and $B_{1,2,r}$). This setup is present in every inspection and repairing loop.
- n is the number of the inspection and repairing stations in the serial production line.
- In each stage, the repairing station machines are labeled $M_{n,r}$ and the merging machines $M_{n,m}$.

There are multiple rules for the production line model. Each buffer has a finite capacity that is characterized by its position. A machine is empty and cannot produce if the upstream buffer is empty and the upstream machine is not producing in this time step. Additionally, the starting machine M_0 is never starved, and the merging machines are only out of parts if both upstream buffers are empty. In contrast, a machine is regarded as blocked if its downstream buffer is filled up. However, machine M_n is never blocked as it is at the end of the production line. Furthermore, after passing through quality inspection and the repairing station, each part can have one of three states: high quality with no repairs required (h_i), high quality after repairing (r_i) and a defective condition (d_i). For this model, it is assumed that no inspection failure is possible, and no parts are scrapped during the process. The resulting production rate is a function of all the buffer and machine parameters in their steady state. This problem will be addressed in detail within Section 2.2.

2.1. *Sensor networks for battery manufacturing*

Battery production is still evolving and has to tackle challenges such as ramping up the capacity as well as integrating new technologies. To correctly analyze and model a battery manufacturing process, it is necessary to gather data during production. Additionally, market fluctuations have to be expected, and thus, flexible manufacturing gains interest as a field of research. However, most conventional production models focus on long-term steady-state manufacturing and cannot describe changing scenarios.[22,23] Due to integrated sensor networks, it is possible to achieve high modeling quality in fast-changing manufacturing environments. Figure 2 shows the setup of a lithium-ion battery production line.

The manufacturing line consists of multiple machines and process steps, starting with the electrode formation (agitation, slitter, drying, coating) and continuing with the assembly (winding, stacking, transfer). Not every process step has to be

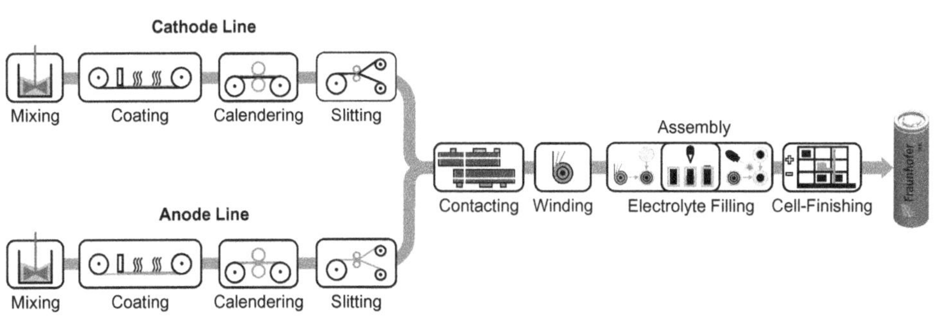

Fig. 2. Detailed model of a lithium-ion battery manufacturing line with crucial transfer and process steps.

monitored. The focus has to be on the stations that mainly influence the product quality. In this case, sensors for multiple parameters are distributed throughout the manufacturing line.

It starts with the agitator, which mixes the battery materials, later coated onto the lithium-ion battery's electrodes. Here, a torque control sensor can monitor the motor and, thus, indirectly measure the viscosity. Additionally, the blend ratio can be determined with a high-speed data logger unit and is stored in an online database. The coating process conventionally struggles with tension fluctuation and applying a uniform coating. In this case, it is suitable to integrate a simple tension controller which can quickly realize a highly accurate tension control. Furthermore, a synchronized speed controller of all axes increases the homogeneity of the applied coating. Identical process step supervision can be employed for the slitter to suppress winding deviation or tightening. One of the most critical quality parameters at the winding station is the positional deviation between the cell and separator. An edge sensor can be installed in order to compensate the position with a photoionization control so that the winding finish is accurate.

For the downstream assembling steps, the stacking process is of utmost importance to the product quality. On the one hand, it is essential to have the best fitting tension control. There are multiple possible solutions, ranging from simple tension detectors to direct monitoring of the unwinding shaft motor's torque. On the other hand, any vibrations while setting the electrodes have to be negated. Here, a servo amplifier can detect and suppress vibration in the device's application arm. The transfer process is often conducted with robots. Therefore, coordination monitoring between the vision system and the robots has to be installed to prevent workpiece damage from possible collisions. Suitable supervision can be conducted by multiple optical sensors coupled with a 2D calibration function. Finally, a data management system should link all production line machines and guarantee wholesome quality control and possible traceability operations.

2.2. *Quality and failure propagation*

In serial battery manufacturing lines with multiple inspections and repairing stations, the quality characteristics are interrelated between the consecutive stages, even if each inspection is independent. Significantly, the upstream throughput is linked to the upstream quality and thus influences the downstream productivity. Here, the interrelation between quality and productivity is evident. The resulting production rate can be determined by combining the quality propagation with the repairing processes. To study the quality in such a serial production line, a flow

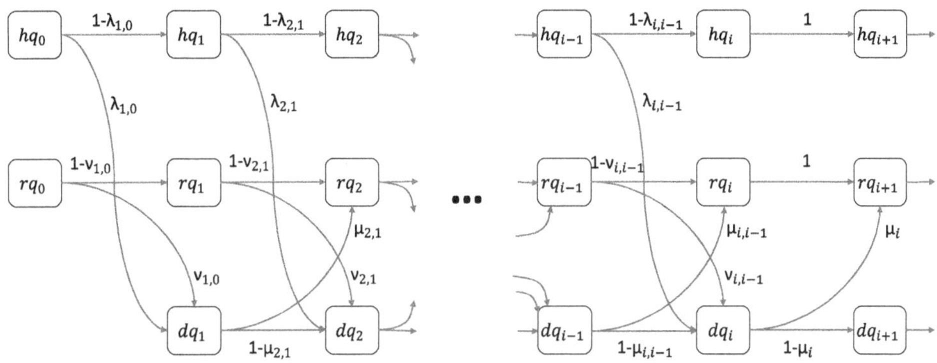

Fig. 3. Quality flow model, highlighting the relations between quality inspections and repairing processes in a serial battery manufacturing line. The left-hand side displays flow models' starting steps and the right-hand side the model behavior after *i* iterations.

model has been developed by Ju *et al.*[6] Based on the flow model, a systematic approach is needed to study the combined influence of quality checks and repair stations on the final product's quality. Figure 3 shows the quality flow model, highlighting the mentioned relations between repairing and quality propagation.

There is a general distinction between a successful repairing and a non-successful attempt. However, in the case of battery production, there has to be distinguishability between an error-free component and restored components that might still have a slightly lower quality. Consequently, the flow model accommodates three possible states for each inspection that have been previously introduced. A good state without repairing (h_i), a failure-free state after repairing (r_i) and a defective state (d_i). Here, it can be seen that each categorized part exhibits different failure probabilities. To trace these reliably, the weighted failure rate Λ is introduced for the ith inspection

$$\Lambda_1 = P(h_0)\lambda_{1,0} + P(r_0)\vartheta_{1,0} \tag{1}$$

$$\Lambda_i = P(h_{i-1})\lambda_{i,i-1} + P(r_{i-1})\vartheta_{i,i-1} + P(d_{i-1})(1-\mu_{i,i-1}) \tag{2}$$

$$\Lambda_{i+1} = P(d_i)(1-\lambda_i), \tag{3}$$

where $\lambda_{i,i-1}$ is the transitional probability from state h_i to state d_{i+1} and $1-\lambda_{i,i-1,i}$ to remain in the conforming state h_{i+1}. Furthermore, $\vartheta_{i,i-1}$ is the transitional probability from state r_i to state d_{i+1} and, respectively, $1-\vartheta_{i,i+1,i}$ to be in f_{i+1}. The last transitional process for a part from d_i to f_{i+1} has the probability $\mu_{i,i-1,i}$ and, respectively, the probability $1-\mu_{i,i+1,i}$ to remain in a defective state d_{i+1}.

Therefore, it is possible to directly obtain the repair probability by solving the balance equations from the presented quality flow model. The repair probability function is as follows:

$$P(h_i)=P(h_{i-1})(1-\lambda_{i,i-1}) \tag{4}$$

$$P(r_i)=P(r_{i-1})(1-\vartheta_{i,i-1})+P(d_{i,i-1})\mu_{i,i-1} \tag{5}$$

$$P(d_i)=P(g_{i-1})\lambda_{i,i-1}+P(r_{i-1})\vartheta_{i,i-1}+P(d_{i-1})(1-\mu_{i-1}) \tag{6}$$

$$P(d_{i+1})=P(d_i)(1-\mu_i) \tag{7}$$

These equations can be expressed in a matrix form to simplify the formulation of the recurrence formula, which is then combined with the weighted failure rate equations. In this way, we receive the probabilities of a part being in a defective state

$$\Lambda_1=P(d_1)=\gamma_1\cdot Y_0 \tag{8}$$

$$\Lambda_i=P(d_i)=\gamma_i\cdot\left(\prod_{j=1}^{i-1}\tau_j\right)\cdot Y_0 \tag{9}$$

$$\Lambda_{i+1}=P(d_{i+1})=(1-\mu_i)\gamma_i\cdot\left(\prod_{j=1}^{i-1}\tau_j\right)\cdot Y_0, \tag{10}$$

where τ_i is the parameter matrix of the failure probabilities. Finally, it has to be noted that $P(d_i)$ represents the weighted failure rate Λ_i in every ith inspection of the production line (cf. Fig. 1). The development of this weighted failure rate model enables the further development of a throughput model as well as the production rate of the considered line.

3. Conclusion

In this section, an integrated model of quality and control has been introduced. It consists of a serial battery manufacturing line with multiple inspection stations and repairing loops. Another part of this section focuses on the integration of smart sensors for battery manufacturing. Here, various production steps are regarded, and fitting sensors are chosen to monitor and regulate the quality. Furthermore, based on this approach, a quality flow model has been proposed to calculate the failure probabilities in the inspection stations. This method can additionally be applied to separate manufacturing lines with similar layouts, such as automotive paint shops or power train assembling. Therefore, this work is the basis for more specific quantitative tools for production engineering in general.

References

1. Dallery Y, David R, Xie X-L. An efficient algorithm for analysis of transfer lines with unreliable machines and finite buffers. IIE Trans. 1988;20(3):280–3.
2. Emmons H, Rabinowitz G. Inspection allocation for multistage deteriorating production systems. IIE Trans. 2002;34(12):1031–41.
3. Han MS, Lim JT, Park DJ. Performance analysis of serial production lines with quality inspection machines. Int J Syst Sci. 1998;29(9);939–51.
4. Inman RR, Blumenfeld DE, Huang N, Li J, Li J. Survey of recent advances on the interface between production system and design and quality. IIE Trans. 2013;45(6): 557–74.
5. Jacobs D, Meerkov SM. Asymptotically reliable serial production lines with a quality-control system. Comput Math Appl. 1991;21(11):85–95.
6. Ju F, Li J, Xiao G, Arinez J. Quality flow model in automotive paint shops. Int J Prod Res. 2013;5:6470–83.
7. Lawless JF, Mackay RJ, Robinson JA. Analysis of variation transmission in manufacturing processes: part I. J Qual Technol. 1999;31(2):131–42.
8. Li J, Meerkov SM. Bottlenecks with respect to due-time performance in pull serial production lines. Proc IEEE Int Conf Rob Autom. 2000;3:2635–40.
9. Li J. Performance analysis of production systems rework loops. IIE Trans. 2004;36(8):755–65.
10. Li J. Modeling and analysis of manufacturing systems with parallel lines. IEEE Trans Autom Control. 2004;49(10):1824–32.
11. Colledani M, Tolio T. Performance evaluation of production systems monitored by statistical process control and off-line inspections. Int J Prod Econ. 2011;120(2):348–67.
12. Chiang SY. Bernoulli serial production lines with quality control devices: theory and application. Math Probl Eng. Article No. 81950. 2006;1–31.
13. Lim JT, Meerkov SM, Top F. Homogeneous, asymptotically reliable serial production lines: theory and a case study. IEEE Trans Autom Control. 1990;35(5):524–34.
14. Liu XG, Buzacott JA. Approximate models of assembly systems with finite inventory banks. Eur J Oper Res. 1990;45(3):143–54.
15. Liu Y, Li J. Split and merge production systems: performance analysis and structural properties. IIE Trans. 2010;42(6):422–4.
16. Arinez J, Biller S, Meerkov SM, Zhang L. Quality/quantity improvement in automotive paint shops: a case study. IEEE Trans Autom Sci Eng. 2010;7(4):755–61.
17. Meerkov SM, Zhang L. Bernoulli production lines with quality-quantity coupling machines: monotonicity properties and bottlenecks. Ann Oper Res. 2011;182(1):119–31.
18. Papadopoulos HT, Heavey C, Browne J. Queueing theory in manufacturing systems analysis and design. London, UK: Chapman & Hall; 1993.
19. Tolo T, Matta A, Gershwin SB. Analysis of two-machine lines with multiple failure modes. IIE Trans. 2002;34(1):51–62.
20. Meerkov SM, Zhang L. Production quality inspection in Bernoulli lines: analysis bottlenecks, and design. Int J Prod Res. 2010;48(16):4745–66.
21. Mhada F, Hajji A, Malhalme R, Gharbi A, Pellerin R. Production control of unreliable manufacturing systems producing defective items. J Qual Maint Eng. 2011;17(3):238–53.

22. Chang Q, Ni J, Bandyopadhyay P, Biller S, Xiao G. Supervisory factory control based on real-time production feedback. ASME Trans J Manuf Sci Eng. 2007;129:653–60.
23. Chang Q, Biller S, Xiao G, Liu J. Transient analysis of downtimes and bottleneck dynamics in serial manufacturing systems. ASME Trans J Prod Sci Eng. 2010;132:051015.

https://doi.org/10.1142/9789811245626_0013

Chapter 12

Method for the Identification of Process Quality Characteristics and Suitable Measurement Systems in Battery Production

Andreas Aichele*, Anselm Lorenzoni† and Jonas Lips‡

Fraunhofer IPA, Nobelstraße 12, Stuttgart, Germany
Coperion GmbH, Theodorstraße 10, 70469 Stuttgart, Germany
**andreas.aichele@ipa.fraunhofer.de*
†anselm.lorenzoni@coperion.com
‡jonas.lips@ipa.fraunhofer.de

Continuous improvement in terms of efficiency and quality is necessary for competitive battery cell production. An essential prerequisite for improvement is the knowledge of cause–effect relationships in production processes. The approaches of digitalization and its industry-specific implementation "Industry 4.0" offers the promising opportunities to recognize these relationships and to use them for improvements, as initial approaches are already showing. However, these approaches can only make a profitable contribution if decisive process variables are recorded correctly. Therefore this article intends to picture existing deficiencies in current research work and additionally outlines possibilities to identify the relevant variables of battery cell production as well as applying suitable measurement systems.

1. Introduction and Motivation

To remain competitive despite increasing production capacities and possible overcapacities, the total production costs of lithium-ion cells need to be reduced. Regarding the relevant parameter "currency per kWh", this can be achieved by increasing the energy density as well as reducing the manufacturing costs. The former can be achieved by improved manufacturing accuracy and cell chemistry, the latter by reducing investment and operating costs, as well as by increasing efficiency through better yield rate and rejection rates.[1,2]

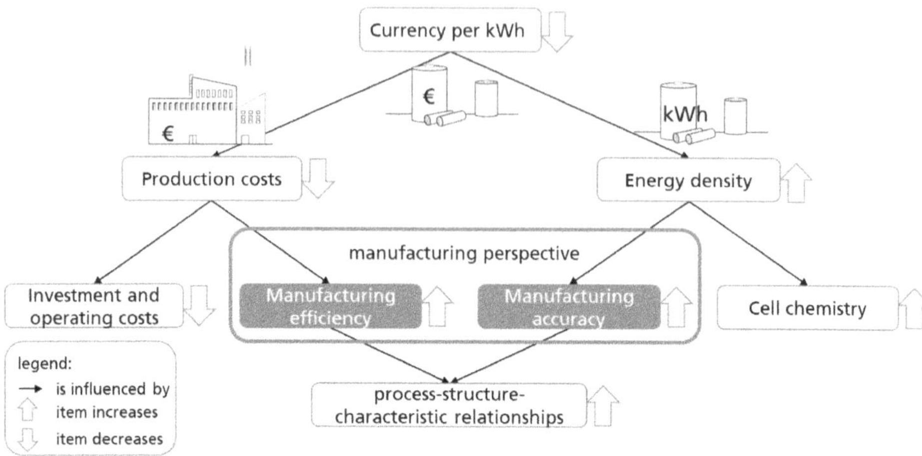

Fig. 1. Influence of manufacturing on competitiveness in the battery market.

From a manufacturing perspective, two issues can be addressed: manufacturing accuracy and increased efficiency — which currently also constitutes the most decisive challenges in battery production. The aim is to increase quality while reducing manufacturing costs.[3] Scrap reduction rates of 10–16%, as found in comparable literature, across the entire process chain of battery production shows the potential for efficiency improvements.[4,5] The challenge is to improve these parameters of a complex process chain which characterizes battery production. It consists of multiple process–structure–characteristic relationships which must be mastered for systematic process improvement to enhance manufacturing accuracy and increase efficiency (see Fig. 1). Regarding the current state of battery production, these improvements still have to be implemented.[3]

Therefore, the knowledge of process–structure–characteristic relationships, also known as cause–effect relationships, of the complete process chain must be gained. This results in the following exemplary points for improving battery cell production:

1. Increasing the quality of the end product.
2. Reduction of rejects.
3. Optimization of the inspection planning.
4. Adaptive process control.

Optimized settings in the processes can be derived from known relationships. This leads to an increasing quality of the final products, as well as a reduction in rejects. Knowledge of the cause–effect relationships also provides transparency of the intermediate product characteristics that lead to scrap. Thereupon the

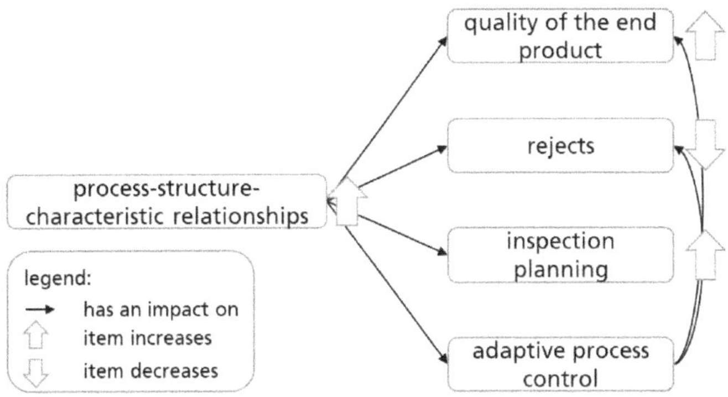

Fig. 2. Potential for improving battery cell production through knowledge of the process–structure–characteristic relationships.

inspection planning can be optimized to detect and eliminate defects at their origin with the help of a suitable measuring infrastructure. Further processing of defective parts can thus be prevented. In addition, knowledge of the cause-and-effect relationships is the basis for adaptive process control. By means of the known relationships, subsequent process steps can be directly influenced for an optimization of the outcome. This contributes to a reduction in rejects and an improvement in the quality of the products (see Fig. 2).

2. Research Issue

To support the acquisition of knowledge about the cause–effect relationships of products and processes, approaches like as Six Sigma, Statistical Design of Experiments and various variations of these have existed for decades. However, due to the complexity of battery cell production as mentioned above, these approaches have not yet been able to achieve any significant added value. One way to improve the understanding of production processes is by digitalization or its industry-related subarea "Industry 4.0". Here, algorithms are used to analyze a large number of data points from production processes and derive optimizations.[6]

Another approach as described by Michaelis is frugal innovation. It is based on a sufficient understanding of the process in order to reduce its complexity. This is achieved by continuous data collection and analysis, which is in line with the idea of Industry 4.0. Both approaches are based on the use of Industry 4.0. Therefore, an increasing information technology networking of production facilities is necessary.[7]

Some research has taken advantage of the possibilities of digitalization and addressed the description of cause–effect relationships in lithium-ion cell manufacturing.[8–12] Relevant process variables were identified, and the manufacturing processes of battery cell production were optimized. Contrary to the approaches known as described before, these "new" approaches do not address how the process variables under consideration become part of the scope of the study and how they are being recorded. This part in particular will be addressed in the presented approach.

According to the known approaches, the procedure for deriving cause-and-effect relationships is always similar: at first, the scope of the investigation and the objective of the investigation must be defined. Based on that, possible influencing variables need to be identified, then the most relevant influencing variables are selected. The selected influencing variables must be recorded in the course of tests or the production of products. For this purpose, it is necessary to select a suitable measuring system in advance. The recorded data is then analyzed and a model is derived which describes the cause–effect relationships. If the

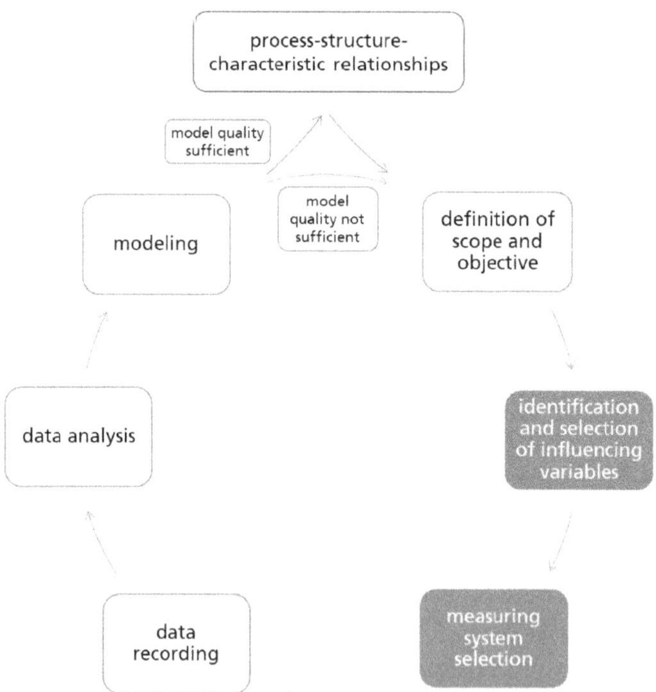

Fig. 3. Procedure for deriving the cause-and-effect relationships and the focus of the approach presented.

achieved model quality is sufficient, i.e. if the model withstands a check for plausibility and correctness, it can be released for further use, and the implementation potentials for improving production mentioned in Section 1 can be addressed. If the model quality is not sufficient, the steps as shown in Fig. 3 must be repeated. The cycle representation makes it clear that for the objective — sufficient model quality — the steps build on each other and therefore the objective depends on each step.

In the approach presented here, special attention is paid to the second and third steps: identification and selection of influencing variables and the measuring system selection.

3. State of the Art

The focus of the literature review results from the key research issues identified in Section 2:

1. Identification and selection of influencing variables.
2. Measuring system selection.

Before discussing the current state of the art of these topics, the variable types in process–structure–charactersitic relationships will be clarified.

3.1. *Variables in process–structure–characteristic relationships*

In order to determine process–structure–characteristic relationships, it is necessary to define what kind of variables characterize the production process. As shown in Fig. 4, processes are generally characterized by output, input, control and disturbance variables. According to *Jung and Wappis*,[13] output variables in the context of production processes are product characteristics, for example, diameter, surface quality or length. Possible input variables are materials, information and services that are used in the process. Control variables are variables influencing the process, which can be adjusted to a specific value. On the contrary, the disturbance variables like humidity cannot generally be set and controlled, or only with great effort (like a drying room for humidity).[13]

Kleppmann[14] uses a slightly different classification of variables to characterize production processes: Instead of output variables, Kleppmann speaks about target variables are being looked at. In addition, influencing variables are seen as the highest level. These are further subdivided into control and disturbance variables. The additional input variables introduced by Wappis and Jung are part of the control variables by Kleppmann.

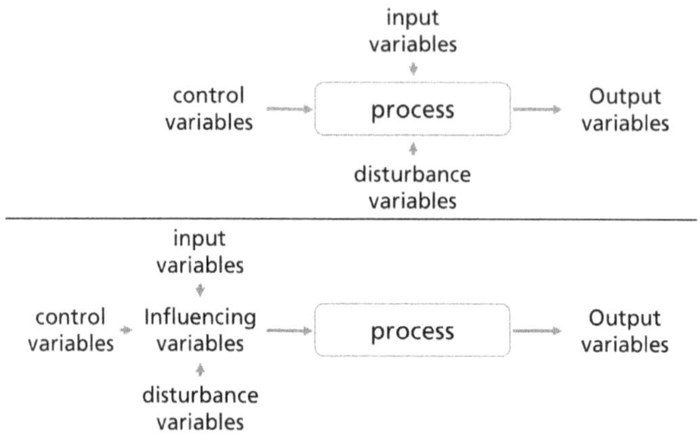

Fig. 4. Different approaches of characterizing variables in a production process.

The terms input, control and disturbance variables can be summarized in the superordinate term "influence variable" is used for the approach discussed here. For a clear distinction between (control) variables that can be set in the process and (input) variables produced by pre-processes, the term is further divided into input variable, control variable and disturbance variable. For variables which represent the process result, the term output variable is used uniformly. The connection is shown in Fig. 4.

3.2. *Identification and selection of process-relevant variables*

For the identification of concrete characterizing variables of the production process, the output variables must be defined first. In production processes, these are product characteristics. These product characteristics which are part of the specification of the product and are given by the design or development. They are defined in product drawings, for example, and are usually subject to tolerances which have to be fulfilled during manufacturing. Based on the product characteristics, the influencing variables are set. In order to determine the influencing factors of the respective product features, brainstorming is generally carried out.[14,13] Collecting ideas via brainstorming intends to stimulate the creative thinking of a team. This process can be supported by various tools:

- Ishikawa diagram,
- Process flowchart,
- influencing factor target variable matrix, and
- Failure Mode and Effects Analysis (FMEA).

These will be explained briefly in the following. The most widely known tool to support brainstorming to find influencing variables is the **Ishikawa diagram**, also called cause–effect diagram. This tool graphically illustrates the relationship between an event and all its influencing factors in a structured way. This form resembles the bones of a fish in an abstract form and is therefore called a "fishbone" diagram.[15] The use of the Ishikawa diagram helps to structure the topics of brainstorming and provides inspiration for further points. In the present application, the event is an output variable produced by the production process. The factors are the variables influencing the output variable. To better structure the influencing variables, five categories are used — the so-called "5 Ms":

- Manpower: what are the causes by humans and people for the effect?
- Machines: what are the causes by equipment, such as machinery, tools, instruments, technology, for the effect?
- Materials: what are the causes by defect or material properties for the effect?
- Methods: what are the causes by regulations or standards for the effect?
- Mother Nature: what are the causes by the environment and context for the effect?[16]

Since a first query usually mentions superordinate influencing variables, it is helpful to further split these up into so-called basic influencing variables. In comparison to superordinate influencing variables, basic influencing variables represent the basic cause. With the basic influencing variables, it is therefore possible to make more precise settings or adjustments.[17]

A further support is provided by the **process flowchart**, which visualizes the individual production process steps and provides a description of the procedure in each step. This helps to identify input and output variables.[14] Therefore, it is necessary to define the starting and end point of the process, identify the individual steps and sort them in the correct order. Afterward, branches and loops have to be defined to connect the process steps.[18] An example of a process flowchart is seen in Fig. 5.

The **influencing factors-target variable-matrix** is suitable for the representation of correlations between influencing variables and output variables. Compared to the Ishikawa diagram, the matrix is showing correlations to more than one output variable. The example in Table 1 represents an influencing factors-target variable-matrix and will be explained in the following.

The output variables listed in the top row are compared in a matrix with the influencing factors listed in the first column. In the matrix, dependencies between influencing factors and product characteristics can now be displayed. In addition

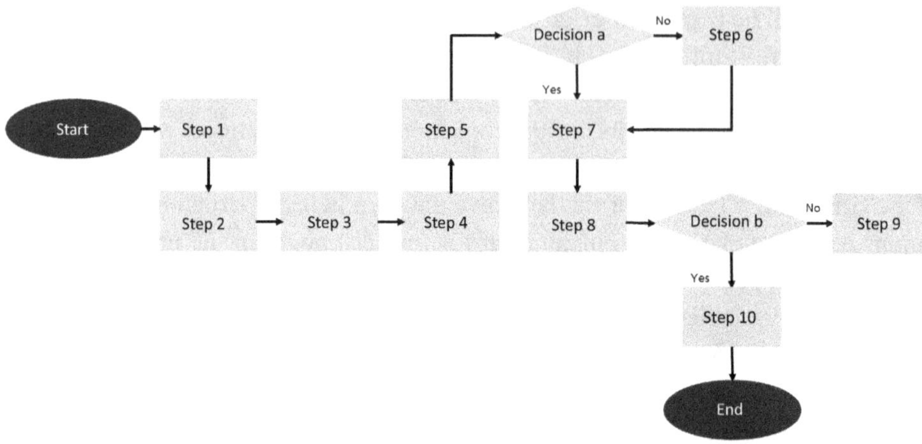

Fig. 5. Example of an process flowchart according to Ref. [18].

Table 1. Example of an influencing factors-target variable-matrix.

Factors	Output Variable 1	Output Variable 2	...
Influencing factor 1	3	0	...
Influencing factor 2	3	2	...
Influencing factor 3	1	3	...
Influencing factor 4	2	2	...
...	...	...	...

to the simple dependency as in the Ishikawa diagram or process flow chart, the strength of the influence of the respective influencing factor on the corresponding target variable can be represented in the influence factors-target variable-matrix by means of a special marking, for example, by numbers.[14,18] The representation of the strength of influence has the advantage compared to the Ishikawa diagram and the process flow chart that a prioritization of the influencing factors can be derived. A prioritization is necessary because despite low-cost sensor technology and a high available computing capacity (compare above), only limited resources are available in the context of Industry 4.0. A selection must therefore be made from a large number of possible influencing factors.

The identification of variables can additionally be supported by a **Failure Mode and Effects Analysis** (FMEA).[18] The actual field of application of FMEA is the preventive risk analysis in the investigation of products and processes.[15] Due to the possibility of using FMEA to analyze causes, it is also suitable for identifying possible influencing factors of a process.[19] However, FMEA has so far mostly been applied in tabular form, although it has been used increasingly in its original field of risk analysis for the last 15 years with visualization in meshes.[20] The advantage of the visualization in meshes compared to the presentation in tabular form is that the number of chain links is not limited. This is particularly necessary for cross-process chain production processes, because influencing variables from preceding process steps via intermediate variables have an effect on subsequent process steps. The advantage of linking chains was already used by Aichele and Landwehr in the context of battery cell manufacturing, but just for the specific step of dry coating.[21]

The already considered tools, process flow diagram and influencing factor-target variable matrix, do not offer the possibility for linking beyond intermediate variables, compared to the Ishikawa diagram. However, the Ishikawa diagram does not allow parallel consideration of several output variables. This means that the effect of process-relevant influencing variables on several output variables cannot be shown by using the Ishikawa diagram. Nevertheless, this is necessary to select the relevant variables of the entire process chain. This deficit of the limited mapping of cross-process chain cause–effect relationships was already recognized and researched decades ago. For the special interrelationships of a process chain with the combination of process- and production-related processes, this was addressed from the mid-1990s for ceramic components.[22] *Westermeier* has provided a detailed overview of the various approaches and has himself developed an approach for complex process chains using the example of li-ion battery cell production. On the basis of a modified process FMEA, experts are questioned about possible cause-and-effect relationships and their responses are then transferred to a matrix representation. This first qualitative evaluation serves to identify particularly relevant variables. The influence of these particularly relevant variables is further investigated by means of experiments and is finally quantified. The focus of these approaches to illustrate cause-and-effect relationships, is a quality-oriented analysis. This means that process relevant influencing factors are determined. Afterwards the product quality is optimized regarding the improvement of product characteristics. This optimization, or parts of it, is based on the performance of tests with subsequent evaluation. Therefore data which reflect the concrete values of the pre-selected relevant quality variables during the test execution is used.[23–27]

For a better understanding of the process, it should be added that the stated approaches do not provide a visualization of the cause–effect relationships besides a table. However, according to Schloske,[28] up to 30% more effect relationships can be identified via network structures. Cause–effect diagrams are an example of such network structures.

3.3. *Measuring systems*

In general, a valid analysis and evaluation of processes is only possible if the relevant variables are recorded in advance in the correct form. The consideration of all relevant variables was addressed earlier in this chapter. Recording in the correct form, by using suitable measuring systems, will be addressed in this chapter.

In the current state of the art, there are a few approaches that deal with the proper usage of measurement systems. They carry out a suitability test, for example, with the aid of a measuring system analysis[29] or test process suitability.[30] The suitability test is a comprehensive investigation with regard to "measurement uncertainty, repeatability, reproducibility, and variations over the measurement range (linearity)".[15] These investigations are carried out following the procurement and installation of the measuring systems. However, any support in the selection of the measuring systems is missing in the presented approaches. Currently, the search for suitable measurement systems is extremely extensive. As an example, an web-based search for the measurement of a specific variable produces a large number of search results. Even evaluating some of the search results requires a lot of time. An alternative search on manufacturers' websites is also time-consuming. First, it must be clarified whether sensors for a specific variable are offered in principle. Then, the characteristic values have to be checked. Depending on the manufacturer, the scope of the stated characteristic values differs and a comparison is also difficult due to the different representations.

3.4. *Summary of the state of the art*

As described in Section 2, there are many approaches in the literature for identifying influencing variables in production processes without having to use already available measurement data. These approaches are not limited to simple processes but are also used in complex process chains. They have also been used in battery cell production in several examples. These approaches are lacking a suitable visualization for complex process chains in battery cell production. As described, this offers the possibility of improving the identification of influencing variables. Furthermore these approaches lack the possibility to justify why certain influencing variables are relevant and others are not. The traceability for the selection of relevant influencing variables is therefore only given to a limited extent. Moreover, the known approaches for determining cause–effect relationships lack any support for selecting appropriate measurement systems.

4. Procedure for Establishing Process Quality Characteristics and Suitable Measurement Systems

As a basis for a prognosis of the quality and a dynamic adjustment of the manufacturing processes, this work develops a procedure for improving the understanding of cause-and-effect relationships for lithium-ion battery cell production. Figure 6 shows the basic procedure as well as the seven steps which are divided into two different areas.

1. Identification of process-relevant factors.
2. Selection of suitable measuring systems.

Therefore, the following chapters are structured in this way.

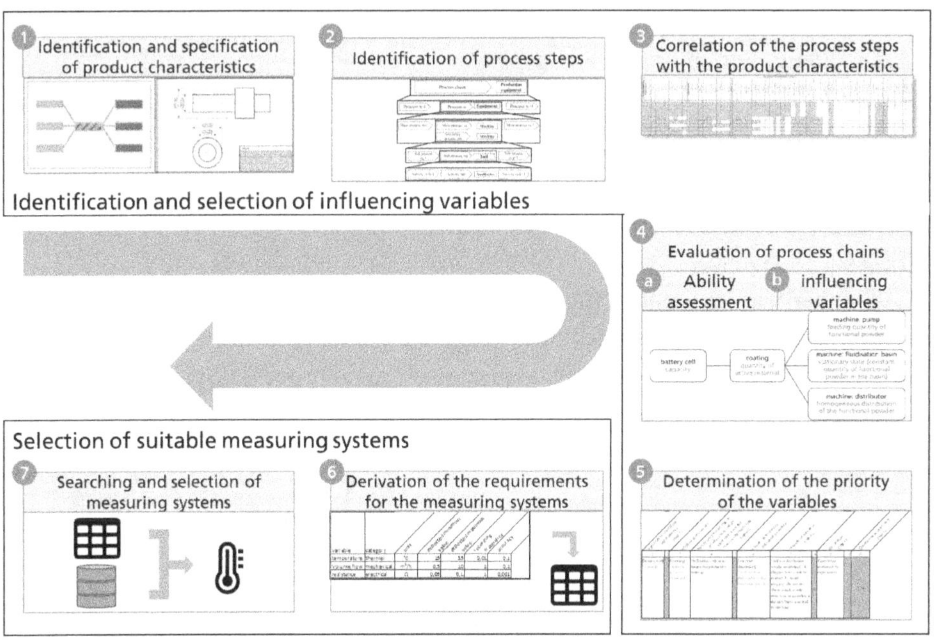

Fig. 6. Procedure for establishing the cause-and-effect relationships.

4.1. *Identification and selection of influencing variables*

The procedure for the identification and selection of influencing variables consists of five steps:

4.1.1. *Identification and specification of product characteristics*

In the first step, it is essential to identify and specify all product characteristics of the cell to be manufactured. Which will be the output variables as stated in Section 2.

Product and construction drawings can be used to support the identification and specification as they include all product characteristics. In contrast to this, the use of a Failure Mode and Effects Analysis (FMEA) is recommended, as product characteristics are determined by means of the functional nets. As a result, conclusions about the influence of product characteristics on the function or key characteristic of the battery cell can be drawn. It is also important to evaluate the significance of the product characteristics with regard to the effects on the key characteristics of a battery cell. This evaluation helps in the further procedure to identify relevant process characteristics. The used evaluation logic is shown in Table 2.

4.1.2. *Identification of process steps*

In a second step, all process steps of the lithium-ion cell production must be identified. For existing production lines, the value stream method is recommended. Alternatively, a process flow diagram is a possible source of information, too. It is important to ensure that the process steps are in the correct processing sequence.

Another way to identify the process steps is to break down the process chain in a discussion with experts, for example in the context of a workshop. The breakdown can be made via the individual processes, main processes, sub-processes and sub-processes up to the activities. To ensure a complete recording, it is advisable

Table 2. Evaluation proposal for the significance of individual product characteristics.

Valuation	Meaning
3	Security for the customer is no longer guaranteed
2	Main functions cannot be guaranteed
1	Sub-functions, desired appearance, service lifetime, etc. cannot be achieved

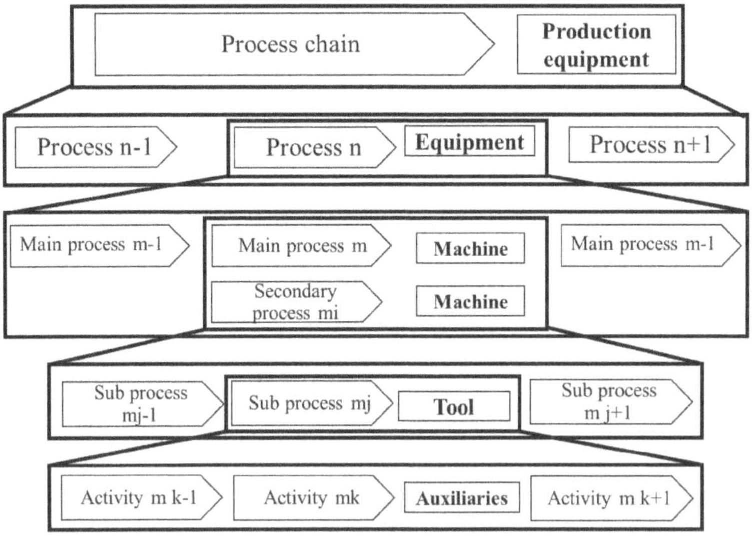

Fig. 7. Breakdown of a process chain to activities.

to carry out both a forward and a backward view. In addition, also to ensure completeness, the breakdown should be considered not only top-down but also bottom-up. In order to support the identification of influencing factors in step 4b, it is recommended to additionally include the respective plants, machines and tools in this step. Figure 7 illustrates this procedure.

4.1.3. *Correlation of process steps with product characteristics*

Based on the identified process steps, the product characteristics can now be correlated with the process steps. A tabular form has shown advantages for this process. This step can be performed either independently by an expert or in a team. The correlation logic in Table 3 can be applied. With this evaluation, it can be determined which product characteristics have to be further evaluated in each specific process step.

4.1.4. *Evaluation of process chains*

The obtained data can then be combined for each process step. Thus, the process owner is aware of the product characteristics he can influence through his process step. In many cases, process owners are not aware of which product characteristics can be changed or affected by their processes and what effect an incorrectly

Table 3. Evaluation proposal for the correlation of process steps with product characteristics.

Valuation	Meaning
3	High influence: the product characteristic is produced in this process step
2	The process step has a medium influence on the product characteristic
1	The process step has no influence on the product characteristic

Table 4. Evaluation proposal for the assessment of the ability of a process step to produce a reliable product characteristic.

Valuation	Meaning
3	The process step has not the ability to produce the product characteristic in a stable manner
2	The process step has the ability to produce the product characteristic most times in a stable manner
1	The process step has the ability to produce the product characteristic every time in a stable manner

manufactured product characteristic can have on the customer. The following steps (Table 4) are suggested to be done in a workshop.

4.1.4.1. Assessment of the ability of a process step to produce the product characteristic in the respective process step

For a differentiated assessment of the relevance of individual product characteristics and their influencing variables, it is also helpful to evaluate to what extent a product characteristic can be produced in a stable, i.e. without major deviations, current state within a process step. Therefore an evaluation according to Table 4 can be applied.

4.1.4.2. Determination of the influencing variables per product characteristic in the respective process step

In order to identify relevant influencing variables for the entire process chain, possible influencing variables must be identified for each product characteristic in the respective process step. Due to the significantly higher number of influencing variables found when using network structures (see Section 2), the use of a process

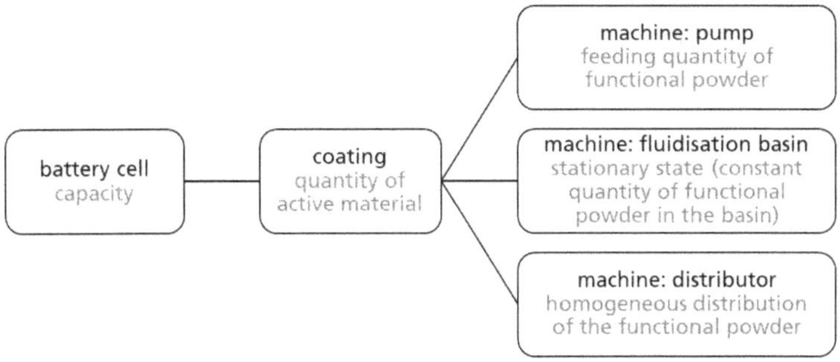

Fig. 8. An example of a network from an FMEA.

FMEA is suggested here (an example of a visualization of a network is shown in Fig. 8).

Additionally this method offers the advantage of an assessment of the strength of the influence of the influencing variable on the product characteristic to be produced, which can be carried out in parallel. Moreover, it offers the possibility to make a further assessment. This is useful for the certainty with which the strength of influence can be assessed. In addition to a purely numerical assessment, a written justification can be provided, which improves the comprehensibility of the assessment. Table 5 shows a proposal for a possible assessment of the strength of the influence and its uncertainty.

4.1.5. *Determination of the priority of the variables considered in the process steps*

The product characteristics and influencing factors can be ranked with the valuation in steps 1, 4a and 4b. Based on the significance in step 1, the process stability in step 4a and the strength of influence combined with the uncertainty of the strength in step 4b, a specific value can be calculated. A possible calculation is shown in Eq. (1) for the product characteristics and Eq. (2) for the influencing variables.

Calculation of the weighting of the product characteristics.

$$W_{PCx} = \sqrt{SxA} \tag{1}$$

Calculation of the weighting of the influencing variables.

$$W_{IVx} = \sqrt[3]{\left(\frac{SxAx\sqrt{IxK}}{3}\right)} \tag{2}$$

Legends:
W: Weight
PC: Product characteristic
IV: influencing variable
x: specific product characteristic or influencing variable
S: Significance (Step 1)
A: Ability of a process step (Step 3)
I: Strength of influence of influencing variable (Step 4a)
K: Knowledge about strength of influence of influencing variable (Step 4b).

An example to document the key elements of steps 1–5 for the further proceeding is shown in Table 6. In the first and second row, it is shown which function or key characteristic of the battery cell or which specific product characteristic has an influence and how the significance is evaluated (Step 1). The ability of the specific process step to produce the product characteristic is evaluated in row 4 and 5 (Step 3). The results from step 4 are shown in columns 6–10. In row 5, the influencing variable(s) of the product characteristic are integrated. The evaluation of the strength of influence and the knowledge about that are integrated in the following row. The result of step 5, the priority of each variable, is shown in row 11.

The different calculation of product characteristics and influencing variables has the advantage that despite different numbers of multipliers, the priority value is always in the range of one and three. Depending on this priority value, a ranking of product characteristics and influencing variables can now be carried out, thus a priority for measuring systems can be determined, which will be produced in the entire measuring chain. However this approach suggested a measuring system with a calculated priority of 1.5 and higher to create a sufficient data base for later evaluations. If there is the possibility to procure further measuring systems due to available resources, this can be done by using the ranking of the calculated priority.

Table 5. Proposal for the evaluation of the strength of influence from variables on product characteristics and the knowledge about them.

Valuation	Evaluation of the Strength of Influence	Evaluation of the Uncertainty of the Strength of Influence
3	High influence	Influence strength is demonstrably neither confirmed nor refuted
2	Medium influence	Experience shows a confirmation of the assessment
1	Low influence	Influence strength is demonstrably proven

Table 6. Documentation of the key elements of steps 1–5 for further proceeding.

Step	No.	Description
Significance of product characteristic (Step 1)	1	[battery cell] ‹capacity›
	2	2
Product characteristic	3	[coating] ‹quantity of active material›
Ability of a process step to produce the product characteristic in the respective process step (Step 4a)	4	Difficulties only arise during the production start-up
	5	2
Process characteristic/ influencing variable	6	[machine: distributor] ‹homogeneous distribution of the functional powder›
Strength of influence (Step 4b)	7	Uniform distribution insofar as ensured. A clogged hose could be critical. To avoid clogging: Do not use fibers which would otherwise accumulate in the distributor and lead to blockage
	8	2
Knowledge (Step 4b)	9	Knowledge confirmed by experiments
	10	1
Determination of priority (Step 5)	11	1,24

4.2. *Selection of suitable measuring systems*

The selection of suitable measuring systems consists of various steps which are presented in the following chapters.

4.2.1. *Derivation of the requirements for the measuring systems*

When working in a team, e.g. in the context of a workshop or step-by-step processing by individual team members, the requirements for the measuring systems have to be derived from the variables prioritized in the previous step. This can be realized with the help of an exploration sheet, which contains all prioritized variables.

At the beginning, the variables are assigned to a specific measurement category. This may be one of the following categories, for example:

- **Mechanical quantities:** Lengths, temporal derivatives, pressure, force, volume flow, vibrations, time, mechanical stress and geometry, shape and weight.

- **Chemical quantities:** Material composition, pH value, gases and particle size.
- **Thermal quantities:** Humidity, dew point, temperature of liquids, air and materials.
- **Radiation quantities:** Laser parameters.
- **Electrical quantities:** Energy, power, current, voltage, resistance and conductivity of liquids.

Depending on the category, various criteria are then defined as requirements. An important basic requirement is the expected measurement range to be recorded, the frequency of data recording and the necessary measurement accuracy and thus the measurement resolution.

Due to the effects of influencing variables on the product characteristics, which are partly unknown at the time of the determination of the requirements, an exact estimation of the necessary measuring accuracy is difficult. Since prices increase strongly with increasing resolution and resources are generally limited, high-resolution measurement systems cannot be used for all quantities. Here, depending on the expected measuring range and the expected sensitivity of the quantity to its target quantities, the necessary resolution could be chosen higher or lower. With a high sensitivity, for example, the measuring system must be able to resolve even in a low percentage range. With a low sensitivity, a coarser resolution is sufficient and thus, under certain circumstances, a more favorable measuring system. Furthermore, possible installation situations in the process must be taken into account: How much space is available, what are the conditions in the room regarding environmental conditions (temperature, humidity, radiation, etc.)? In addition to the requirements of the specific measurement, which could be further supplemented, the requirements of higher-level systems must be taken into account in the context of digitalization. These can be interfaces or the maximum number of data to be processed, which would require additional logic for processing the data.

4.2.2. *Searching and selection of measuring systems*

Based on step 6 with the defined requirements for each variable, suitable measurement systems can be selected. In order to make the research and subsequent selection of measuring instruments as simple as possible, it is advisable to have a database available. This database should contain measuring instruments from different relevant manufacturers in order to have a sufficient selection. In addition, the requirements defined in step 6 for each measuring instrument should be integrated and defined in the database. By assigning the measuring instruments available in the database to the categories mentioned in step 6 (mechanical quantities, chemical quantities, thermal quantities, radiation quantities, electrical quantities), it is possible to have suitable measuring instruments suggested directly on the basis of the information from step 6. These can be used as a basis for a more specific selection (e.g. with regard to price, other connectivity to the system).

5. Conclusion

In the presented chapter, an approach was introduced which supports the digital quality management in battery production. In the introduction, the necessity of quality improvement within the process chain of the battery production was explained. Furthermore, the specific challenges in this complex process chain were presented. To master these challenges, it is essential to use the possibilities of digitalization and artificial intelligence. A sufficient data basis has to be created for this. By digitally recording the relevant variables of battery production. It has to be ensured that this process is carried out in a form that allows a correct evaluation by artificial intelligence. These two points have been realized in the presented approach. By the systematic use of suitable tools, a suitable prioritization of the essential variables of battery production is possible. With the subsequent derivation of the requirements for the metrological recording of the variables and the generation of suggestions for suitable measuring systems, it is ensured that the resulting data basis and the following analysis improves the quality of battery production.

References

1. Kuepper D, Kuhlmann K, Wolf S, Pieper C, Xu G, Ahmad J. The future of battery production in electric vehicles; 2018 Sep 11 [cited 2019 Jul 9]. Available from: https://www.bcg.com/publications/2018/future-battery-production-electric-vehicles
2. Wood DL, Li J, Daniel C. Prospects for reducing the processing cost of lithium ion batteries. J Power Sources. 2015;275:234–42.
3. Michaelis S. Roadmap Batterie-Produktionsmittel 2030. Update 2018. Frankfurt am Main: VDMA Batterieproduktion; 2018.
4. Brodd RJ, Helou C. Cost comparison of producing high-performance Li-ion batteries in the U.S. and in China. J Power Sources. 2013;231:293–300. Available from: https://www.sciencedirect.com/science/article/abs/pii/S0378775312018940
5. Hakimian A, Kamarthi S, Erbis S, Abraham KM, Cullinane TP, Isaacs JA. Economic analysis of CNT lithium-ion battery manufacturing. Environ Sci Nano. 2015;2(5):463–76.
6. Roland Berger Strategie Consultants / BDI. Die digitale Transformation der Industrie; 2015.
7. Drossel W-G, Brecher C, Bauernhansl T, Gumbsch P, Hompel M ten, Miehe R, *et al.* Biointelligenz: Eine neue Perspektive für nachhaltige industrielle Wertschöpfung. Stuttgart: Fraunhofer Verlag; 2019.

8. Thiede S, Turetskyy A, Kwade A, Kara S, Herrmann C. Data mining in battery production chains towards multi-criterial quality prediction. CIRP Ann. 2019;68(1):463–6.
9. Schnell J, Nentwich C, Endres F, Kollenda A, Distel F, Knoche T, *et al.* Data mining in lithium-ion battery cell production. J Power Sources. 2019;413:360–6.
10. Kornas T, Daub R, Karamat MZ, Thiede S, Herrmann C. Data-and expert-driven analysis of cause-effect relationships in the production of lithium-ion batteries. In: 2019 IEEE 15th International Conference on Automation Science and Engineering (CASE). IEEE; 2019. pp. 380–5. Vancouver, BC, Canada.
11. Kornas T, Knak E, Daub R, Bührer U, Lienemann C, Heimes H, *et al.* A Multivariate KPI-based method for quality assurance in lithium-ion-battery production. Procedia CIRP. 2019;81(3):75–80.
12. Kornas T, Wittmann D, Daub R, Meyer O, Weihs C, Thiede S, *et al.* Multi-criteria optimization in the production of lithium-ion batteries. Procedia Manuf. 2020;43:720–7.
13. Wappis J, Jung B. Null-Fehler-Management: Umsetzung von Six Sigma; [extra mit kostenlosem E-book. 4., überarb. und erw. Aufl. München: Hanser; 2013. (Praxisreihe Qualitätswissen).
14. Kleppmann W. Versuchsplanung: Produkte und Prozesse optimieren. 9., überarbeitete Auflage. München: Hanser; 2016. (Praxisreihe Qualitätswissen). Available from: https://www.hanser-elibrary.com/doi/book/10.3139/9783446447172
15. Schmitt R, Pfeifer T. Qualitätsmanagement: Strategien–Methoden–Techniken. 5., aktualisierte Auflage. München: Hanser; 2015. Available from: https://www.hanser-elibrary.com/isbn/9783446440821
16. The Ishikawa Diagram: identify problems and take action. 1. Aufl. s.l.: 50 Minutes; 2015. (Management & Marketingv.5).
17. Winz G. Qualitätsmanagement für Wirtschaftsingenieure. Qualitätsmethoden, Projektplanung, Kommunikation. München: Hanser; 2016.
18. Meran R, John A, Staudter C, Roenpage O, Lunau S. Six Sigma+Lean Toolset: Mindset zur erfolgreichen Umsetzung von Verbesserungsprojekten. 3. Aufl. 2012. Berlin, Heidelberg: Springer; 2012. (SpringerLink Bücher).
19. Medina-Oliva G, Iung B, Barberá L, Viveros P, Ruin T. Root cause analysis to identify physical causes; 2012.
20. Produkt- und Prozess-FMEA: Sicherung der Qualität in der Prozesslandschaft; Allgemeines, Risikoanalysen, Methoden, Vorgehensmodelle; 2006. Band 4.
21. Aichele A, Landwehr I. Identification and selection of parameters in battery cell production for analyses of process improvement using FMEA using the example of dry coating. In: LabFactory TUB/B, editor. 3rd International Battery Production Conference. 2020 Nov 2–4, Virtuell; 2020. 1 Poster.
22. Westermeier M. Qualitätsorientierte Analyse komplexer Prozessketten am Beispiel der Herstellung von Batteriezellen. München: Herbert Utz Verlag; 2016. (Forschungsberichte IWB).
23. Mayers B. Prozeß- und Produktoptimierung mit Hilfe der statistischen Versuchsmethodik [Zugl.: Aachen, Techn. Hochsch., Diss., 1997]. Als Ms. gedr. Aachen: Shaker; 1997. (Berichte aus der Produktionstechnik; vol 97,9).
24. Molitor M. Prozesskettenorientierte Qualitätstechnik: Dargelegt am Beispiel der Getriebe-Serienfertigung [Zugl.: Aachen, Techn. Hochsch., Habil.-Schr., 1995]. Als Ms. gedr. Aachen: Shaker; 1997. (Berichte aus der Produktionstechnik; vol 97,4).
25. Steins DJ. Entwicklung einer Systematik zur qualitätsgerechten Optimierung komplexer Produktionssysteme. Aachen: Shaker; 2000. (Berichte aus der Produktionstechnik).
26. Eichgrün K. Prozesssicherheit in fertigungstechnischen Prozessketten: Systemanalyse, ganzheitliche Gestaltung und Führung. Kaiserslautern: Universität, Lehrstuhl für Fertigungstechnik und Betriebsorganisation, 2003.

27. Schäfer L. Analyse und Gestaltung fertigungstechnischer Prozessketten: Konzept zur datenbasierten Ermittlung qualitätswirksamer Einfluss-Ursache-Wirkzusammenhänge und zur Ableitung von Maßnahmen zur Prozesssicherung [Zugl.: Kaiserslautern, Univ., Diss., 2003]. Kaiserslautern.
28. Stuttgarter Produktionsakademie, Schloske A, Mannuß O, editors. Prozess-FMEA, Besondere Merkmale und Control-Plan : Integriert und durchgängig erstellen. Seminar, 2019 Sep 19, Stuttgart; 2019.
29. Chrysler Group LLC, Ford Motor Company, General Motors Corporation, Automotive Industry Action Group. MSA–Measurement Systems Analysis [Reference Manual]. 4th Ed., Second Printing 2017. Southfield, Mich.: Automotive Industry Action Group; 2010.
30. Verband der Automobilindustrie e.V. Prüfprozesseignung. Berlin; Juli 2011. VDA-Band 5.

https://doi.org/10.1142/9789811245626_0014

Chapter 13

Cross-Process Stabilization and Optimization of Battery Cell Production Through Digital Networking of Process Stations

Pablo Mayer*, Michael Oberle† and Sascha Gärtner‡
Plus 10 GmbH, Nobelstraße 15, 70569 Stuttgart, Germany
**pablo.mayer@plus10.de*
†michael.oberle@ipa.fraunhofer.de
‡sascha.gaertner@ipa.fraunhofer.de

This chapter deals with the method of development to overcome the existing challenges in Li-ion battery cell manufacturing. In particular, the complex interlinking and interdependence of several process steps make it difficult to set up a stable manufacturing process in Li-ion battery production. To overcome these challenges, special methods for process stabilization and optimization were developed. In order to control the complexity of the processes, there is a great focus on combining process knowledge in interdisciplinary teams. Furthermore, the method relies on different statistical tools which are used in different phases during the optimization. For validation, the developed method is used to stabilize a single process.

1. Introduction

The majority of the value chain for battery system production takes place entirely outside Europe. Apart from Tesla, this market is claimed almost completely by Asian manufacturers.[1] Forecasts show that Tesla will also gain an increasing market share in the production of battery systems in the coming years.[2] Meanwhile, society's approval of electromobility is growing, while acceptance of combustion engines is declining. This trend is underlined by the increasing number of electric cars registered in recent years.[3]

Therefore, the German government would like to use economic stimulus packages to focus on promoting forward-looking areas such as alternative drive concepts and digitalization.[4] In the context of electromobility, especially in battery cell production, technology leadership lies with companies outside Germany.[1] The German automotive industry lacks expertise in this regard, as for years the focus remained on the conventional drive strategy of the internal combustion engine.

German automotive suppliers are experiencing an indirect decline in orders for the production of components for combustion engines as a result of this trend.[5] At the same time, setting up production facilities for battery cell manufacturing is seen as a major risk in view of the high investment costs and lack of technical expertise in this area.[6]

Battery technology can be seen as a key technology for the competitiveness of companies in the field of electrification of transport systems. In order to build up a market advantage over competing manufacturers, a resource-saving and cost-effective high-performance battery system is necessary. By using current technologies of digitalization as well as process knowledge, Europe can take an important step toward technology leadership.[7] However, knowledge gaps in battery cell production must first be closed and optimization potentials exploited. Process analyses in this context have shown that the start-up phase of battery cell production in particular offers great potential for optimization. The aim is therefore to develop and validate a method to accelerate the start-up phase of battery cell production. This is being done using digitalization measures and is intended to help medium-sized companies in particular to overcome the challenges in battery cell production.

The practical activities in the context of this article are implemented in the so-called center for battery cell manufacturing (ZDB, for short). The ZDB researches and develops strategies for the sustainable production and the life cycle-optimized usage of battery cells. Multiple individual battery cell production processes are available for this purpose in a laboratory environment.[8]

2. Li-Ion Cells and Method Development — Theoretical Background

This section provides an overview of fundamental topics that have an influence on the development of the overall concept. At the beginning, the principle of a lithium battery cell and its manufacturing process are explained. This is followed by an introduction to the topic of Industry 4.0 and current technologies of machine-to-machine communication.

2.1. *Electrochemical storage systems*

The generic term battery covers all electrochemical storage cells that enable a self-sufficient power supply. A distinction is made between primary (non-rechargeable) and secondary (rechargeable) cells. Both systems use carbon (graphite) and toxic heavy metals such as lead, cadmium, mercury, nickel or lithium as chemical starting materials.[9] This chapter focuses in particular on cylindrical lithium-ion cell systems.

Li-ion accumulators are considered state of the art in the use of electric batteries in vehicles.[10] The structure consists of a total of two electrodes, a separator and a high-purity aqueous electrolyte. The electrolyte is usually a solution of a lithium salt in an organic solvent. The separator is located between the two electrodes and isolates them from each other.[9] A protective passivation or top layer (solid electrolyte interface — SEI) is formed on the electrode surfaces by a reaction with the lithium. This layer enables the exchange of lithium ions between the electrodes through the separator layer. Depending on the direction of flow, this results in a discharging or charging process.[11]

In summary, the accumulator consists of a housing, positive electrode, negative electrode, separator and an electrolyte. Depending on the application, different types of accumulators are used. The most common types are button cells, pouch cells, cylindrical cells and prismatic cells.[11]

Individual cells are usually connected in series or parallel in a so-called module. The combination of several modules results in the battery pack. The module housing provides protection against external influences, primarily mechanical loads. In addition, there are electrical, mechanical and thermal components. The battery management system is used to monitor the condition of the battery and to enable connection to the overall system.[12]

Basically, the manufacturing process of a lithium-ion cell can be divided into the steps of electrode production, cell assembly, and formation and aging. The entire process takes place under strictly controlled clean room conditions. The operation of the rooms, appropriate equipment and automated machines that meet the requirement for high cleanliness in production is related with high costs.[13]

At the beginning of the manufacturing process is the electrode production which consists of further sub-steps. In addition to the two active materials for the anode and cathode, electrical conductors, binders and additives are mixed together during the mixing process. With the aid of solvents, a slurry-like mixture is then produced in the plant. The mixing process affects the subsequent quality of the cell and must therefore be carried out with extreme care.[13]

The slurry is then applied to the respective electrode material. Cell capacity can be influenced by the layer thickness in this process step. The applied mass is now dried together with the carrier material using air jets with specific temperature profiles. Afterwards solvent is extracted from the carrier film by the heat. The so-called calendar is an arrangement of rollers arranged one above the other to roll the electrodes to the correct thickness. Without adjusting the process parameters, significant influence can be exerted here on the performance and quality indicators.[13] Knife or laser cutting systems then cut the electrode material to the appropriate width. This process is called slitting. Finally, the electrodes are dried in a vacuum oven for 24 to 48 hours. This removes the residual moisture from the coil.[11] In the next process step, the electrodes are wound around a core together with the separator strip. Contact plates are then attached to the end faces and the cover assembly is welded on. To accelerate the filling of the cell with the electrolyte, it is placed under a vacuum. Finally, the cell is placed in a housing and sealed.[13] The last block of the manufacturing process is the cell finishing, which again consists of several individual processes. First, the cell is slowly charged. During each charging cycle, the loading current is successively increased. This contributes to the complete formation of the SEI layer and is called forming.[11] The subsequent aging process involves storing the battery cells for up to three weeks at normal or high temperature. Due to the increased space required for the cells, storage takes place in so-called storage towers. If no significant changes in the cell properties have occurred after storage, it can be classified as a good part. The final end-of-line test uses measured values to categorize the produced cells and assigns a quality class to each of them.[13]

2.2. *Methods of process stabilization and optimization*

The term process optimization covers the improvement of a process with regard to criteria such as quality, costs, speed, efficiency and effectiveness. Process stabilization, on the contrary, requires productions to detect and react to deviations. A stable process is defined by the ability to deliver a product within the specified tolerances.[14] The following paragraphs give a brief overview of statistic tools that are applied in this context.

Statistical process control follows the principle of continuous quality improvement. For this purpose, the method makes use of various statistical procedures and tools. A common example is a quality control chart used to achieve a satisfactory quality and to minimize scatter in the process. When using tolerance zones, the aim should always be to achieve target-oriented utilization, rather than utilization of the entire tolerance zone. Samples are used in the quality control chart as a simple way of detecting deviations in the process with regard to a quality characteristic.[15]

Statistical design of experiments (DoE) is a branch of applied statistics and deals with the planning, execution, analysis and interpretation of controlled tests, also called experiments. Through this method, the effect of influencing parameters on the overall result (output) can be determined and analyzed. In an iterative process, the process parameters can then be adjusted so that a stable process is established.[16]

Failure Mode and Effects Analysis (FMEA) is used to uncover potential sources of error in products or processes at an early stage in the planning phase. The relevance of these sources of error must be assessed in order to initiate suitable measures if necessary.[17]

Advanced Process Control (APC for short) links several components, including Statistical Process Control (SPC for short), Run-to-Run (R2R for short), and Fault Detection and Classification (FDC for short) to form an overall system. The method solves the handling of complex processes and originates from semiconductor manufacturing, where quality and throughput are of very high relevance.[18]

Data mining describes the collection and identification of patterns in large datasets. In total, the process steps consist of the selection of data, preprocessing, transformation (conversion), analysis using various methods and interpretation of the result. Data mining methods can provide information about the influence of process parameters. Using a sufficiently large dataset, correlations between quality and influencing variables can be determined, for example. These findings can in turn be used profitably in the setting of a stable process or in the statistical DoE. If APC is applied, large amounts of data from different sources of different data types are combined. This results in a high complexity so that data mining methods, especially neural networks, are often the only way to gain insights from the data.[19]

2.3. *Machine-to-machine communication*

At the heart of the fourth industrial revolution are the Cyber-Physical Systems (CPSs). These systems combine information and software components together with mechanical and electronic components to communicate as a system using a data infrastructure. Therefore, an organization of cyber physical systems is created by interconnecting embedded systems through wired or wireless communication networks.

Special requirements are placed on CPSs in the production environment. For example, the devices must be network compatible and have suitable interfaces for communication with peripheral devices. Furthermore, many applications, such as controllable automation systems in the production environment, require real-time capability via bus systems.[20]

In addition to the physical product, machines and software form a flexible network in which all participants are networked with one another. According to the Reference Architectural Model Industry 4.0 (RAMI 4.0), this enables communication between all participants regardless of their hierarchy level.[a] This communication not only needs to be enabled by syntactical interoperability but also by enabling semantic interoperability. In order to enable both, OPC Unified Architecture (UA) published a platform-independent architecture in 2008 that can be used for machine-to-machine (M2M) communication allowing the semantic modeling of data.[21]

3. Method Development

This section covers the development of the method for stabilizing and optimizing individual and overall processes in battery cell production.

The manufacturing process of a battery cell represents a complex interaction of individual sub-steps. Any deviations and influencing factors in the process can lead to a reduced quality of the end product. For this reason, the cross-process dependencies and their origins or interdependencies must be recorded and subsequently evaluated. The focus for the method in this chapter lies on the process steps of assembling, welding and filling.

In targeted method development, a fundamental distinction is made between stabilization and optimization. Furthermore, a distinction is made between operating phases of production plants, the individual process and the higher-level process as well as the core processes and their knowledge carriers.

A single process is understood to be a single operation or process step, for example, the electrolyte filling of a battery cell. The overall process is therefore the production of the complete battery cell. There are various standardized models for further description. This chapter uses the models of the ISA-88 process model (see Fig. 1) in further course.[22]

In addition to the general knowledge of the process flow, the specific process knowledge of the employees and machine operators plays a decisive role in the optimization and stabilization of a process. In addition to the business understanding (objectives of the project), which is essentially developed from an economic and strategic point of view, operational knowledge carriers are irreplaceable.

[a]The terminology in the production context follows the standards IEC62264 and IEC61512 which define reference models for batch control and the integration of company control systems.

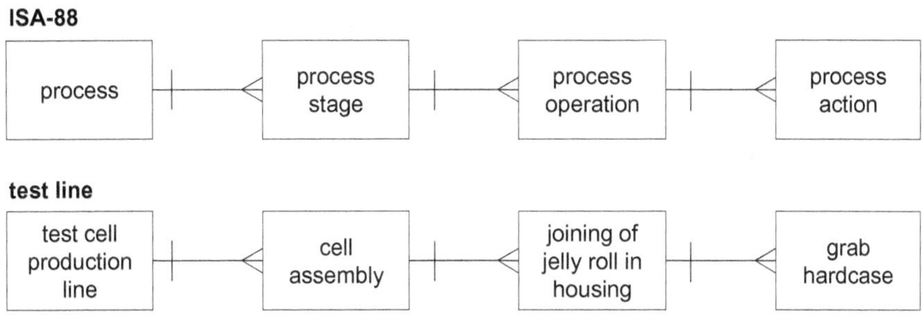

Fig. 1. Process model (top) and application on the test cell production line (bottom).

A distinction can be made between process knowledge, data knowledge and analytical understanding.

Process knowledge

- *Qualification*: These individuals have knowledge of the actual process (chemical, physical, technical); ideally, they are developers of the processes or have years of experience in directly or indirectly related processes.
- *Job titles*: For example, process engineer, chemist, materials scientist and physicist.
- *Role*: Among other things, identification of parameters, properties or test criteria that are meaningful or relevant. Qualitative assessments regarding product and process.
- *Example based on mixing process*: Person with strong experience in mixing processes, raw material properties and ideally high experience in mixing processes of electrode materials.

Data knowledge

- *Characteristics*: Person who knows the implementation of the process or the automation from a technical point of view, i.e. how, at what time and where in the machine which processes are carried out.
- *Job titles*: For example, process engineer, control engineer, programmer and automation engineer.
- *Role*: Generation of measured values by sensors, by means of design and implementation of test systems. In addition, the person is an expert in the control and regulation technology and can thus identify and modify or even generate the corresponding variables and data.

- *Example based on the mixing process*: Control technician who has programmed the mixing process and can extract the appropriate actual and setpoint values from the machine controls, such as viscosity.

Analytical understanding

- *Characteristics*: Has the ability and knowledge to analyze complex data and identify relationships from the same. Proficient user of state-of-the-art analysis tools and procedures. Basic knowledge of data storage and databases.
- *Job titles*: For example, data scientist, mathematician, computer scientist, statistician and physicist.
- *Role*: Working closely with the process and data scientists, the analysis scientist performs the data preparation and the actual analysis. For the interpretation of the results as well as for the modeling, the input of process and data knowledge is again necessary.
- *Example based on the mixing process*: Data scientist who can use the viscosity data as well as temperature and mixing time to draw conclusions about the choice of the most suitable agitator.

The mix of specialists from different areas thus forms the basis for the next steps and is valid for both individual and overall processes.

3.1. *Development of a method for individual processes*

The method for process stabilization and optimization can be divided into three phases: ramp-up, stabilization and optimization. In each phase, important subtasks are performed in order to ultimately master the complex process. In all three phases, teams of experts from the intersections of the domains are absolutely necessary.

3.1.1. *Phase 1: Ramp-up — Individual process*

First, all knowledge carriers (process knowledge, data knowledge and analytical understanding) are involved in the first phase, the ramp-up of an individual process, so that the strongest possible exchange of knowledge between the process experts is created.

In detail, the approach is defined, possible features are identified and test plans are set up and executed. A wide variety of tools are used to visualize processes, but control programs and measurement technology can also be used. The overall goal is to derive relevant process parameters to enable a stable process. The approach is defined as follows:

- *Procedure*:
  (a) Build up process understanding.
  (b) Process feature identification and recording.
  (c) Set up experimental plan and execute experiments.
  (d) Evaluation and adjustments.
- *Tools*:
  (a) Process visualizations, flow charts, technical drawings, moderation techniques.
  (b) Control programs, measurement protocols, flow charts, analog documentation, measurement techniques, circuit diagrams, data loggers.
  (c) DoE software tools, modeling software (for example UML (Unified Modeling Language)), databases.
  (d) Presentation tools, moderation techniques.
- *Goal/Result/Outcome*:
  (a) Intrinsically stable process within tolerances.
  (b) Reduced number of variables or process parameters.
  (c) Derived SPC parameters.
  (d) Graphical presentation of the results as a basis for discussion.

3.1.2. *Phase 2: Stabilization — Individual process*

The knowledge gained is then used in phase 2 to reduce the variables required for stable operation of a process to the bare minimum and to implement them in an SPC (Figure 2). This allows to minimize the scatter of a process and to set

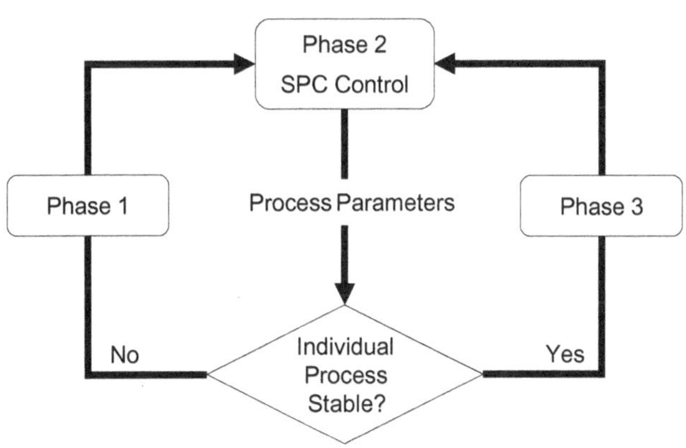

Fig. 2. Relationships between the different phases for the method individual process.[23]

fluctuating influencing parameters correctly so that a stable process is created. In practice, this is often an iterative process.[23]

3.1.3. *Phase 3: Optimization — Individual process*

Basically, phase 3 is a closed process optimization for a single process, i.e. only parameters of the single process are considered. This explicitly excludes parameters of upstream and downstream processes. Examples of approaches for optimizing the individual process can include the use of optimized tools or sensors for process parameter recording, new recipe parameters or modified process sequences.

An example is the implementation of an optimization on the mixing process with the aim of shortening the mixing times. After understanding the process, a mechanical optimization of the dissolver is carried out, for example, the reworking of the flanks on the dissolver. For this purpose, the viscosity change of the slurry paste is measured over time, which is recorded by a viscometer. Based on these and other parameters such as temperature, speed and start-up ramps, tests can be carried out to establish optimized parameters.

3.2. *Cross-process method*

Instead of the individual process analysis, the focus of the stabilization and optimization measures is now on the overall process. The differentiated approach to the overall process analysis differs by focusing on the influences of processes on each other.

3.2.1. *Phase 1: Ramp-up — Cross-process*

The number of influencing parameters increases compared to the individual processes, and interactions between the processes make the identification of important characteristics and the development of process knowledge even more difficult. The process variables to be identified in Phase 1 are characterized by the data having a direct impact on downstream processes, the final or intermediate product, and direct quality parameters of the final product. Therefore, it is recommended to use the same team of experts for the analysis of the single and the overall process.

- *Procedure*:
  - (a) Build process understanding.
  - (b) Identify dependencies of the processes.
  - (c) Process characteristics and recording.

(d) Set up experimental plan and execution experiments.
(e) Evaluation and adjustments.

- *Tools*:
  (a) Process visualizations, flow charts, technical drawings, analog documentation.
  (b) Flow charts and process visualizations.
  (c) PLC programs (programmable logic controller), measurement protocols, measurement techniques, circuit diagrams.
  (d) DoE software tools, UML, Databases.
  (e) Presentation and moderation techniques.
- *Goal / Result / Outcome*:
  (a) Intrinsically stable cross-process.
  (b) Dependencies between individual processes.
  (c) Derived parameters for APC.
  (d) Graphical presentation of the results as a basis for discussion.

3.2.2. *Phase 2: Stabilization — Cross-process*

High complexity in the manufacturing process, as well as pronounced coupling effects of multiple input parameters on the final process results, make the use of a classical single variable feedback system impossible.

A concept of upstream and downstream adaptations to process parameters would be required for stabilizing the cross-process. An APC system with feedforward and feedback loops (see Fig. 3) can automatically determine the optimal recipe for each individual product. Based on the characteristics of the individual product itself and the machine's tool condition, the optimal recipe is determined.[24]

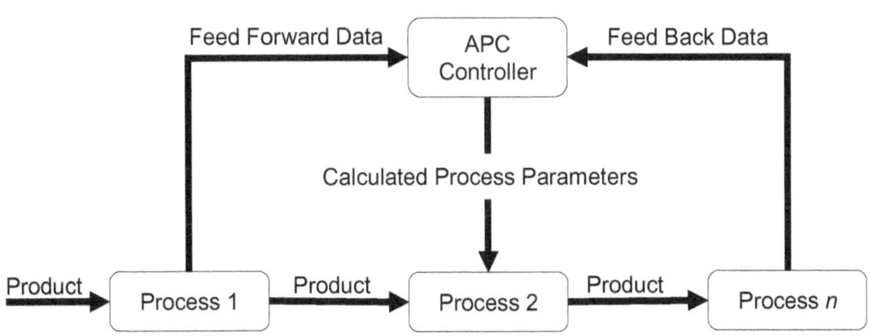

Fig. 3. Mode of action APC Feedforward and Feedback Loop.[25]

The APC system uses measurement techniques, process models and sophisticated control algorithms to provide dynamic fine-tuning of intermediate process targets that improve the final product.[22] Once the process has been stabilized, the APC system can be used to selectively shift the scatter band within the tolerance field. For example, it is possible to compensate for tool wear on a machine while maintaining quality. In this context, the APC controller can be used for processes with a run-to-run (R2R) characteristic. It provides a discrete controller that is parameterized for each run in advance. The run itself can correspond to a batch (batch processing or batch production) or to a single product.[25] Measured values of the current process step can be used to adjust the process parameters for the subsequent workpieces. In this way, process fluctuations, such as those caused by tool wear, can be compensated.

In contrast to the SPC approach from single process stabilization, with higher-level stabilization, reactions are stabilized on characteristics of output or intermediate products from phase 1. By using information across individual processes, it is possible to compensate for deviating process results in upstream and downstream processes.

Furthermore, the added value of upstream processes is not lost, especially in the phase of "ramping up" a line after a successful start-up. Ultimately, this results in a minimization of scrap.[25]

3.2.3. *Phase 3: Optimization — Cross-process*

In phase 3, the highest complexity is represented by the cross-process optimization of already stable process chains (after phase 2). Under certain conditions, it would also be possible to proceed with a statistical design of experiments (similar to phase 3 optimization of individual processes). However, overarching process experts are again involved and suitable means are used to visualize complex interrelationships. The experimental effort must be reduced to a minimum and be in proportion to the expected improvement.

Depending on the complexity of the interrelationships and the processes themselves, such projects are not economical due to the personnel capacity and the associated costs. One approach here can be an Evolutionary Operation (EVOP), a strategy for optimizing ongoing production processes according to statistical methods. However, these approaches are limited in the number of parameters and are only conditionally suitable for the analysis of multidimensional data.

As an alternative approach, data mining approaches can be used, such as pattern feature recognition, association, regression or sensitivity analyses. The

Table 1. Comparison of the two methods developed for individual and cross-processes.

Phase	Individual Process	Cross-Process Method
Phase 1	• Individual process is in the test and ramp-up phase • Use statistical experimental design • Stable process as a result • Extraction of key features	• Total process is in the test and ramp-up phase • Use statistical experimental design • Stable process chain as a result • Identify interactions
Phase 2	• Increase stability within tolerances • Use SPC method in individual process	• Increase stability within tolerances • Use APC method in total process
Phase 3	• Testing of new parameters • Optimization as a result • Suitable for carrying out series of experiments	• Optimization of complete process chain (data mining) • Suitable for carrying out series of experiments

selection of the methods depends, among other things, on the application and the available data. Prerequisites for data mining approaches in general are that data are available in sufficient quantity and have been recorded over a correspondingly long period of time. For a statistically valid statement, completeness and sufficient quality (e.g. sampling rate for time-based sensor data) are mandatory.

The developed method offers approaches for the optimization and stabilization of individual processes and comprehensive processes. In the following chapter, parts of the developed method are applied in practical tests on plant components for the production of cylindrical Li-ion batteries. Table 1 provides an overall view of the proposed methods for process stabilization and optimization.

4. Process Understanding and System Analysis

In this section, the necessary process understanding of the ZDB Li-ion cell production line is built up in detail, relevant characteristics are identified and subsequently the planning for characteristic acquisition is carried out.

4.1. *System boundaries for method testing*

Within the considered cell production, system boundaries are necessary to clearly define the framework for the analysis and tests. The entire process chain is shown in Fig. 4. In further course, the developed method for individual processes is applied to the process steps of assembly (No. 5 & 6). For process step No. 11, the filling of the electrolyte, a theoretical consideration is also carried out.

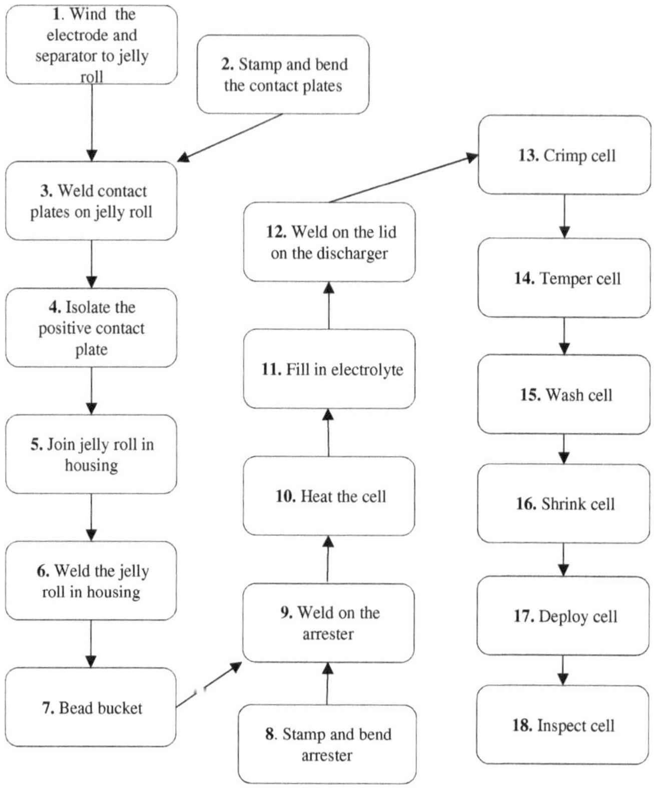

Fig. 4. Process flow for the production of a Li-ion cell at the test production line.

4.2. *Structuring of the exemplary Li-ion cell production*

For simplified representation, the entire chain of manufacturing under consideration is mapped into a hierarchical model (Fig. 5 according to ANSI/ISA88). The process model contains subordinate process sections (PA), which in turn contain process operations (PQ). Thereby, it is possible that arbitrarily many process operations with remaining many parameters exist.

The pyramid abstracts the chain at different levels of granularity. The structure provides a consistent naming of processes and variables, so that an unambiguous assignment is possible.

Table 2 shows the correct assignment of the process steps within the cell production considered. As an example, the assembly process step is mentioned here, which in turn comprises several process operations. A machine, controller or location does not necessarily have to be assigned to a process section.

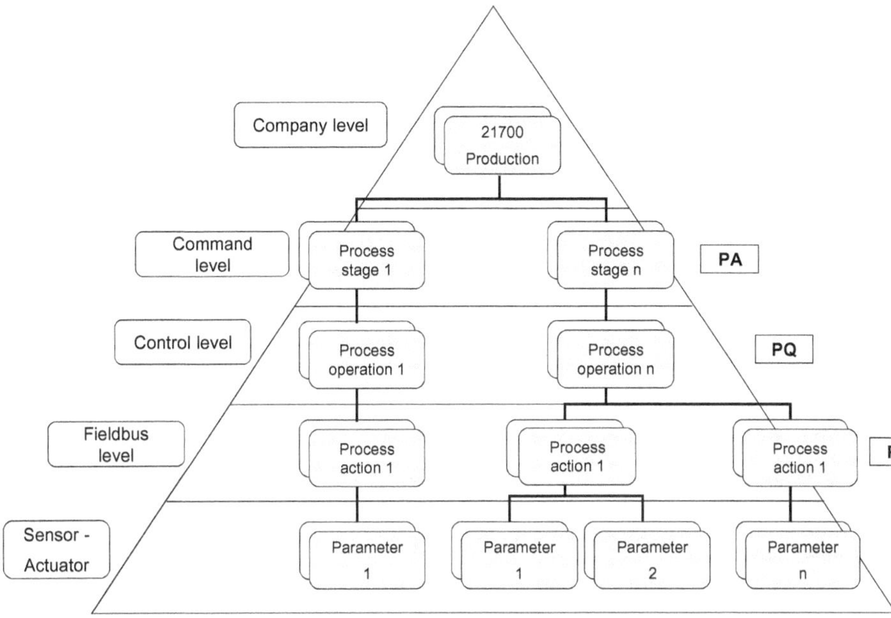

Fig. 5. Physical model of the test cell production line with communication levels.

Table 2. Overview of the process stages within Li-ion test cell production.

Process Step [PA]	Description
PA010	Mixing
PA020	Coating
PA030	Drying
PA040	Calendaring
PA050	Cutting
PA060	Vacuum drying
PA070	Winding
PA080	Assembly
PA090	Electrolyte filling
PA100	Cell Sealing
PA110	Formation
PA120	Aging
PA130	End of line control

4.3. *Analysis and feature identification for PA082 and PA082.1*

The assembly process section is made up of a total of six individual process operations. Input materials are coils including contact plates and the bucket. The output

Table 3. Process operations within the assembly process.

Process Operation [PQ]	Description
PA080	Assembly
PQ081	Contacting
PQ082	Mounting
PQ082.1	Weld on jelly roll
PQ083	Beading
PQ085	Bake out cell

product is therefore the assembled cell with the exception of the electrolyte. Filling takes place in the downstream process section.

To provide a better overview, Table 3 lists the individual process operations within the assembly.

Process operation PQ082 is split into two operations due to the following circumstances. The welding device is integrated in the robot cell for assembly.

In addition, the welding station is in the same safety circuit and part of the bus system of the main controller for assembly. The last feature is the interrelated process of assembly and resistance welding, which are coupled via a SCARA robot arm.

In order to accurately describe the actions performed by the process operations, they are classified according to VDI standards. The standard distinguishes between joining, handling, checking, adjusting and special operations. These categories cover all actions occurring in manufacturing and production, so that a clear assignment can be made.

These indices allow a clear classification in the data model and provide information about the component under consideration, the production status and the type of process step.

The introduced categories also allow a direct classification of the process steps and can thus, for example, enable targeted searches for measurement results (category I) or other categories. This approach can be especially helpful for manual or partially automated analyses with a lot of data and saves time compared to searching in unstructured data. This procedure is used as an example for the assembly process (see Table 4). The third column contains the categories of the process steps which allow an unambiguous assignment.

The focus for the exemplary implementation of single process stabilization is on the welding process [PA080]. The detailed description is listed in Table 5. For example, the abbreviation "H_CELL" stands for a cell handling step, while "I_CELL" stands for a cell inspection step.

Table 4. Overview of assembly process steps [PQ082].

Process Step [PS]	Description	Category
PA080PQ082.1PSxxx	Assembly	
PS010	Identification tray	I_IC
PS020	Measure jelly roll	I_JR
PS030	Grab jelly roll tray	H_JR
...	...	...
PS180	Drop cell in tray	H_CELL

Table 5. Process steps "welding on jelly roll in housing" [PQ082.1].

Process Step	Description	Category
PA080PQ082.1PSxxx	Weld on jelly roll	
PS010	Measure electrode 1	I_WEL
PS020	Measure unwelded cell	I_CELL
PS030	Translation axis AP	H_CELL
PS040	Weld on jelly roll	P_CELL
PS050	Translation axis GS	H_CELL
PS060	Measure welded cell	I_CELL
PS070	Resistance measuring	I_CELL
PS080	Measure electrode 2	I_WEL

5. Technical Implementation and Modeling of Individual Process

This section describes the necessary technical framework for carrying out the subsequent tests. In addition to the architecture for the IT and control systems, a data model for the assembly process section is described. Finally, the modeling of both use cases follows.

5.1. *Design of IT and control technology architecture*

Basically, the control system consists of the main components of a multicore controller (PLC), robot controller KUKA KR C2, a high-frequency inverter system PrimusHFI and an EtherCAT bus system.

On the field bus side, the converted bus signal data of the decentralized peripherals are communicated to the PLC via the master. The connection between the control and IT environment is implemented via OPC UA. The Beckhoff PLC acts as the server and the IT environment as the client.

The Manufacturing Operations Management System (MOMS) maps the IT environment with its micro services. With its services, this system facilitates processes and operations in production. In addition, the MOMS provides the central database and associated services such as test planning or parameter management.

5.2. *Data model for process section PA082*

The basic structure of the data model is derived from the systematic analysis of the processes from Section 4 and is aligned with the ISA 95 standard. The modeling is done directly in the software UaModeler,[26] a modeling tool for OPC-UA.

In the modeling, each process section is assigned its own folder structure (Object Folder Type). In this, in turn, the assigned operations and process steps are mapped in subfolder structures. The process variables are stored in the corresponding hierarchy level according to their characteristics.

Depending on the hierarchy level, process variables are already created in the process section object (for example, global variables of the machine control such as the machine status). The associated process parameters, such as the current flow time in milliseconds (PS040), are located within the individual process steps.

5.3. *Process modeling — Individual process PA082*

The use cases are modeled by using graphical methods (e.g. cause–effect diagram). On the basis of the findings, the individual process of "welding a jelly roll onto a cup" is now examined in more detail for the subsequent test. To determine the objective of the experiment, the target variables are defined at the beginning.

For this purpose, the following four criteria are selected: quality welding process, geometry, optics and strength. The points mentioned are based on ISO 17677-1:2019[27] for resistance spot projection and roller seam welding. The evaluation takes place according to a school grading system (grade 1–6).

The cause–effect diagram records all influencing variables and evaluates them according to their influence. Expert discussions and research in technical literature show that the factors welding current, welding time and electrode force are the main influencers of the process.

The welding current is defined by the charge per time, which is initiated by two directly opposing electrodes. The value is set and measured via the welding inverter. The welding time is defined by the period of time during which the welding current flows continuously. It is also set and measured via the welding inverter. The electrode force is the force acting on the workpiece via the electrodes. This is set by a proportional control valve and transmitted by means of a spring system.

On the welding head, there is an upper electrode holder which is rigidly connected to the pneumatic traversing axis and can thus perform a vertical movement. The electrode on the holder moves with a defined force onto the components to be welded. The welding inverter then sets the required current. The materials to be welded are melted and joined by the locally generated heat. A multi-stage factorial plan is selected as the experimental plan. The final setup of the multilevel factorial experimental design is performed with the support of Minitab software.[28]

6. Experimental Procedure and Evaluation

The test is carried out directly in the assembly and welding cell on site (see Fig. 6). Strict attention is paid to always maintaining the same sequence during the welding test. The following list shows the operations carried out:

(1) Removal of beaker from carrier system (randomized numbering).
(2) Cleaning of the housing with alcohol.
(3) Cleaning of contact plate material.
(4) Placing the objects to be welded in the resistance welding machine.
(5) Setting of parameters.
    (a) Welding current and time on the inverter system.
    (b) Force via pneumatic pressure manometer.
(6) Starting the recording.
(7) Initiation of the welding process and visual observation.
(8) Removal of the housing.

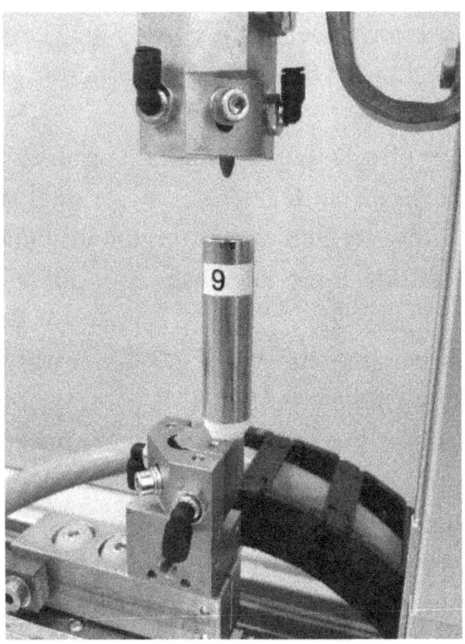

Fig. 6. Housing in resistance welding machine before welding process.

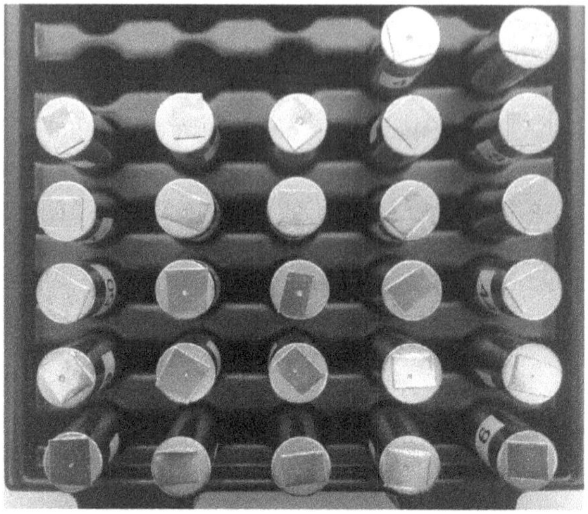

Fig. 7. Detailed view of the housings produced in the resistance welding process with clearly visible welding lens.

After conducting the tests, all collected data were transferred to the DoE software and the evaluation was performed based on the criteria.

During the test execution, strong weld spatter was observed at currents of 3.5 kA. Likewise, increased force had to be applied to detach the adhering cup from the electrode. Significant differences were found in the size of the weld lenses. Pronounced welding lenses indicate excessive heat input during the welding process and are rated poorly accordingly (see Fig. 7). Assessment is confirmed by the evaluation of the recorded data. The higher the welding current, the poorer the evaluation of the weld.

Figure 8 shows the effects on the target variable. It becomes clear that an increase in the welding current leads to a poorer quality of the welding seam. The welding duration, however, has only a minor influence in the present case. Furthermore, as the electrode force decreases, the quality of the welding seam improves.

From DoE, an ideal welding current of 2.5 kA, a welding time of 16 ms and an electrode force of 70 N are determined. This parameter combination shows the best results. The cells' inner view shows no discoloration or damage to the cup from the inside. From this, it can be concluded that neither high thermal input nor extremely high forces acted during the welding process.

Further test welds with identical welding parameters to those identified in the DoE are therefore considered to be very suitable and are used as default or basic parameters in the start-up phase of the assembly and welding cell.

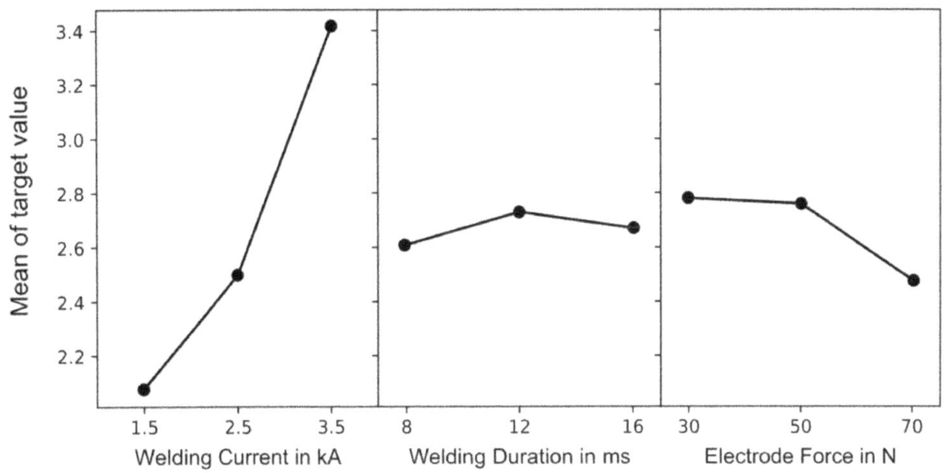

Fig. 8. Diagram showing the main influencing variables on the welding process (evaluation according to grading system).

7. Theoretical Contemplation of Cross-Process Method

This section includes the analysis of the process as well as the modeling of the process section. At the time of writing, the plant is still under construction. For this reason, only a theoretical consideration is made for this part of the plant.

7.1. *Analysis and feature identification for PA090*

The PA090 process section of electrolyte filling follows assembly. After completion of the plant, the method can be tested and validated in further work.

The filling system is housed together with the laser welding system in a gas-tight glove box. Despite being housed together, the filling of the cell with the electrolyte is to be regarded as a single process step. Electrolyte filling is composed of these process steps as follows. For clarity, the categories of each process step are determined and shown in Table 6.

7.2. *Process modeling — Cross-method PA090*

Due to the high complexity and the interdependencies, only a theoretical examination of sub-processes can be carried out within the scope of this work. The aim of this analysis is to identify and model the cross-process interactions of the process section assembly (PA080), in particular the process operation assembly (PQ082), and electrolyte filling. The identified characteristics will then be used to develop a concept for cross-process stabilization.

Table 6. Process steps "electrolyte filling" [PQ091].

Process Step	Description	Category
PA090PQ091PSxxx	Electrolyte filling	
PS010	Identification cell & carrier	I_WEL
PS020	Handling cell	I_CELL
PS030	Tare weight of cell	H_CELL
PS040	Translation axis vacuum chamber	P_CELL
PS050	Vacuumize cell	H_CELL
PS060	Electrolyte filling	I_CELL
PS070	Resistance measuring	I_CELL
PS080	Measure electrode 2	I_WEL
PS090	Handling cell	H_CELL

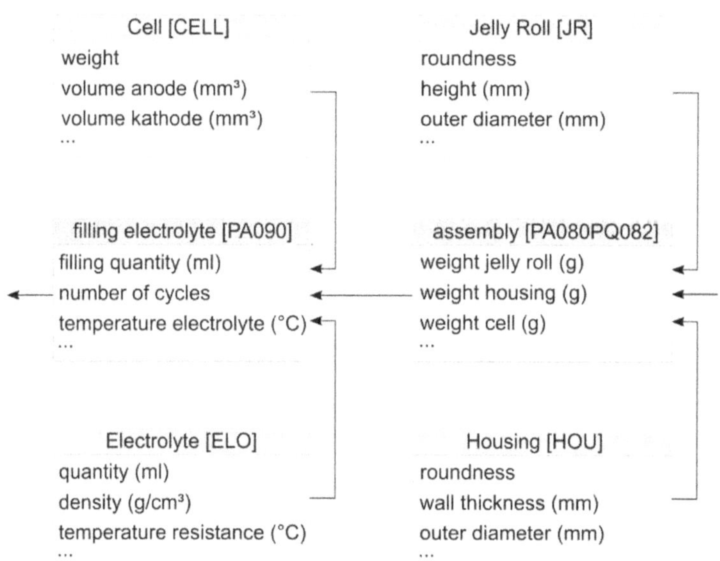

Fig. 9. Interactions between assembly [PQ082] and electrolyte filling [PA090].

In the first step, all relationships of the process operation [PQ082] to the process section [PA090] are visualized in an influence diagram and brought into dependence. Figure 9 shows the section of the interrelationship analysis between [PQ082] and [PA090]. Feeder parts to each process are shown in the top and bottom rows. The actual process can be found in the middle row. In each listing are the key process parameters of the Li-ion cell that have a direct impact on lifetime, performance and safety.

One approach to cross-process stabilization is to adjust the filling parameters of the electrolyte depending on each individual cell, which requires the following parameters, among others:

- *PA080PQ082*:
  - Weight of the housing, jelly roll and cell.
  - Dimensions of jelly roll and housing.
- *PA090PQ091*:
  - Fill parameters (quantity, cycles, fill time).
- *Cell parameters*:
  - Calculated cavity volume.
- *Set up experimental plan and carry out experiments*:
  - Use DoE Software Tools, UML, Databases.
- *Evaluation and adjustments*:
  - Use presentation and moderation techniques.
  - Penetration speed electrolyte.

Mathematical models such as the Washburn equation can be used to determine the theoretical penetration speed of the electrolyte in the porous materials from the cell materials, electrolyte properties and cavity volume. The porous materials can be divided into the separator and electrode materials (anode as well as cathode).[29] The microstructure of the electrodes is strongly dependent on the compression in the calendar during electrode fabrication and is therefore very well suited for cross-process stabilization via APC. Faster and more stable wetting of the porous materials improves electrolyte wetting, which has a positive effect on the performance and quality of the battery.

8. Summary and Future Research

A new pilot line for cell manufacturing, such as the one at the Fraunhofer Campus in Stuttgart, provides an excellent framework for testing new production processes and manufacturing environments.

Based on previous research work, a method for the stable operation and optimization of a test series line was developed. It was shown how complex the individual processes for the production of cylindrical Li-ion battery cells are. With the developed method, it is possible to shorten the ramp-up phases and at the same time to enable a high process stability as well as transparency along the entire process chain.

After the necessary basics for the understanding of the topic were worked out, methods were explained, which are used for process stabilization and optimization. The following chapter deals with the explicit procedure for solving the problem. In addition to the analysis of the actual processes, detailed process knowledge is built up.

Furthermore, in the chapter a building concept was developed, which runs over the different start-up phases of a production, from ramp-up to optimization. When considering individual processes, statistical methods are used to initially stabilize the process so that optimization can take place downstream.

Section 4 provides a systematic system analysis based on existing industry standards and norms. Section 5 then describes the technical implementation, from control technology and information technology to the final modeling of the processes.

Finally, the method is applied and validated using the resistance welding process as a practical example. First, the most important process parameters are identified and then varied within suitable limits in a multi-stage, factorial experimental plan. By later executing the plan with subsequent evaluation, the ideal process parameters can be determined.

The results of the practical application now provide the basis for the automated process and the implementation of the stabilization concept using SPC. However, further steps are still missing before the method can be implemented on the entire test production line. By forming expert groups at the individual process level, the theoretical foundations can be worked out. After stabilizing the process, this team has the task of transferring the knowledge to the next phase, for cross-process stabilization.

After data structures and IT connections have been implemented, the complex modeling of interactions must take place. Among other things, this poses organizational challenges. On the one hand, the large number of people involved requires project management, and on the other hand, a suitable infrastructure must be created so that tools and software are available on a sufficient scale.

Approaches such as the use of an intelligent workpiece carrier to store and directly transmit relevant process variables are also conceivable. Likewise, specific functions should be added to the micro services on the IT platform. This can further reduce the initial effort of modeling and acquiring process knowledge.

The work also shows that statistical experimental design increasingly reaches its limits as the number of influencing parameters increases. In the future, artificial intelligence methods could be applied to analyze large amounts of data and uncover new potentials.

References

1. Statista. Elektroautobatterien–größte Hersteller nach Absatz H1 2018. Statista; 2021 [cited 2021 Mar 5].
2. Statista. Batterieproduktion E-Autos Marktanteil Hersteller weltweit 2022. Statista; 2021 [cited 2021 Mar 5]. Available from: https://de.statista.com/statistik/daten/studie/1113139/umfrage/prognose-des-marktanteils-der-hersteller-von-elektroautobatterien-nach-laendern/
3. Statista. Zulassungszahlen von Elektroautos 2021. Statista; 2021 [cited 2021 Mar 5]. Available from: https://de.statista.com/statistik/daten/studie/244000/umfrage/neuzulassungen-von-elektroautos-in-deutschland/
4. Bundesfinanzministerium. Eckpunkte des Konjunkturprogramms; 2021 [cited 2021 Mar 5]. Available from: https://www.bundesfinanzministerium.de/Content/DE/Standardartikel/Themen/Schlaglichter/Konjunkturpaket/2020-06-03-eckpunktepapier.html
5. Ludwigsburg24. Stuttgarter Mahle-Konzern plant Abbau von fast 8000 Arbeitsplätzen. AG Group24 Ludwigsburger Nachrichten UG; 2020 Sep 17 [cited 2021 Mar 5]. Available from: https://www.ludwigsburg24.com/stuttgarter-mahle-konzern-plant-abbau-von-fast-8000-arbeitsplaetzen/
6. Frankfurter Allgemeine Zeitung. Wasser Marsch! Frankfurter Allgemeine Zeitung; 2020 Feb 13 [cited 2021 Mar 5]. Available from: https://www.faz.net/aktuell/stil/quarterly/die-rolle-des-wasserstoff-im-verkehr-der-zukunft-16622466.html
7. Krause E. LB1. thesen-zur-industriellen-batteriezellfertigung-in-deutschland-und-europa; [cited 2021 Mar 5]. Available from: https://www.bmwi.de/Redaktion/DE/Downloads/S-T/thesen-zur-industriellen-batteriezellfertigung-in-deutschland-und-europa.pdf?__blob=publicationFile&v=5
8. S-TEC. S-TEC–Zentrum für Digitalisierte Batteriezellenproduktion; 2020 [cited 2021 Mar 5]. Available from: https://s-tec.de/zentren/zentrum-fuer-digitalisierte-batteriezellenproduktion/
9. Kurzweil P, Dietlmeier OK. Elektrochemische Speicher: Superkondensatoren, Batterien, Elektrolyse-Wasserstoff, rechtliche Grundlagen. 1. Aufl. 2015. Wiesbaden: Springer Vieweg; 2015.
10. Kampker A, Vallée D, Schnettler A. Elektromobilität: Grundlagen einer Zukunftstechnologie. Berlin, Heidelberg: Springer; 2013. Available from: https://publications.rwth-aachen.de/record/210661?ln=de
11. Korthauer R, editor. Handbuch Lithium-Ionen-Batterien. Berlin, Heidelberg, s.l.: Springer Berlin Heidelberg; 2013. Available from: https://link.springer.com/book/10.1007%2F978-3-642-30653-2
12. Heimes HH, Kampker A, Wessel S, Kehrer M, Michaelis S, Rahimzei E. Montageprozess eines Batteriemoduls und -packs. 3. Auflage, [revidierte Ausgabe]. Aachen, Frankfurt am Main: PEM der RWTH Aachen; VDMA; 2018.
13. Heimes HH, Kampker A, Lienemann C, Locke M, Offermanns C, Michaelis S, *et al.* Produktionsprozess einer Lithium-Ionen-Batteriezelle /Autoren Dr.-Ing. Dipl.-Wirt.-Ing. Heiner Hans Heimes, Prof. Dr.-Ing. Achim Kampker, Christoph Lienemann, Marc Locke, Christian Offermanns [und weitere]. 3. Auflage, [revidierte Ausgabe]. Aachen, Frankfurt am Main: PEM der RWTH Aachen; VDMA; 2018.
14. Becker T. Prozesse in Produktion und Supply Chain optimieren. 3., neu bearbeitete und erweiterte Auflage. Berlin: Springer Vieweg; 2018.
15. Geiger W, Kotte W. Handbuch Qualität: Grundlagen und Elemente des Qualitätsmanagements: Systeme, Perspektiven. 5., vollständig überarbeitete und erweiterte Auflage. Wiesbaden: Friedr. Vieweg & Sohn Verlag | GWV Fachverlage GmbH Wiesbaden; 2008. (Praxis und Studium).

16. Dietrich E, Schulze A. Statistische Verfahren zur Maschinen- und Prozessqualifikation. 7., aktualisierte Auflage. München: Carl Hanser Verlag; 2014.
17. REFA.de. FMEA; 2021 [cited 2021 Mar 5]. Available from: https://refa.de/service/refa-lexikon/fmea
18. Partners O. How to implement Advanced Process Control (APC) in a manufacturing facility to improve process capability and product quality. https://objectpartners.com/2013/06/14/how-to-implement-advanced-process-control-apc-in-a-manufacturing-facility-to-improve-process-capability-and-product-quality/
19. Otte R, Wippermann B, Otte V. Von Data Mining bis Big Data: Handbuch für die industrielle Praxis. München: Hanser Carl; 2018.
20. Vogel-Heuser B, Bauernhansl T, Hompel M ten, editors. Handbuch Industrie 4.0: Bd. 1: Produktion. 2., erweiterte und bearbeitete Auflage. Berlin: Springer Vieweg; 2017. (Springer Reference Technik).
21. Wiki–Cyber-Physical Production System (CPPS); 2021 [cited 2021 Mar 5]. Available from: https://i40.iosb.fraunhofer.de/Cyber-Physical%20Production%20System%20(CPPS)
22. Enterprise-control system integration: models and terminology. Research Triangle Park, N.C.: International Society of Automation (ISA); 2000.
23. Kleppmann W. Versuchsplanung: Produkte und Prozesse optimieren. 10., überarbeitete Auflage. München: Hanser; 2020. (Praxisreihe Qualität).
24. Kübler K, Verl A, Riedel O, Oberle M. Simulation-assisted run-to-run control for battery manufacturing in a cloud environment: 21–23 Nov. 2017. Piscataway, NJ: IEEE; 2017. Available from: http://ieeexplore.ieee.org/servlet/opac?punumber=8168204
25. Philipp S. Ansätze zur Prozessstabilisierung in der Fertigung von Batterie-Rundzellen. Stuttgart; University of Stuttgart (university publication — not publically available) 2016.
26. UaModeler "Turns Design into Code"–Unified Automation; 2021 [cited 2021 Mar 6]. Available from: https://www.unified-automation.com/products/development-tools/uamodeler.html?gclid=CjwKCAiAiML-BRAAEiwAuWVggjeAjWCBzzCGrio-Cxlvow8Fg0pGG4qHNEiKivg1vsmImrrCUsdbsBoCCIsQAvD_BwE
27. DIN EN ISO 17677-1:2019-06, Widerstandsschweißen_- Begriffe_- Teil_1: Punkt-, Buckel- und Rollennahtschweißen (ISO_17677-1:2019); Dreisprachige Fassung EN_ISO_17677-1:2019. Berlin: Beuth Verlag GmbH.
28. Softwarepaket für Statistik und Datenanalyse. Minitab; 2021 [cited 2021 Mar 6]. Available from: https://www.minitab.com/de-de/products/minitab/
29. Knoche T, Surek F, Reinhart G. A process model for the electrolyte filling of lithium-ion batteries. Procedia CIRP. 2016;41:405–10.

https://doi.org/10.1142/9789811245626_0015

Chapter 14

Analyzing and Maximizing Line Productivity — Bottleneck Identification and Predictive Maintenance

Thomas Hilzbrich*, Brandon K. Sai† and Lennard Sielaff‡
Fraunhofer IPA, Nobelstraße 12, 70569 Stuttgart, Germany
**thomas.Hilzbrich@ipa.fraunhofer.de*
†brandon.sai@ipa.fraunhofer.de
‡lennard.sielaff@ipa.fraunhofer.de

The production of battery cells is made in interlinked production lines, which have a high number of parameters that influence the resulting product. Consequently, material properties and process parameters have an influence on the effectiveness and output, so the optimization becomes a challenging task. The optimization of productivity itself can be achieved by three main optimization goals: increasing equipment availability, boosting performance and reducing quality losses.

Since quality may be subjectively defined, this chapter focuses on solutions to increase availability and performance. In fact, the approach of root cause analysis and predictive maintenance is presented. In this case, root cause analysis focuses on performance optimization and predictive maintenance on the availability of the production system.

In the following, both topics will be elaborated in the form of a literature review and a conceptual presentation. Afterward, the application of both topics will be discussed or illustrated with examples. Finally, predictive maintenance as well as root cause analysis will be discussed and an outlook on both applications in battery cell production will be given.

1. Introduction

Battery cells are produced in interlinked manufacturing processes, which have a high number of parameters influencing the resulting product. Material properties and process parameter have an influence on the effectiveness and output of the manufacturing processes. Due to the interlinkage of the manufacturing processes,[1] the individual processes have an impact on each other, causing waiting or blocked conditions as well as quality issues.

According to the VDMA the overall equipment effectiveness in battery cell manufacturing is around 90%.[1] The effectiveness directly affects the economic efficiency of production, so that the improvement of the effectiveness is a continuous target in manufacturing.

To improve the effectiveness, the causes of effectiveness losses in manufacturing (downtimes, performance losses or quality issues) need to be analyzed. Due to the high number of process parameters, the cause of an effectiveness loss can be manifold and not trivial to identify. In an interlinked manufacturing process, an effectiveness loss, which occurred in a specific process step, can have its root cause in a prior manufacturing process, meaning that a direct linkage between cause and effect is not easily possible.

To demonstrate the bottleneck identification and the concept to identify components where predictive maintenance would benefit the most, an example battery cell assembly, including the production steps from separating until filling, is shown in Fig. 1. The example is based on the study of the PEM of the RWTH Aachen[2] and the battery production equipment road map published by the VDMA.[1] The quantities shown in Table 1 for the individual production steps for a small series and a large series production are based on the VDMA road map. The additional assumptions required (number of sheets in one cell and electrolyte volume within one cell) were selected in such a way that continuous production with simultaneous operation of all production steps is possible. The cutting speeds and number of machines required to separate the plates needed to be adjusted

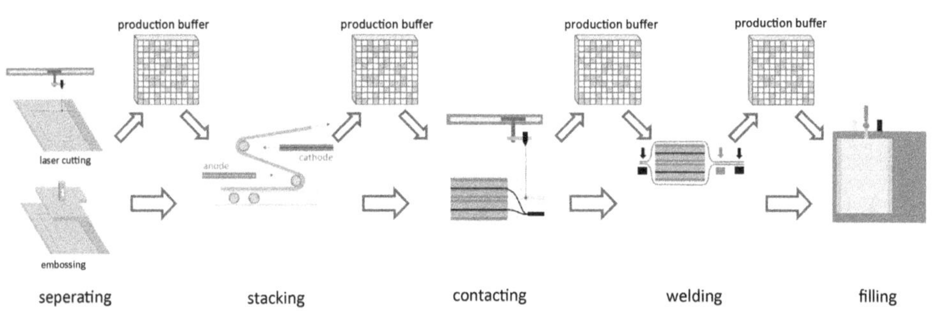

Fig. 1. Assembly steps for battery cell used to discuss the concepts as pictured and described in the study of the PEM of the RWTH Aachen[2] and the battery production equipment road map published by the VDMA.[1]

Table 1. Number of machines and maximum number of cells for each production step at a small-scale and large-scale battery assembly. Targets and numbers based on the production technology road map from the VDMA.[1]

	Production Step									
Small Scale Target: 0.2 cells/sec Large Scale Target: 5 cells/sec	Separation		Stacking		Contacting		Welding		Filling	
	Small	Large	Small	Large	Small	Large	Small	Large	Small	Large
Number of production machines for each process	4	10	4	5	1	4	1	1	3	4
Number of interacting equipment for each machine	20	60	4	4	6	12	1	1	7	15
Maximum cells per second for each machine	0.06	0.54	0.07	1	0.75	6	0.2	2	0.32	2.08
Maximum cells per second for the process	0.24	5.36	0.27	5	0.75	24	0.2	6	0.32	8.33

from the technology road map to fit with the assumption of 40 plates for each battery cell.

In this example, only battery cell assembly is considered, since different technologies and different degrees of interlinking are possible within the individual steps for this part of the process. Additionally, both the technology and the possibility of intermediate storage are the main factors for bottleneck detection as well as the selection of the maintenance strategy. Therefore, the assembly steps of separating, stacking, contacting, welding and filling are depicted, summarized and used in the demonstration of the concepts. The material handling within and between the stations as well as supporting equipment will not be considered.

Within one process step, the assembly is achieved on different machines of the same type, which enable parallel production lines. In most cases, a line has several components that can process the battery cells of the line either in parallel or in series (interacting equipment). Based on the configuration of the units and the auxiliary technologies, a continuous process as well as intermediate buffering can take place between the process steps, which influences the impact of bottlenecks and the optimal maintenance strategy.

The separation of the plates can be done either with a laser cutter or by embossing from the coil. While the cutting edge of the tool is subject to heavy wear during embossing, and corresponding maintenance strategies must be selected, laser cutting causes hardly any wear on the production equipment. The stacking of anodes and cathodes with the separator and the subsequent contacting form the inner workings of the battery cell. Welding closes the cell so that it can be filled with electrolyte in the last step considered. Based on our defined production scenario, Table 1 shows the number of machines, number of interacting components for each machine (handling of the cells) and the calculated maximum numbers of battery cells (or material for one battery cell) for each machine and all machines combined for each production step within a small-scale assembly and a large-scale assembly. The differentiation between small and large production capacity and the production numbers and machines are based upon the production technology road map from the VDMA.[1] Differences between small and large scale are the result of different number of machines as well as the different technology used. The calculated maximums can be used to identify bottlenecks and can be used to choose maintenance strategies accordingly.

2. Root Cause Analysis of Downtimes and Performance Losses

Multiple approaches to measure and analyze effectiveness in manufacturing are used in industry. For instance, Production data acquisition (PDA) can be used to document errors of production equipment and statistical methods to aggregate them. Often, PDA relies on manual input from production personnel, which reports particular errors with naming the assumed cause and the used measure to solve the error. In addition, methods like statistical process control (SPC) are used to monitor certain production parameters.

These established methods often only take into account a specific part of the whole interlinked manufacturing process and do not explicitly include prior and following processes in the analysis. Furthermore, most methods only use a small number on input variables, so that these focus on analyzing the effect of an error and not the real cause.

In research, data mining and machine learning methods are used to identify error patterns that help explain the cause of the error. By using a database including a high number of features describing the manufacturing process, the methods focus on identifying multivariate relationships.[3] Several open research areas exist here, to bring these methods into broad industry application. For instance, detailed sensor data are often lacking, which are needed to identify root causes of errors on a detailed level. In addition, for classification of a cause, a labeling is often needed, which requires manual input from a human, which can be resource-intense.

2.1. Literature review

In the literature, only a few contributions specifically discuss the analysis of effectiveness losses. The main focus of these is the analysis of quality issues in manufacturing of battery cells. Thiede *et al.* adapt a multivariate regression to predict quality parameters of lithium-ion cells based on sampled set parameters and measurements.[4] Schnell *et al.* compare classification methods for prediction of battery cell capacity prior to the formation process.[5] The electrode manufacturing process and the electrolytes' filling process were identified as having the main impact on the battery cell quality. With both approaches, battery cells with quality issues can be identified during manufacturing and can be discharged early. In addition, the learned models (regression, classification) can be used to understand the underlying patterns resulting in quality issues.

Many approaches of effectiveness loss analysis in the literature can be used for multiple manufacturing processes. An important criterion for the transferability of a method is adaptability for continuous or discrete manufacturing processes; here, Niggemann and Frey introduced a model for anomaly detection in manufacturing processes.[6] Based on sensor data, the manufacturing process is modeled as a finite-state machine, which maps the sequential manufacturing process steps. It also includes the optimal (minimum) time of transitions. The model can be used to detect time deviations in the manufacturing process, which can represent anomalies.

Laxmann *et al.* developed a method to identify time-based relations between errors in manufacturing processes.[7] If a specific event frequently occurs after an individual error, a time-based relation between these events can be detected.

2.2. *Concept*

In order to improve the economic efficiency, continuous optimization of battery cell manufacturing needs to be established. Expert- and data-based approaches, which analyze sole manufacturing processes or parameters, are often not sufficient to identify the root cause of an effectiveness loss as multiple parameters and processes can have an impact.

The subsequent concept focuses on identifying root causes of downtimes in battery cell production and can been seen as a hybrid of both, a data-based approach enriched by expert knowledge. The concept is subdivided into multiple building blocks, which are displayed in Fig. 2.

To identify the root causes of errors, detailed data of the manufacturing processes are needed. For instance, if an error is caused by a malfunction of a sensor, data of this sensor are needed, which include the situation of the malfunction. Instead of using aggregated manufacturing data, which are available, for instance,

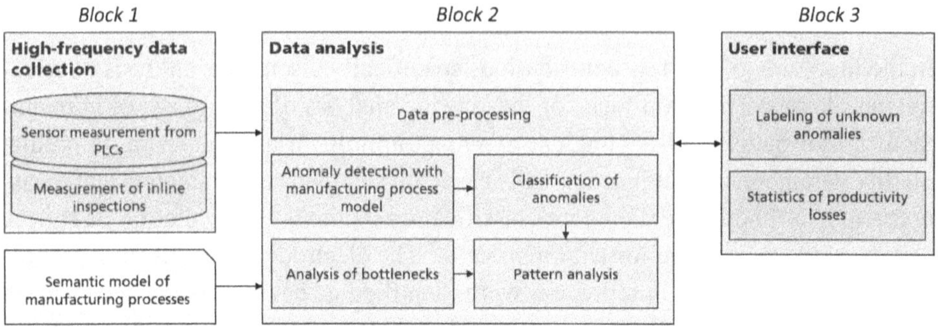

Fig. 2. Building blocks of root cause analysis of downtimes.

in Manufacturing Execution Systems (MESs), sensor and actuator data from programmable logic controllers (PLCs) of the manufacturing processes can be used (Block 1). For the collection of PLC data, the sampling rate has to be defined based on the alteration rate of the measured signals, aiming to measure all signal value changes in order to completely describe the manufacturing process. For the high-frequency collection of machine data, various software solutions are available on the market, which are able to read data from PLCs via proprietary and open protocols, for instance, the iba-Analyzer.[1]

For the analysis of the behavior of a manufacturing process, a descriptive model of it is needed, which includes the structure of the process (sequence of process steps shown in Fig. 3) as well as a semantic description of the signals used for analysis, for instance, process step 9 can be identified as welding.

The collected data can be analyzed in different ways and aggregations (Block 2). Detection of anomalies in a production process is done using a model of the process behavior. The process of production equipment can be subdivided in multiple production steps, which occur in a sequential order. A production step can be described as a state by a specific signal vector. The transition from one state in another can be annotated with a transition time, describing the optimal duration of the transition. Thus, the optimal, time-wise behavior of a production process can be described. The model can be described automatically using a batch of signal data, which includes a situation of optimal production. Also, the variation of set parameters has to be taken into account. With the model, deviations of the optimal process behavior can be detected and isolated to a specific process step.

[1]Available at https://www.iba-ag.com/en/software (last accessed 28-04-2021).

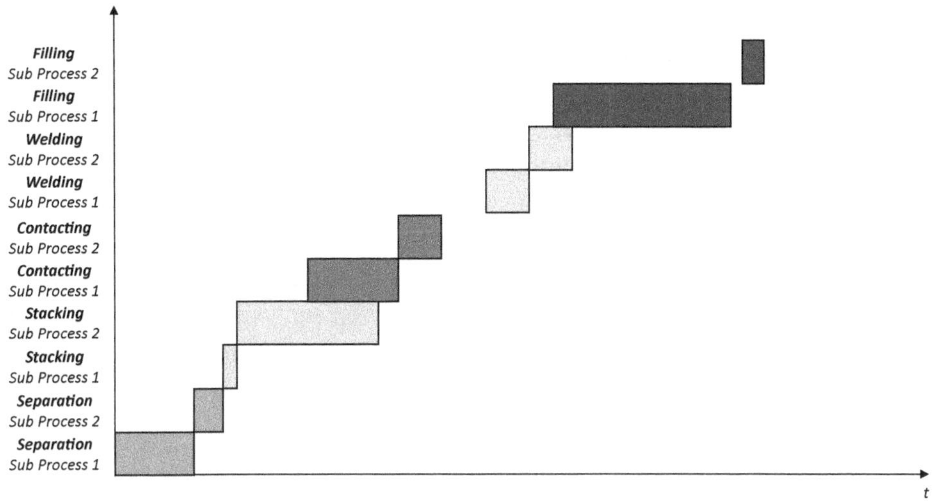

Fig. 3. Abstract process model in form of a gantt chart.

To determine the cause of an error, a classification based on the signal data and the process model can be performed. The signal data can be specified by the time of occurrence of the process step deviation and the signals involved in the specific process step. For the classification, a labeling of the error classes, which describe them semantically, is needed. In general, this has to be performed by a human (Block 3).

While the anomaly detection and classification identify deviations on a detailed level, the bottleneck analysis focuses on analyzing the interdependencies between subsequent manufacturing processes. By analyzing waiting and blocked conditions of manufacturing processes, the impact of anomalies on prior and following processes can be determined.

Based on the results of bottleneck analysis as well as anomaly detection and classification, overall patterns can be identified, for instance, in regard to time-based relations between anomalies and general parameters (like day or air temperature).

2.3. *Efficient modeling with transfer learning approaches*

To generate the models needed for analysis, the manual effort can be quite high as these often have to be adapted to individual manufacturing equipment and technologies. For an efficient use of data-based analysis methods, transfer learning approaches can be used to transfer models and their parameters to similar production equipment. The basic idea is to reuse a once generated model for similar production equipment, assuming that the behavior is similar and similar errors will occur there. Therefore, an evaluation of the similarity of production equipment based on semantic description is necessary, determining the level of transferability. Based on that, an existing model or parts of it can be used as an initial starting point for a new production process.

Using machine data, a production process can be described on a detailed level and by that, anomalies can be identified and analyzed to determine the root cause. The identified root causes can be used to define measures for optimization. In particular, in order to be able to use machine data-based analysis methods on a larger scale, transfer learning can be a promising approach to transfer models to similar production systems and thus reduce the effort required for model creation.

2.4. *Application of bottleneck analysis*

For the application, a rough grasp of the object, at least a classification, is necessary, because there are differences regarding the bottleneck, there is a static bottleneck and a dynamic bottleneck.

The simplest production systems in terms of bottleneck detection and capacity management are static systems. Within these systems, there is no random fluctuation. There are no failures, conversions or other changes; each time, the system behaves exactly as before. Due to variations in cycle times, failures with a random Mean Time Between Failure (MTBF) and a random Mean Time To Repair (MTTR), the assumption of a static system does not match the observation represented by the data.

Hence, in reality, there are no static production systems and the assumption is made to a dynamic system. The behavior of any real system is subject to countless random events, which make the system different every single time. These random events create different probability functions, describing how the randomness behaves. Therefore, bottlenecks shift in dynamic systems, i.e. the bottleneck may change over time.

If we take the information from Table 2, it is possible that every other production step except the static bottleneck "*Welding*" has a failure and the bottleneck shifts. It does not matter if the failing process step is located before or after the static bottleneck. The location would only affect the shifting speed.

However, before defining the dynamic bottleneck and working with it, a common misunderstanding regarding bottlenecks needs to be clarified. As stated before, the intuitive assumption that a certain process is identified as a constant bottleneck cannot be taken as correct. This is due to the fact that dynamic bottlenecks are the system constraint only for a certain period of time. In effect, there

Table 2. Number of machines and maximum number of cells for each production step at a small-scale and large-scale battery assembly. Targets and numbers based on the production technology road map from the VDMA[1] with the bottleneck process highlighted.

Small Scale Target: 0.2 cells/sec Large Scale Target: 5 cells/sec	Production Step									
	Separation		Stacking		Contacting		Welding		Filling	
Production Scale	Small	Large	Small	Large	Small	Large	Small	Large	Small	Large
Number of production machines for each process	4	10	4	5	1	4	1	3	1	4
Number of interacting equipment for each machine	20	60	4	4	6	12	1	1	7	15
Maximum cells per second for each machine	0.60	0.54	0.07	1	0.75	6	0.2	2	0.32	2.08
Maximum cells per second for the process	0.24	5.36	0.27	5	0.75	21	0.2	6	0.32	8.33

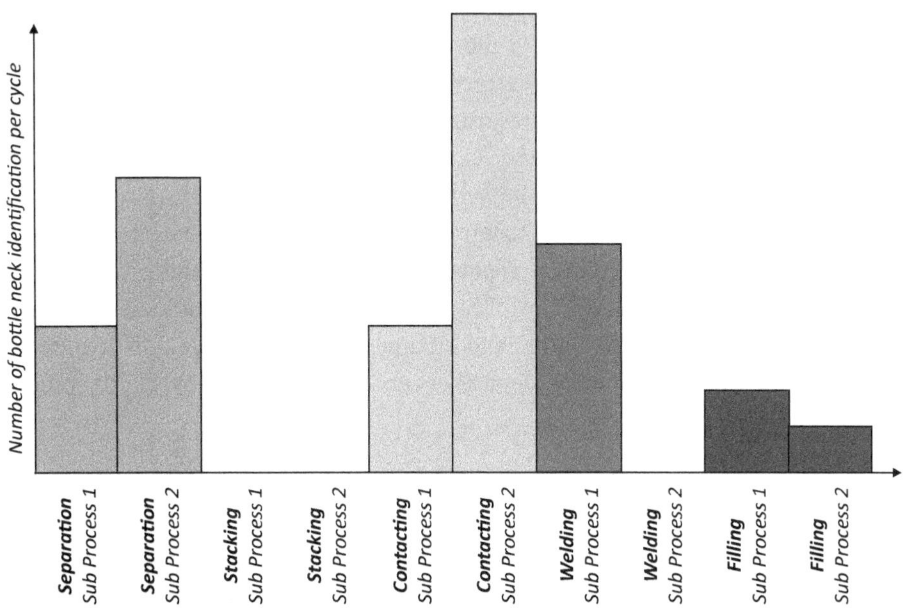

Fig. 4. Histogram of temporary bottleneck identifications of processes.

are only temporary bottlenecks. However, in order to be able to make a general statement about the bottleneck of production systems, a comprehensive examination is necessary as shown in Fig. 4.

Therefore, long-term bottlenecks can be determined based on the duration of the processes that represented the current bottleneck. The process that was the most frequent current bottleneck is also the biggest long-term bottleneck, although it may not necessarily be the only long-term bottleneck.

3. Predictive Maintenance

Mastering the plant technologies required is one of the central tasks of battery cell manufacturing line maintenance. The aim of maintenance is to ensure plant availability while minimizing the costs involved.[8] Particularly challenging for maintenance are, in addition to the large plant scope, the innovative production steps and the new technologies used for this purpose, for whose maintenance there is still little experience available.[5]

Given this uncertainty regarding the technologies used, the decision for a maintenance strategy is difficult. Predetermining strategies such as fixed maintenance intervals and also monitoring and, if necessary, making forecasts of plant wear and tear generally lead to higher plant availability, but they also incur higher costs than reactive maintenance strategies.[9] This means that the strategy decision includes, among other things, weighing up the required availability, technical possibilities, probability of failure and the direct and indirect costs. Bengtsson and Lundström also argued in their case study that new developments in maintenance, such as predictive maintenance, are just another tool for maintenance and should not be used as a general solution.[10]

In order to select maintenance strategies for battery cell manufacturing lines and to be able to take advantage of the benefits of predictive maintenance, both a technical and an economic analysis and evaluation of the machines and plants must be carried out before they are introduced. Due to the innovative technologies, the strong interdependencies among each other and little previous knowledge, this requires the creation of a maintenance concept that enables the employees to proceed systematically.

3.1. *Literature review*

The publications on maintenance in battery cell manufacturing in most cases indicate a need for further research. Pfleging, for example, examined the use of nanosecond lasers in 2018 for cutting electrodes for lithium-ion battery production and concluded that both the system and its maintenance are still too new for series production and that a proven concept is needed first.[11] Even in publications that do not deal with individual technologies or process steps, but with battery cell manufacturing as a whole, there is little specific information on the subject of maintenance. In their case study, Pei *et al.* examine production planning in a battery cell manufacturing facility and consider only aggregated values for maintenance costs.[12] In the publication of Thiede *et al.*, in which the prediction of product quality in a battery cell manufacturing facility is examined in a data-driven manner, neither the maintenance of the equipment nor the selected strategies are considered,[5] although many defects in a finished lithium-ion battery can be attributed to the equipment and its condition at the time of production.[13]

The fact that maintenance plays an important role in battery cell manufacturing lines, despite little consideration in the literature, can be indirectly seen in the costs per hour of maintenance staff in a comparison of US lithium-ion battery cell manufacturing with Chinese production. The cost per hour worked for a maintenance employee of the battery cell manufacturing facilities is the same in both countries, while the cost per hour worked for a maintenance employee responsible for the building in China is only one-fifth as high as in the USA. The authors explain the equality of costs for the maintenance staff responsible for the production facilities with the high complexity of the tasks and the specialists required.[14] From this, it can be seen that maintenance is very important for battery cell manufacturing (compared to building maintenance, for example).

Outside of battery cell manufacturing, the topic of maintenance is driven by technological innovations and the integration into workflows. New monitoring methods and possibilities of predicting wear are regularly published.[15,16] In their outlook, Roy *et al.* also state that the future of maintenance is not only the future of technical developments but also the intelligent use of these.[17] The combination of new approaches such as predictive maintenance and existing strategies leads to better results overall. This is confirmed by the overviews of various predictive maintenance operations by Sakib and Wuest[18] and the case study by Bengtsson and Lundström.[10]

3.2. *Concept*

As pointed out in the problem description, the biggest challenge for maintenance in battery cell manufacturing is the novelty of the technologies and equipment and the resulting lack of experience, which is necessary for an appropriate maintenance strategy selection. At this point, the developed concept for the technical and economic selection of suitable equipment and the implementation of predictive maintenance in battery cell manufacturing starts.

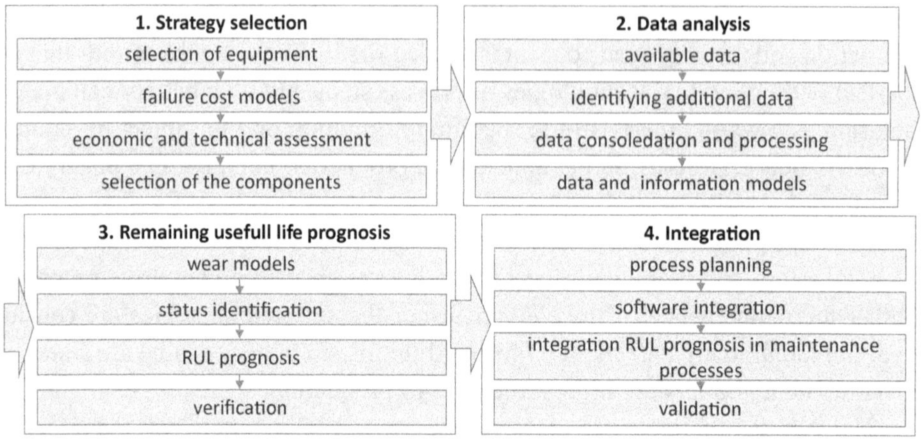

Fig. 5. Concept for implementing predictive maintenance into a battery cell manufacturing.

The concept shown in Fig. 5, which is divided into four phases, first describes how the systems, and in even more detail the individual components to be monitored, are selected for the economically sensible introduction of Predictive Maintenance. The second step describes how existing data are analyzed to determine whether further information is needed for the predictive models. Based on this, the concept describes how to choose the necessary data and how to create information models. In the third phase, the wear models are created and, based on this, the forecasts for the possible remaining useful life (RUL) will be calculated. Once the forecasts are known, the final step of the concept describes how the new information can be integrated into the maintenance processes of battery cell manufacturing.

For the selection of the components to be monitored, first of all, the systems must be identified for which the effort for the implementation of predictive maintenance is also economically worthwhile. For this purpose, all battery cell manufacturing facilities are first recorded using a checklist in order to identify those facilities whose failure will have critical consequences. Failure cost models are created for the plants identified in this way, which describe how much the unforeseen failure of these plants will cost depending on the duration of the failure. Example failure costs models are shown in Fig. 7. With the help of this information, malfunction modes and effects analyses (MMEAs) are carried out, in which the plants are technically analyzed on the basis of their components and the failure probability is offset against the previously determined failure costs. This enables the risk to be estimated and thus compared with the additional costs that would be caused by the introduction of predictive maintenance. In combination with the

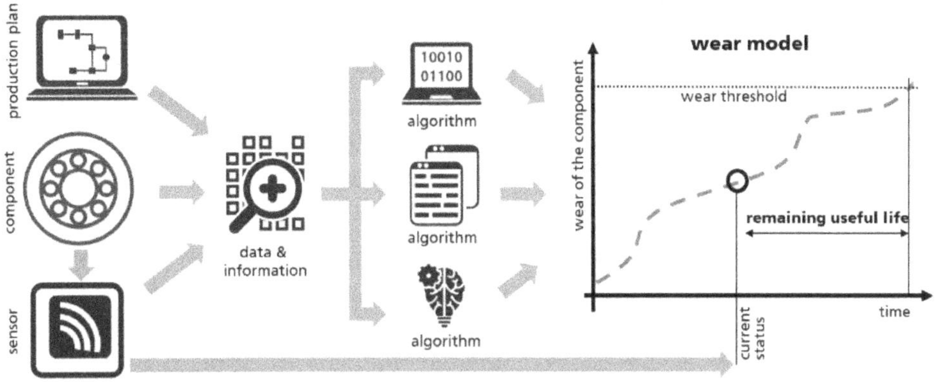

Fig. 6. Relationship of component, data/information algorithms and RUL.

technical analysis of the components during the MMEAs, it is thus possible to identify those components of the plant for which the introduction of monitoring and prediction of the remaining useful life is economically viable.

Once the components have been identified, the existing data on these components are analyzed. Different methods, like using the Ishikawa diagram or employing a gap analysis, are used to identify missing information for the determination of the condition and for the possibility of predicting wear. The information and data identified in this way, as well as the processing regulations developed (data acquisition, pre-processing, storage, processing and documentation), are recorded either by interfacing with existing machines or by deploying additional sensors and pre-processing equipment and then transforming the information into data and information models.

Once the information and data are available and the future production plan is known, the wear models are developed using various algorithms. Combined with the current information on the individual components, these wear models can be used to make forecasts of the remaining useful life as visualized in Fig. 6. These forecasts can be verified using historical data and during operation.

The final step in the concept for selecting suitable components and applying predictive maintenance involves integrating the forecasts into the existing processes of battery cell manufacturing line maintenance. The existing processes will be extended by the possibility of automated condition monitoring and failure prediction to ensure the availability of the battery cell manufacturing facilities despite the new technologies. A fully digitalized battery cell manufacturing would also allow for smart maintenance scheduling with respect to the predicted wear and other relevant parameters.

3.3. *Exemplary identification of machines and components for Predictive Maintenance*

If the concept for identifying the components for Predictive Maintenance is applied to the example described above, the critical components must first be determined. One conceivable process step is separation by embossing, as the cutting edges are subject to considerable wear. However, a closer look at the process figures shows that, for example, four embossing machines work in parallel in small-scale production and each of them can emboss material for 0.06 cells per second. The overall target of the line is 0.2 cells per second. If one machine fails and the other three machines are running at full capacity at the same time, this would result in a minus of 0.02 cells per second.

If there is a buffer of 30 cells (or material for 30 cells) between each process step, it would take 25 minutes for the buffer to be used up for the next process step. Since stacking and contacting are not the bottlenecks in the process chain, these machines can run empty without causing an interruption in output. Only when the shortfall in quantity arrives at the welding stage can the missing quantity not be made up again without additional measures. From this point on, 0.02 cells per second less than planned are produced. In the event of a machine failure in the embossing process, the maintenance employee has a maximum of 75 minutes to repair the machine and reintegrate it into the process. The machines further down the production line are able to process the backlog more quickly and refill the buffers. The relationships between downtime of a machine and lost battery cells are shown for small-scale production in Fig. 7 with initial buffers of 0, 30 and 60 cells.

The production step filling is in line after the bottleneck and is able to fill 0.32 cells per second. If there was a defect at the filling machine, the number of missed battery cells would be as shown in Fig. 7, but after the defect was repaired and the machine is fully running it could make up all battery cells that are buffered after the welding. Therefore, the production loss would be linear with 12 cells per minute, but would be reduced after the machine is running based on the buffer size.

From these production loss functions, it is clear that the contacting, welding and filling machines are the most critical from a maintenance point of view. These process steps only rely on one machine where the separation and the stacking steps employ multiple machines. If there is a defect in one machine, the other can mostly, in case of the separation, or fully, in case of the stacking, make up for the loss.

After identifying the critical machines in the process, these machine need to be examined for possible malfunctions and their root cause. If wear and tear

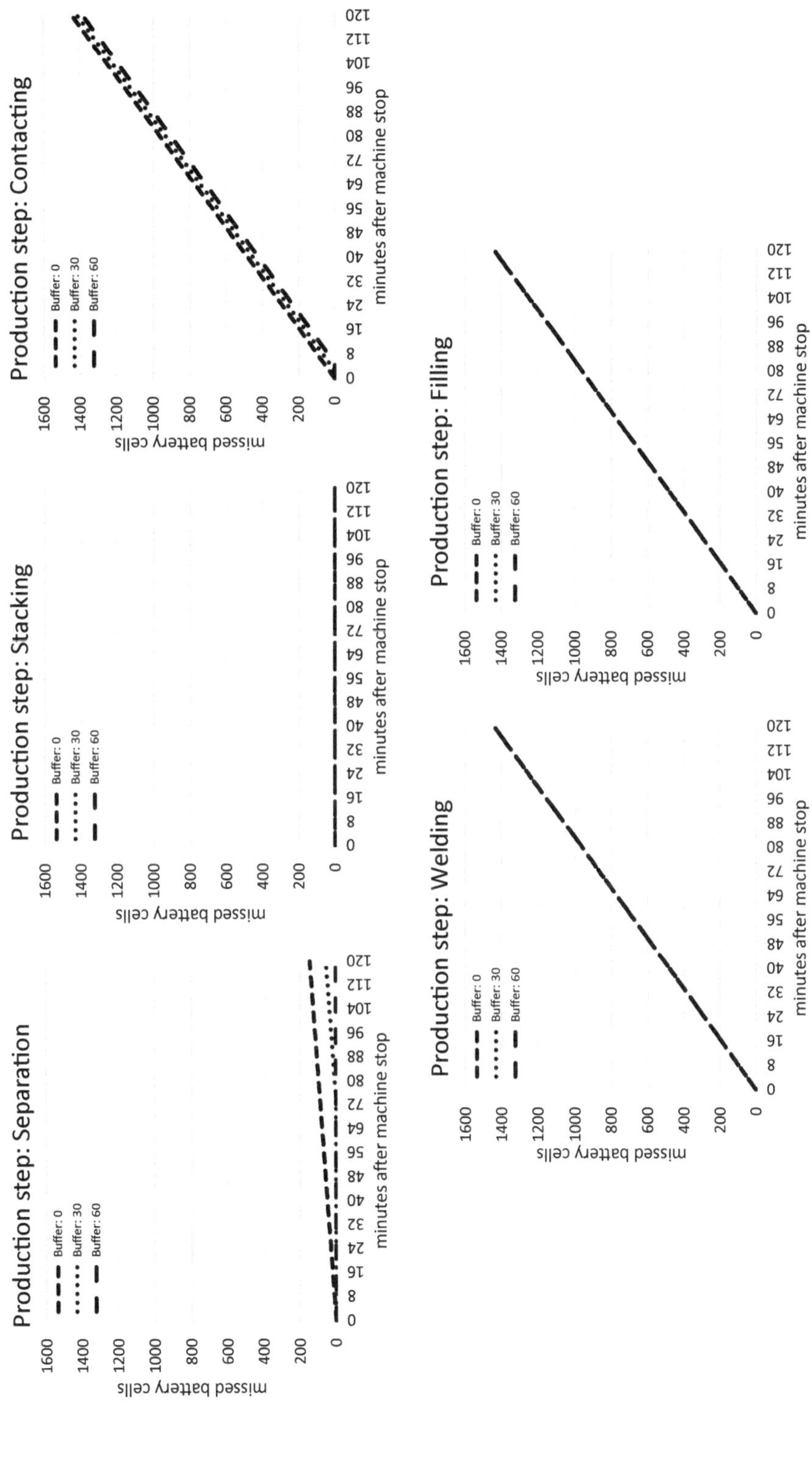

Fig. 7. Battery cells lost for a machine failure in the indicated process step as a function of the buffer size between process steps.

presents itself as a root cause, the wear mechanisms will be determined and a technical and economic analysis will be performed to identify if predictive maintenance will be used on the component.

3.4. *Exemplary integration of predictive maintenance into the maintenance processes*

Integrating predictive maintenance into the processes of a maintenance department enables direct and indirect costs to be saved and production times to be enhanced through better planning opportunities. If a failure is predicted, it is possible to react before damage with possible consequential impact occurs. Thus, the scope of the maintenance action can be reduced, which frees both production time and maintenance resources. In addition, with prior information about the needed action, the procedure of the measure can be significantly improved. For example, the required maintenance can either be planned into other downtimes, or further planned tasks such as regular maintenance of other components can be performed at the same time. This can reduce the overall downtime of the machine. In addition, the buffer levels can be specifically increased before a planned task is carried out to not interrupt the production line. Furthermore, it is possible to coordinate the procurement of spare parts in such a way that unnecessary storage costs are avoided and all parts are still available in time before the required use. According to the same principle, the availability of tools and qualified personnel for the respective operation can be ensured.

To be able to use these advantages, it is necessary to analyze and record the individual processes and dependencies and to integrate them into the maintenance planning software used. The interfaces, especially for production planning, must work in both directions (planning of maintenance and planning of production must be able to be done simultaneously for both systems) or at least allow the maintenance system to provide input so that possible necessary operations can be taken into account in production planning.

4. Summary and Discussion

Concerning root cause analysis, fully digitalized battery cell manufacturing is a specific form of interlinked production lines and can therefore benefit from optimization methods for interlinked production lines. Here, it is important to ensure that the identification of bottlenecks is performed with each production cycle, after which a general statement can be derived. By using predictive maintenance, wear-related faults in the selected systems can be eliminated, thus significantly reducing the costs caused by unforeseen disruptions in battery cell manufacturing. Within the framework of the presented concept, the resources required for this purpose are compared with the expected cost savings and individual faults are evaluated for their possible economic impact. In the context of a fully digitalized battery cell manufacturing, smart maintenance scheduling makes it possible to reduce production downtime due to systematically preparing for maintenance events by clustering and rearranging to minimize the impact on production.

References

1. Maiser E, Michaelis S, Müller D. Roadmap Batterie-Produktionsmittel 2030: Update 2016. Frankfurt am Main: VDMA Verlag; 2016.
2. Heimes HH, Kampker A, Lienemann C, Locke M, Offermanns C, Michaelis S, *et al.* Produktionsprozess einer Lithium-Ionen-Batteriezelle /Autoren Dr.-Ing. Dipl.-Wirt.-Ing. Heiner Hans Heimes, Prof. Dr.-Ing. Achim Kampker, Christoph Lienemann, Marc Locke, Christian Offermanns [und weitere]. 3. Auflage, [revidierte Ausgabe]. Aachen, Frankfurt am Main: PEM der RWTH Aachen; VDMA; 2018.
3. Wuest T, Weimer D, Irgens C, Thoben K-D. Machine learning in manufacturing: advantages, challenges, and applications. Prod Manuf Res. 2016;4(1):23–45.
4. Thiede S, Turetskyy A, Kwade A, Kara S, Herrmann C. Data mining in battery production chains towards multi-criterial quality prediction. CIRP Ann. 2019;68(1):463–6.
5. Schnell J, Nentwich C, Endres F, Kollenda A, Distel F, Knoche T, *et al.* Data mining in lithium-ion battery cell production. J Power Sources. 2019;413:360–6.
6. Niggemann, O., Frey, C. Data-driven anomaly detection in cyber-physical production systems. at-Automatisierungstechnik 2015;63(10):821–32.
7. Laxman S, Shadid B, Sastry PS, Unnikrishnan KP. Temporal data mining for root-cause analysis of machine faults in automotive assembly lines; 2009 Apr 29.
8. Strunz M. Instandhaltung: Grundlagen–Strategien–Werkstätten. Berlin: Springer Vieweg; 2012.
9. Swanson L. Linking maintenance strategies to performance. Int J Prod Econ. 2001;70(3): 237–44.
10. Bengtsson M, Lundström G. On the importance of combining "the new" with "the old" — One important prerequisite for maintenance in Industry 4.0. Procedia Manuf. 2018;25:118–25.
11. Pfleging W. A review of laser electrode processing for development and manufacturing of lithium-ion batteries. Nanophotonics. 2018;7(3):549–73.
12. Pei W, Ma X, Deng W, Chen X, Sun H, Li D. Industrial multi-energy and production management scheme in cyber-physical environments: a case study in a battery manufacturing plant. IET cyber-phys syst. 2019;4(1):13–21.
13. Wu Y, Saxena S, Xing Y, Wang Y, Li C, Yung W, *et al.* Analysis of manufacturing-induced defects and structural deformations in lithium-ion batteries using computed tomography. Energies. 2018;11(4):925.
14. Brodd RJ, Helou C. Cost comparison of producing high-performance Li-ion batteries in the U.S. and in China. J Power Sources. 2013;231:293–300.
15. Lughofer E, Sayed-Mouchaweh M. Prologue: predictive maintenance in dynamic systems. In: Lughofer E, Sayed-Mouchaweh M, editors. Predictive maintenance in dynamic systems. Cham: Springer International Publishing; 2019. pp. 1–23.

16. Lorenzoni A, Kempf M, Mannuß O. Degradation model constructed with the aid of dynamic Bayesian networks. Cogent Eng. 2017;4(1):1395786.
17. Roy R, Stark R, Tracht K, Takata S, Mori M. Continuous maintenance and the future — Foundations and technological challenges. CIRP Ann. 2016;65(2):667–88.
18. Sakib N, Wuest T. Challenges and opportunities of condition-based Predictive Maintenance: a review. Procedia CIRP. 2018;78:267–72.

Part V

Life-Cycle Optimized Usage of Smart Battery Products

https://doi.org/10.1142/9789811245626_0016

Chapter 15

A Battery Circular Economy — Definition, Significance and End-of-Life Strategies

Jan Koller*, Oliver Oechsle† and Caroline Hellmich‡
Fraunhofer IPA Universitaetsstr. 9, 95447 Bayreuth, Germany
**jan.koller@ipa.fraunhofer.de*
†oliver.oechsle@ipa.fraunhofer.de
‡caroline.hellmich@ipa.fraunhofer.de

This chapter first addresses the importance of electric vehicles in the transportation sector to achieve climate targets. In this context, the circular economy as a contrast to the current linear economy is introduced. This is followed by an overview of possible End-of-Life strategies for technical products in the waste hierarchy context. Due to the emerging volume growth, the limited lifetime of the battery and the valuable and limited raw materials a battery contains, it is essential to close the loop for battery systems in terms of product and material flows. Forecasts predict that the battery circular economy will generate an estimated $10 billion in revenue opportunities in 2030. The End-of-Life strategies suitable for lithium-ion battery systems, such as reuse, repurpose, remanufacturing and recycling, are then compared and examined in more detail concerning their technical implementation and industrial feasibility. Finally, current issues to establish a circular economy for battery systems are highlighted.

1. Introduction

To achieve the proposed and more ambitious goal of reducing greenhouse gas emissions in the European Union by 55% in 2030 compared to 1990 levels, and to be climate neutral by 2050, emissions in the transport sector must be reduced significantly.[1] Road transport, which currently accounts for one-fifth of the EU's greenhouse gas emissions, plays an essential role in this context.[2] Light-duty vehicles (cars and vans) are responsible for around 15% of total CO_2 emissions in the EU.[3] Therefore, on the one hand, the European Commission attempts to reduce greenhouse gas emissions by regulating the permitted emissions per kilometer for new vehicles.[4] On the other hand, the replacement of vehicles with internal combustion engines by electric vehicles in private transport is intended to further reduce greenhouse gas emissions.[5] According to studies, electric vehicles show a better climate balance after 50–80,000 driven kilometers when compared to conventional vehicles with combustion engines.[6,7] Supported by direct subsidies and policy approaches, the number of electric vehicles will continue to increase, and in the long term will account for the vast majority of the passenger car fleet.[8] It is estimated that electric vehicles will amount to more than 30% of new sales in 2030, with the number of electric vehicles worldwide likely exceeding 300 million.[9]

However, there are also issues associated with the use of lithium-ion battery systems, e.g. its limited lifespan.[10] According to studies, the average calendar life of lithium-ion battery systems is around eight to ten years.[11–13] Assuming a technologically progressive development of battery systems, the average useful life might increase up to ca. 12 years in 2030, according to experts.[14] Therefore, around four million battery systems of electric vehicles are expected to be at their End of Life (EoL) in 2030.[9] Those battery systems contain many valuable, limited materials, such as lithium, cobalt, nickel and manganese.[14] Consequently, the battery system is the most expensive electric vehicle component with a cost share of more than 30%.[10,15] Due to the increasing importance of electric vehicles and the expected volume growth of battery systems, closing the loop in terms of product and material flows is of central importance.[10,13] Forecasts predict that the battery circular economy will generate an estimated $10 billion in revenue opportunities in 2030.[9]

Against this background, the basic concept of a circular economy is examined in more detail in the next section. The initial focus is on technical products and applicable EoL strategies. See, among others, Ellen MacArthur Foundation,[16] for an overview of EoL strategies for biological materials.

2. A Battery Circular Economy — Definition and Significance

The economic and industrial system currently in place is facing challenges to move beyond the linear consumption system. The Ellen MacArthur Foundation describes this linearity as a "take-make-dispose" pattern, which has been in place since the beginning of industrialization.[16] Well-known evidence proving these circumstances is delivered every year by the Earth Overshoot Day. This day marks the date when humanity's demand for ecological resources in a given year exceeds what the Earth can regenerate in that year.[17] In the linear consumption system, virgin materials are harvested and deployed in manufacturing to create an output, e.g. a product sold to a customer who disposes it when its useful life ends.[16] A linear economy is the result of inexpensive and plentiful resource supplies. Hence, in linear processes, mindfulness for environmental impacts is often absent, and take-back strategies are undeveloped.

In this context, the circular economy concept is introduced, aiming to decouple economic growth from poor resource use. Businesses in a circular economy aim to create value by managing resources already in the markets.[18] Attempts to define the concept of a circular economy have been made by many authors and institutions.[16,19–21] Geissdoerfer *et al.* reviewed the various definitions in a systematic literature review and suggested the following generic definition of a circular economy:

> *A regenerative system in which resource input and waste, emission, and energy leakage are minimised by slowing, closing, and narrowing material and energy loops. This can be achieved through long-lasting design, maintenance, repair, reuse, remanufacturing, refurbishing, and recycling.*[22]

Therefore, a circular economy aims to eliminate both the depletion of natural raw materials as well as the final disposal of waste and recyclable materials, e.g. by landfilling.[23] Instead, resources that have already been used should be kept in the cycle for as long as possible. This can be achieved by closing material loops and cross-industry linking, in which waste can be reused as input material for further processes, if possible, without the reduction of product or material quality.[16] This central aspect of the circular economy can be summarized as cascade use or cascading systems, which retain the inherent value of products and improve resource efficiency.[24]

Cascading systems use raw materials or products made from them as long and as often as possible. Only after all technological, economic and ecological possibilities have been exhausted should an energetic recovery be considered. Therefore, so-called use cascades are utilized, from higher to lower levels of value creation.

The cascade use can be illustrated by the Butterfly Diagram of the Ellen MacArthur Foundation, one of the most prominent representations of material flows within a circular economy alongside the *Comet Circle* from Ricoh.[16,25] The *Butterfly Diagram* represents a value circle and the continuous flow of technical and biological materials within.

In the biological cycle, food and biologically based materials — like wood and cotton — are designed to lead back into the cycle by processes such as anaerobic digestion and composting. The biological cycle regenerates living systems, e.g. soil, to provide renewable resources for the economy.

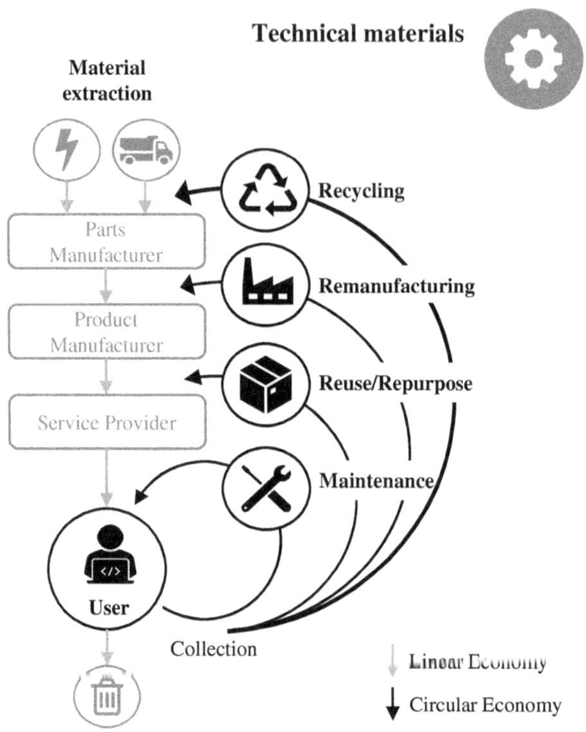

Fig. 1. Section of "Butterfly diagram" for technical materials, according to Ref. [16].

The cycle for technical materials is shown in Fig. 1, with the linear consumption system on the left side and the different options to close the loop to shift to a circular economy on the right side. The displayed End-of-Life strategies, also referred to as value retention processes, are used to extend or ensure the completeness of product life and create new value streams by saving the embedded input.[18,24] The precious input includes the embodied value of product function, material and geometric design, as well as the invested energy and labor input.[16,24] In summary, circularity can reduce the current virgin materials' price volatility, ameliorate resource performance along the entire material chain and simplify handling supply shortages.[18] However, the magnitude of a gradual transition from a linear consumption system to a circular system should not be underestimated.[26]

In the European Union, cascade use is legally embedded by the waste hierarchy in directive 2008/98/EC, which obliges member states to follow a set of prioritizations concerning waste management. As shown in Fig. 1, the priority is the prevention of waste. For generated waste, the options in order of preference are the preparation for reuse, followed by recycling, other forms of recovery, such as energy recovery, and disposal.[27] This directive is amended by the directive (EU)

2018/851/EC, which states that reuse and recycling targets should increase to improve resource efficiency and implementation of circular economy.[28]

There are specific guidelines that apply for industrial batteries or accumulators used in any type of electric vehicle. For example, the directive 2006/66/EC prohibits the landfill of batteries and accumulators, defines minimum collection rates of 45% for batteries and accumulators as of 2016 and specifies minimum recycling efficiencies. Lithium-ion batteries belong to the category "other waste batteries and accumulators". Therefore, the minimum recycling efficiency is 50% by average weight.[30] In addition, the Commission Regulation (EU) No. 493/2012 defines how to calculate the recycling efficiency.[31]

As shown in the waste hierarchy, recycling is only one option toward a circular economy (see Fig. 2). To close the loop, several EoL strategies can be used. These EoL strategies can be categorized and structured to various criteria, for example, concerning the degree of disassembly of a product or the result in the form of product or material recycling.[32] Potting *et al.* differentiate between a total of ten different EoL strategies within the *9R Framework*, as shown in Fig. 3.[33]

Considering the *9R Framework*, the EoL strategies can further be grouped into three categories: "smarter product use and manufacture", "extend lifespan of product and its parts" and "useful application of materials".[33] In contrast, the work of Parker *et al.* includes a total of 15 EoL strategies, while other authors consider only a few, e.g. two to three EoL strategies.[34,35] For an overview of the number of used EoL strategies in a total of almost 70 publications, see Reike *et al.*[36]

There are various reasons why different numbers of EoL strategies are considered depending on the literature used. On the one hand, they include the partly non-uniform definitions of the terms used to describe the EoL strategies and the use of synonyms and, therefore, a partly different interpretation of the terms. On the other hand, the number of EoL strategies considered varies due to the degree

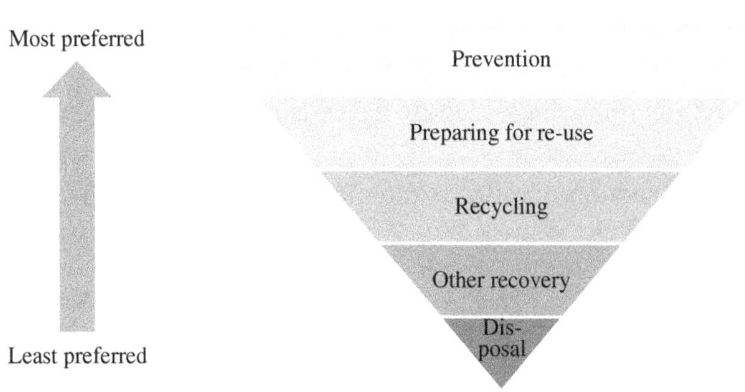

Fig. 2. Waste hierarchy, according to Ref. [27, 29].

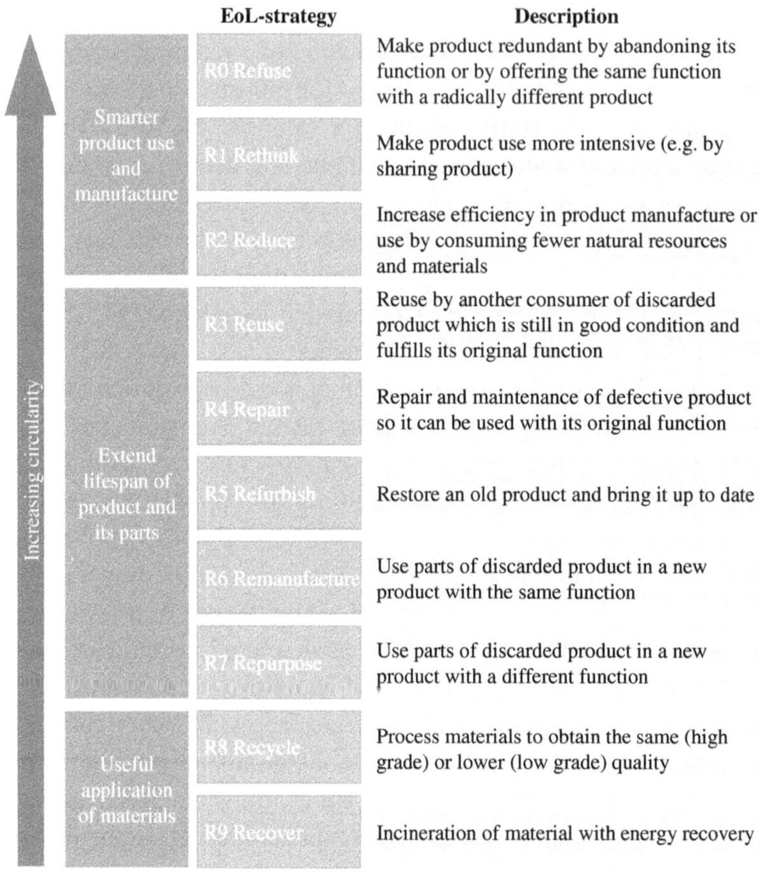

Fig. 3. 9R Framework, according to Refs. [33, 37].

of abstraction used as a basis for the analysis. Regardless of the number of considered EoL strategies, they are usually arranged in a hierarchy that prioritizes their use.[37] The number of EoL strategies that can be used for a product is not uniformly defined and varies widely depending on the study.[36] This also applies to battery systems of electric vehicles whose cascade use will be of particular importance in the future.

In the following section, the EoL strategies for battery systems are described in general terms.

3. End-of-Life Strategies for Battery Systems

Regardless of the selected EoL strategy, the battery system is removed from the electric vehicle in the first step. The battery system is usually attached to the

Table 1. Selected parameters of master, transaction and status data, according to.[39,40,41]

Master Data	Transaction Data	Status Data
• Cell chemistry • Cell design • Number of cells per module • Number of modules • Battery management information • Thermal management information • ...	• Number of cycles • Calendrical service life • Extended producer responsibility holder • Already performed EoL strategies • ...	• State of Health (SoH) • Internal resistance • State of Charge (SoC) • Depth of Discharge (DoD) • Operating temperature • ...

vehicle floor and can be removed using special tools.[38] Subsequently, all available data of the lithium-ion battery system are collected. A data structure that contains master data, transaction data and status data is useful to obtain the necessary information. Master data include general information about the structure of the battery system. Status data record the battery system's current health status along its entire life cycle. In contrast, transaction data provide information about the usage history of the battery system.[39,40] Table 1 provides an overview of the selected parameters that the master, transaction and status data consist of.[39,40,41]

After the information has been obtained, the battery system's decisive condition assessment is performed, taking into account the data and parameters described.[39,41] These and additional information, e.g. market data, serve as decision support for selecting a suitable EoL strategy for the lithium-ion battery system. This decision is of high importance since a battery system with an SoH of 80% still has a residual monetary value of about 70–75% of a new battery system.[10,42]

The applicability of EoL strategies depends strongly on the individual product. Therefore, the EoL strategies defined by Potting *et al.* in the *9R Framework*, see Fig. 3, are further limited concerning their applicability to battery systems.[32,33]

The system boundaries of the analysis are battery systems that have already been produced and installed in an electric vehicle. The manufacturing process of battery systems and their installation in an electric vehicle are not considered, although there is also a potential for increasing resource efficiency in this area. Therefore, the EoL strategies in the "smarter product use and manufacture" category of the *9R Framework* are out of scope. Other EoL strategies can only be applied to a limited extent and are not part of this analysis.

Repair of defective products is also out of scope since it is usually a more artisanal and not an industrial process.[43] Refurbish, sometimes referred to as

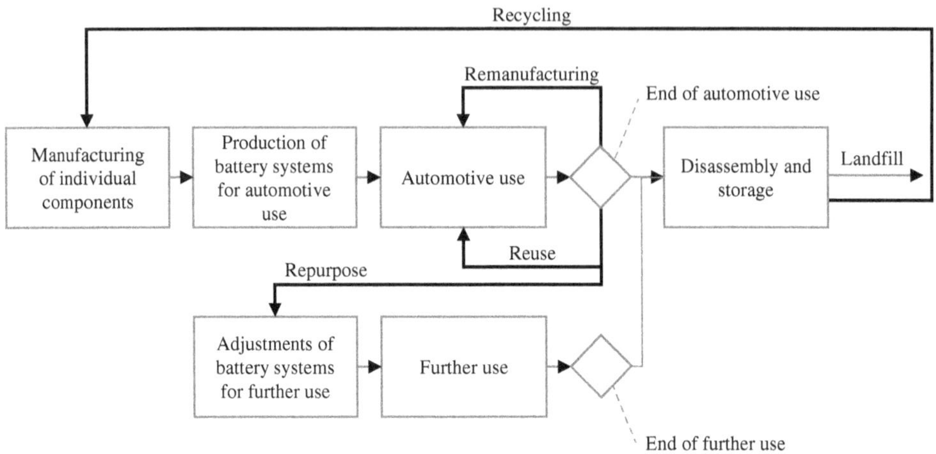

Fig. 4. EoL strategies for battery systems according to Ref. [32].

reconditioning, is not considered separately but included as a form of remanufacturing. The two EoL strategies only differ in detail, e.g. lower costs of refurbishing compared to remanufacturing, and are difficult to distinguish from one another.[44] The EoL strategy recovery is also not considered as this largely corresponds to the incineration of material for energy recovery, which can only indirectly generate input for downstream processes.

Figure 4 gives an overview of the considered EoL strategies suitable for battery systems in an industrial context. These strategies are the reuse of battery systems in an automotive application, the repurpose in other, e.g. stationary applications, the remanufacturing to a same-as-new condition, and the material recycling to, e.g. manufacture new battery cells.

3.1. *Reuse*

Reuse as an EoL strategy for Li-ion battery systems is applied when the removed battery system is used again for the identical application, i.e., when the condition of the removed battery system is as good as new.[32] In this case, the battery system can be used in an electric vehicle again without major reprocessing steps.[39] Although reuse as a term is already mentioned by various Original Equipment Manufacturers (OEM) as a deployed EoL strategy, it is usually not reuse but rather repurpose that is performed in this context.[45]

Reuse of battery systems for their original purpose has not yet been carried out on an industrial scale. In practice, this EoL strategy's potential applications are limited since battery systems are usually removed from an electric vehicle because they have fallen below a certain SoH threshold and are not suitable for reuse in another electric vehicle.

3.2. *Repurpose*

If battery systems are no longer suitable for reuse in electric vehicles, repurposing, i.e., further use in another application, is usually pursued.[32,46] In the first step, the battery systems are disassembled down to module level so that the battery management system (BMS) can be replaced in the next step.[40] The BMS usually differs depending on the application and must be designed specifically for the particular second-life application.[46,47] The new BMS is then used to reconfigure the cells.[47] All other components are cleaned, and wear parts, such as screws and seals, are replaced.[46] The battery system can then be rebuilt using the application-distinctive specification.[47]

It is also possible for a larger number of battery modules from several battery systems to be assembled to meet the second-life application requirements.[40] The battery system is then stored and transported to the second-life application site where it is being installed. These repurposed battery systems are often stationary applications for decentralized energy storage of renewable energy sources.[40,46] Many automotive OEMs already apply this EoL strategy on used battery systems.[48–51]

3.3. *Remanufacturing*

In general, remanufacturing consists of five process steps: disassembly, cleaning, inspection and sorting, reconditioning and replenishment as well as reassembly.[43,52] First, the battery system must be disassembled.[45,47] Whether the battery system is disassembled on cell or module level depends, e.g. on the number of modules or cells, the associated disassembly effort, as well as the cost–benefit ratio in sum. Subsequently, all battery cells are tested since the decrease of the battery system's total capacity is not equivalent to the degradation of each cell. This is the case because each cell has a different capacity and experiences a different depth of discharge with each cycle, which affects aging differently.[53] This also leads to temperature inhomogeneity within the battery system, which increases the likelihood of battery failure. Thus, the loss of capacity of one cell can lead to a significant loss of capacity of the whole battery system.[46] The battery cells that have not failed and still show the required charge characteristics can therefore be used again.[46,47] About 15% of the battery cells have to be replaced,

Fig. 5. Different aging of cells/modules in battery systems (left), new composition (middle) and examples for a future use (right), according to Ref. [52].

either with new battery cells or functional battery cells from other disassembled battery systems of the same type.[42,47,53] Since the behavior of the battery module is based on the properties of the weakest cell, battery cells with the same characteristic values should be linked together.[10] Based on the condition, the used battery systems can be reassembled for different applications at their EoL, see Figs. 4 and 5.[42,46,47,54] The replaced battery cells unsuitable for further use in battery systems are sent for recycling.[47]

Another form of battery system remanufacturing involves the periodic replacement of 5–30% of the battery cells, e.g. after 2,000 cycles, which can bring the battery system back to an SoH of close to 100%.[55]

Once the selection of battery cells is determined, the next step is to replace all wear parts, e.g. screws.[43] All other components are cleaned, if necessary, and evaluated in the next step.[10,41,42] A positive evaluation of the components allows them to be used again.[10] Otherwise, the corresponding component is replaced either by a new spare part or by a functioning part from another battery system.[56] The replaced components are in turn sent for recycling. All functional battery cells are balanced and then reassembled with other components into a battery system according to OEM specifications.[10,47] Afterward, final testing takes place according to OEM specifications.[10,56] The battery system is then stored, transported and installed in the vehicle.[32]

Although Daimler AG, as an Automotive OEM, is already performing remanufacturing of battery systems on the cell level, this is the exception.[57] Especially the non-automated disassembly of battery systems currently presents a challenge. Additionally, remanufacturing is often not applicable due to the design of the battery systems, e.g. glued cells.[57,58] Currently, research projects in Germany, for example, are working on solutions to overcome these challenges.[59–61]

All battery systems, which cannot be reused, repurposed or remanufactured, should be recycled either because of economic or technical reasons. The same is true if an EoL strategy has already been applied on a battery system.[32]

3.4. *Recycling*

About 15% of the removed battery systems have to be recycled because they are not suitable for other EoL strategies due to their condition.[42,47,53] The remaining 85% of the battery systems could theoretically be preserved by means of a higher-value EoL strategy.[40] Furthermore, battery systems that have already been reused, repurposed or remanufactured are also recycled.[32] Thus, it should be noted that, despite the existence of higher-value EoL strategies, recycling should be applied at some point.[62]

Recycling of lithium-ion battery systems can be achieved by different processes. Therefore, the main processes of hydrometallurgy, pyrometallurgy, mechanical treatments or a combination of these processes are used, see Fig. 6.[10,40,63]

These specific recycling processes are preceded and followed by general process steps. In the first step, the battery systems are disassembled entirely down to the module or cell level in preparation for the actual recycling process.[64] Afterward, the resulting materials are divided into different fractions, such as aluminum, copper and plastic.[65] The material fractions are collected and transported to appropriately specialized recycling companies.[64]

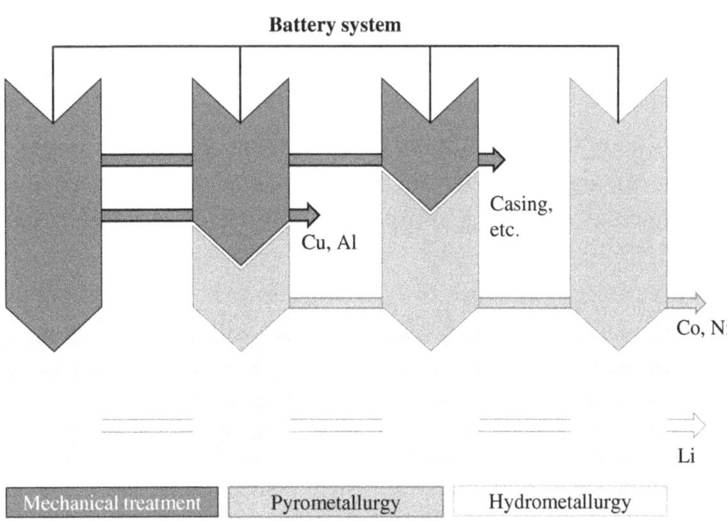

Fig. 6. Basic processes for battery recycling according to Ref. [63].

An essential material fraction, which accumulates during disassembling of the battery system, is the battery cell itself. The value of the battery cells is about 2/3 of the entire battery system, which is why their recycling is of particular importance.[10] Before the battery cells can be fed into the actual recycling process, they must be deactivated to ensure safety during further processing. The deactivation of the battery cell is mostly achieved by crushing or shredding in an inert gas.[62] Deactivation is followed by the actual recycling process. Recycling thus represents the final EoL strategy of the lithium-ion battery system. A detailed overview of material recovery processes, industrial processing routes and the currently installed recycling capacity is provided in Chapter 19 ("Battery Recycling — Material Recovery Processes").

4. Summary and Conclusion

The presented EoL strategies make it possible to close product and material loops in a cascading system, and thus enable a circular economy.

A battery circular economy offers both economic and ecological potential that needs to be tapped. By establishing a battery circular economy, it is on the one hand possible to reduce the material use of rare and valuable raw materials and therefore reduce costs. On the other hand, EoL strategies strengthen automotive supply chains' resilience and sustainability.[66]

Many of these EoL strategies are still in the growth phase and explored for an industrial scale. The EU policy objective of reducing greenhouse gas emissions from the transport sector supports increasing new registrations of electric and hybrid vehicles in Europe.[67] Due to the time lag between the production and sale of new battery systems and the return at the end of the product's life, significant volume increases are on the future horizon. To manage these volumes and close the loop, it is necessary to consider the EoL strategies' requirements already in the product development phase, e.g. through simplified disassembly capability for remanufacturing of battery systems.

There are various challenges to overcome to close the loop entirely. These include, in particular, the following aspects, which will be discussed in more detail in the next chapters:

- The interconnection of all stakeholders involved in the battery life cycle through digitalization and the corresponding data exchange along the value chain.
- Characterization of battery cells with suitable cell assessment processes to assess which battery cells are still useable for applications with lower requirements, e.g. stationary battery storage systems.

- Optimization of the disassembly processes as the basis of many EoL strategies against the background of the expected increase in battery systems at their EoL.
- Ensuring efficient material recovery processes on an industrial scale by combining different forms of recycling to handle high volumes.
- The development and establishment of suitable business models for reintroducing reassembled batteries into a second use phase.

References

1. Proposal for a Regulation of the European Parliament and of the Council establishing the framework for achieving climate neutrality and amending Regulation (EU) 2018/1999 (European Climate Law); 2020.
2. European Environment Agency. Greenhouse gas emissions from transport in Europe; 2020 [cited 2021 Feb 4]. Available from: https://www.eea.europa.eu/data-and-maps/indicators/transport-emissions-of-greenhouse-gases/transport-emissions-of-greenhouse-gases-12
3. European Commission. Road transport: reducing CO2 emissions from vehicles; 2020 [cited 2021 Mar 17]. Available from: https://ec.europa.eu/clima/policies/transport/vehicles_en
4. Regulation (EU) 2019/631 of the European Parliament and of the Council–of 17 April 2019 — Setting CO2 emission performance standards for new passenger cars and for new light commercial vehicles, and repealing Regulations (EC) No 443/2009 and (EU) No 510 / 2011; 2019. (vol. 62).
5. Circular Economy Initiative Deutschland. Resource-efficient battery life cycles: Driving electric mobility with the circular economy. München: Acatech–National Academy of Science and Engineering; 2020.
6. Helms H, Kämper C, Biemann K, Lambrecht U, Jöhrens J. Klimabilanz von Elektroautos: Einflussfaktoren und Verbesserungspotenzial. Berlin: Agora Verkehrswende; 2019.
7. Regett A, Mauch W, Wagner U. Klimabilanz von Elektrofahrzeugen–Ein Plädoyer für mehr Sachlichkeit. Forschungsstelle für Energiewirtschaft (FfE) e.V.; 2019.
8. International Energy Agency. Global EV outlook 2020: Entering the decade of electric drive? 2020.
9. Niese N, Pieper C, Arora A, Xie A. The case for a circular economy in electric vehicle batteries. Boston Consulting Group; 2020. Online available: https://www.bcg.com/de-de/publications/2020/case-for-circular-economy-in-electric-vehicle-batteries
10. Kampker A, Vallée D, Schnettler A, editors. Elektromobilität: Grundlagen einer Zukunftstechnologie. 2. Auflage. Berlin: Springer Vieweg; 2018. Available from: https://link.springer.com/book/10.1007%2F978-3-662-53137-2
11. Skeete J-P, Wells P, Dong X, Heidrich O, Harper G. Beyond the EVent horizon: Battery waste, recycling, and sustainability in the United Kingdom electric vehicle transition. Energy Res Social Sci. 2020;69(2):101581.
12. Kurdve M, Zackrisson M, Johansson MI, Ebin B, Harlin U. Considerations when modelling EV battery circularity systems. Batteries. 2019;5(2):40.
13. Drabik E, Rizos V. Prospects for electric vehicle batteries in a circular economy. Brussels, Belgium: CEPS; 2008. (No 2018/5).
14. Karabelli D, Kiemel S, Singh S, Koller J, Ehrenberger S, Miehe R, *et al.* Tackling xEV battery chemistry in view of raw material supply shortfalls. Front Energy Res. 2020;8:2873.
15. Kampker A, Vallée D, Schnettler A. Elektromobilität: Grundlagen einer Zukunftstechnologie. Berlin, Heidelberg: Springer Vieweg; 2013. Available from: http://dx.doi.org/10.1007/978-3-642-31986-0

16. Ellen MacArthur Foundation. Towards the circular economy vol 1: Economic and business rationale for an accelerated transition; 2013.
17. Global Footprint Network. Overshoot day; 2021 [cited 2021 Jan 30]. Available from: https://www.overshootday.org
18. Lacy P, Rutqvist J. Waste to wealth: The circular economy advantage. Basingstoke: Palgrave Macmillan; 2015.
19. Geng Y, Doberstein B. Developing the circular economy in China: Challenges and opportunities for achieving 'leapfrog development'. Int J Sustainable Dev World Ecol. 2008;15(3):231–9.
20. Webster K. The circular economy: a wealth of flows. 2nd Ed. Cowes, Isle of Wight, UK: Ellen MacArthur Foundation Publishing; 2017.
21. Yuan Z, Bi J, Moriguichi Y. The circular economy: A new development strategy in China. J Ind Ecol, 2006;10(1–2):4–8.
22. Geissdoerfer M, Savaget P, Bocken NMP, Hultink EJ. The circular economy — A new sustainability paradigm? J Cleaner Prod. 2017;143:757–68.
23. Haas W, Krausmann F, Wiedenhofer D, Heinz M. How circular is the global economy? An assessment of material flows, waste production, and recycling in the European Union and the world in 2005. J Ind Ecol. 2015;19(5):765–77.
24. Russell J, Nasr N. Value-retention processes within the circular economy. In: Nasr N, editor. Remanufacturing in the circular economy: operations, engineering and logistics. Beverly, MA, Hoboken, NJ: Scrivener Publishing, Wiley; 2019. p. 1–29.
25. Ricoh. The Ricoh Group contributes to the development of a sustainable society based on the Comet Circle concept; 2021 [cited 2021 Feb 1]. Available from: https://www.ricoh.com/sustainability/environment/management/concept.html
26. Larsson M. Circular business models: developing a sustainable future. Cham: Springer International Publishing; 2018.
27. Directive 2008/98/EC of the European Parliament and of the Council of 19 November 2008 on waste and repealing certain Directives (Text with EEA relevance): 2008/98/EC; 2008.
28. Directive (EU) 2018/851 of the European Parliament and of the Council of 30 May 2018 amending Directive 2008/98/EC on waste; 2018 (vol. 61).
29. Hernández Parrodi JC, Höllen D, Pomberger R. Potential and main technological challenges for material and energy recovery from fine fractions of landfill mining: a critical review. Detritus. 2018;(3):19–29. https://digital.detritusjournal.com/articles/potential-and-main-technological-challenges-for-material-and-energy-recovery-from-fine-fractions-of-landfill-mining-a-critical-review/139.
30. Directive 2006/66/EC of the European Parliament and of the Council of 6 September 2006 on batteries and accumulators and waste batteries and accumulators and repealing Directive 91/157/EEC; 2006 (vol. 49).
31. Commission Regulation (EU) No 493/2012 of 11 June 2012 laying down, pursuant to Directive 2006/66/EC of the European Parliament and of the Council, detailed rules regarding the calculation of recycling efficiencies of the recycling processes of waste batteries and accumulators; 2012 (vol. 55).
32. Becker J, Beverungen D, Winter M, Menne S, editors. Umwidmung und Weiterverwendung von Traktionsbatterien: Szenarien, Dienstleistungen und Entscheidungsunterstützung. Wiesbaden: Springer Fachmedien Wiesbaden; 2019.
33. Potting J, Hekkert M, Worrell E, Hanemaaijer A. Circular economy: Measuring innovation in the product chain; Policy Report. PBL Netherlands Environmental Assessment Agency; 2017.
34. Diener DL, Tillman A-M. Component end-of-life management: Exploring opportunities and related benefits of remanufacturing and functional recycling. Resour Conserv Recycl. 2015;102(3):80–93.

35. Parker D, Riley K, Robinson S, Symington H, Tewson J, Jansson K, *et al.* Remanufacturing market: For horizon 2020, grant agreement No 645984, November 2015 Study; 2015.
36. Reike D, Vermeulen WJV, Witjes S. The circular economy: new or Refurbished as CE 3.0? — Exploring controversies in the conceptualization of the circular economy through a focus on history and resource value retention options. Resour Conserv Recycl. 2018;135: 246–64.
37. Kirchherr J, Reike D, Hekkert M. Conceptualizing the circular economy: an analysis of 114 definitions. Resour Conserv Recycl. 2017;127:221–32.
38. Kampker A. Elektromobilproduktion. Berlin: Springer Vieweg; 2014.
39. Beverungen D, Fabry C, Ganz W, Matzner M, Satzger G, editors. Dienstleistungsinnovationen für Elektromobilität: Märkte, Geschäftsmodelle, Kooperationen. Stuttgart: Fraunhofer-Verl.; 2015.
40. Baltac S, Slater S. Batteries on wheels: the role of battery electric cars in the EU power system and beyond. Element Energy; 2019. See: https://www.transportenvironment.org/wp-content/uploads/2021/07/2019_06_Element_Energy_Batteries_on_wheels_Public_report.pdf
41. Zhang H, Liu W, Dong Y, Zhang H, Chen H. A method for pre-determining the optimal remanufacturing point of lithium ion batteries. Procedia CIRP. 2014;15:218–22.
42. Kampker A, Heimes HH, Ordung M, Lienemann C, Hollah A, Sarovic N. Evaluation of a remanufacturing for lithium ion batteries from electric cars. Int J Mech Mech Eng. 2016;10(12):1929–35.
43. Steinhilper R. Produktrecycling: Vielfachnutzen durch Mehrfachnutzung. Stuttgart: Fraunhofer-IRB-Verl.; 1999.
44. British Standards Institution. Design for manufacture, assembly, disassembly and end-of-life processing (MADE): Part 2: Terms and definitions: British Standards Institution; 2009.
45. Bittner P. Zweites Leben für gebrauchte Akkus: Volkswagen AG; 2021 [cited 2021 Feb 1]. Available from: https://www.volkswagen.de/de/elektrofahrzeuge/elektromobilitaet-erleben/elektroauto-technologie/zweites-leben-fuer-gebrauchte-akkus.html
46. Treffer F. Abschlussbericht zum Verbundvorhaben: Entwicklung eines realisierbaren Recyclingkonzepts für die Hochleistungsbatterien zukünftiger Elektrofahrzeuge (LiBRi). Hanau; Umicore AG & Co.KG; 2011.
47. Standridge CR, Corneal L. Remanufacturing, repurposing, and recycling of post-vehicle-application lithium-ion batteries. Mineta National Transit Research Consortium; 2014. Online available: https://transweb.sjsu.edu/sites/default/files/1137-post-vehicle-Li-Ion-recycling.pdf
48. Still B, Busse J. Audi und EnBW kooperieren bei Batteriespeichern: Gemeinsame Pressemitteilung; 2020 [cited 2021 Feb 26]. Available from: https://www.enbw.com/unternehmen/presse/second-life-batterien.html
49. Stoker L. Renault looks to drive UK storage boom with second-life Powervault deal; 2017 [cited 2021 Feb 19]. Available from: https://www.energy-storage.news/news/renault-looks-to-drive-uk-storage-boom-with-second-life-powervault-deal
50. BMW Group. BMW i Batterien werden als "Second Life Batteries" flexible Speicher für erneuerbare Energien und sichern die Stabilität des Stromnetzes. Hamburg; 2014 [cited 2021 Feb 18]. BMW AG. Available from: https://www.press.bmwgroup.com/deutschland/article/detail/T0193200DE/bmw-i-batterien-werden-als-"second-life-batteries"-flexible-speicher-fuer-erneuerbare-energien-und-sichern-die-stabilitaet-des-stromnetzes
51. Chubu Electric Power Co., Inc., Toyota Motor Corporation. Chubu electric power and Toyota to commence electrified vehicle battery reuse and recycling verification project; 2018 [cited 2021 Mar 15]. Available from: https://global.toyota/en/newsroom/corporate/20929916.html
52. Charter M, Gray C. Remanufacturing and product design. Int J Prod Develop. 2008;6(3/4):375.

53. Kampker A, Wessel S, Fiedler F, Maltoni F. Battery pack remanufacturing process up to cell level with sorting and repurposing of battery cells. J Remanuf. 2020;11:1–23. https://link.springer.com/article/10.1007/s13243-020-00088-6
54. Maltoni F. Challenges and outlook in the remanufacturing of automotive Li-Ion Batteries. In: To elevate automotive remanufacturing by innovation and automation; Fernand J. Weiland. 2019. pp. 135–156.
55. Mathew M, Kong QH, McGrory J, Fowler M. Simulation of lithium ion battery replacement in a battery pack for application in electric vehicles. J Power Sources. 2017;349:94–104.
56. Paterson DAP, Kao C-C, Ijomah WL, Windmill JFC. Incorporating remanufacturing into the end-of-life vehicles directive: current presence and the waste problem. J Remanuf. 2018;8 (1–2):23–37.
57. Daimler AG. Patentierte Frischzellenkur; 2020 [cited 2021 Mar 17]. Available from: https://www.daimler.com/nachhaltigkeit/lebenszyklus-batterie/nutzung-charging/remanufacturing-reportage.html
58. Dürr Systems AG. Lösungen die verbinden. Innovative Verfahren für die Batterieverklebung; 2019.
59. Bundesministerium für Bildung und Forschung. Karliczek: Deutschland ist heute wieder "Hotspot": 100 Millionen Euro für die Batterieforschung: BMBF fördert vier neue Batterie-Kompetenzcluster; 2020 [cited 2020 Aug 8]. Available from: https://www.bmbf.de/bmbf/shareddocs/pressemitteilungen/de/karliczek-deutschland-ist-heute-wieder-hotspot.html
60. University of Bayreuth. Powerful research contribution of the University of Bayreuth to federal battery cluster network. Bayreuth; 2020 [cited 2021 Mar 17]. University of Bayreuth. Available from: https://uni-bayreuth.de/pressemitteilung/batterieforschung
61. Schäfer P. Projekt forscht an automatisiertem Batterierecycling; 2020 [cited 2021 Feb 7] Available from: https://www.springerprofessional.de/batterie/recycling/projekt-forscht-an-automatisiertem-batterierecycling/17649832
62. Harper G, Sommerville R, Kendrick E, Driscoll L, Slater P, Stolkin R, *et al.* Recycling lithium-ion batteries from electric vehicles. Nature. 2019;575(7781):75–86.
63. Hanisch C, Haselrieder W, Kwade A. Recycling von Lithium-Ionen-Batterien–das Projekt LithoRec. In: Recycling und Rohstoffe. Neuruppin: TK-Verl.; 2012.
64. Buchert M, Sutter J. Stand und Perspektiven des Recyclings von Lithium-Ionen-Batterien aus der Elektromobilität. Darmstadt, Germany: Öko-Institut e.V.; 2020.
65. Foster M, Isely P, Standridge CR, Hasan MM. Feasibility assessment of remanufacturing, repurposing, and recycling of end of vehicle application lithium-ion batteries. J Ind Eng Manage. 2014;7(3):698–715.
66. Baars J, Domenech T, Bleischwitz R, Melin HE, Heidrich O. Circular economy strategies for electric vehicle batteries reduce reliance on raw materials. Nat Sustainable. 2021;4(1):71–9.
67. European Environment Agency. New registrations of electric vehicles in Europe; 2021 [cited 2021 Mar 17]. Available from: https://www.eea.europa.eu/ims/new-registrations-of-electric-vehicles

https://doi.org/10.1142/9789811245626_0017

Chapter 16

Digitalization — A Driver for Circular Economy Solutions

Soumya Singh*, Christian Schneider† and Sabri Baazouzi‡
Fraunhofer Institute for Manufacturing Engineering and Automation IPA
Nobelstraße 12, 70569 Stuttgart, Germany
**soumya.singh@ipa.fraunhofer.de*
†christian.schneider@ipa.fraunhofer.de
‡sabri.baazouzi@ipa.fraunhofer.de

One of the main agendas of the European Green Deal (March 2020) is a circular economy action plan (CEAP). It introduces legislative and non-legislative measures along the entire life cycle of products to reduce the pressure on natural resources, encourage sustainable consumption and to ensure that the resources used are kept in the EU economy for as long as possible. The circular economy perspective for the battery sector in particular has great potential to combine socio-economic and economic effects by increasing resource productivity and safely reusing materials at the end of their service life. However, digitalization or extensive use of digital technologies and processes is necessary to enable an information-driven circular economy. Battery-specific circular economy measures includes end-of-life tracking, automated dismantling, remanufacturing and recycling, thus promoting a cost-effective, higher-quality product and material (re)use. Comprehensive digitalization can help to generate business model-relevant data about the life cycle of the batteries and share it between relevant actors to maximize corporate and societal benefits. Moreover, information-driven circular economy implementation results in increased transparency in product tracking and connectivity of downstream value creation paths.

1. Introduction

In a circular economy, multiple stakeholders are involved, such as raw material suppliers, OEMs, users, disassemblers and recyclers, who perform their roles across the different phases of product life cycle. Thus, a circular economy is necessarily a geographically networked environment that involves various organizations across the product life cycle. A networking concept that takes these two dimensions into account is associated with several challenges such as information variety, data security, ensuring stakeholder-oriented availability, data storage and data management. This chapter focuses on the digitalization opportunities to improve the circularity of battery systems. Thereby, the different life cycles are considered to track the evolution of battery properties at different system levels during the whole life cycle, which involves nine phases starting with the refinement process of raw materials until recovery by remanufacturing, repurposing or recycling (see Fig. 1). At the end-of-life phase, detailed information, especially

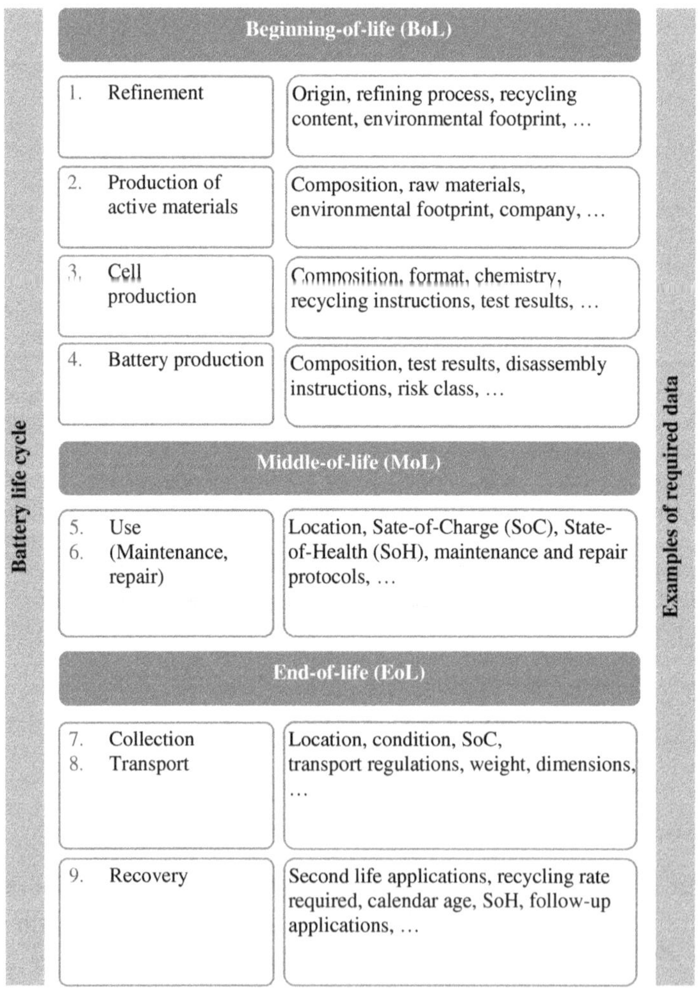

Fig. 1. Battery lifecycle phases and their associated product data.

about the availability, location and condition of products, can improve the recovery efficiency by enhancing the planning capability, enabling optimal selection of end-of-life strategies and optimizing the end-of-life operations such as testing and disassembly. In Fig. 1, we identified exemplary relevant data for circularity along the life cycle in reference to Ref. [29]. Nowadays, transparent and holistic tracking of these data is not possible, resulting in information deficit, which complicates the establishment of highly efficient circular economies. Wils et Berg (Quelle) classified these information deficits in circular economy are classified in four categories: (A) Limited information availability: This concerns certain information related to the quality and quantity of the returns. Consequently, it becomes more challenging to allocate the end-of-life assets to the optimal follow-up applications. (B) Increased transaction and research costs: The information deficits cause high costs, for example, to find out the quality and composition of products. This complicates determining prices and results in a lack of transparency in markets. (C) Distorted perception by users: Second use products and secondary materials are mostly seen as being of lower value than new products and primary materials. (D) Technological deficits: Circularity is mostly not seen as a competitive advantage and is made more difficult due to not being considered during the design phase.

2. Digitalization — Definition and Significance

Battery life cycle management or battery life cycle data management (BLDM) (components of PLM and PLDM) plays a key role in digitalization for circular economy. Today, the need for product life cycle data has become a reality across organizations, as vast amounts of data are created, managed and utilized across functions. Mounting product complexity due to increasing electrical, mechanical, embedded software and service elements has gradually created a demand for structured product data management. In this section, the technologies that enable battery life cycle data management, especially at EoL, are determined and further discussed.

According to Tezri *et al.*[1] product lifecycle management (PLM) is an integrator of tools and technologies that streamlines the flow of information through the various stages of the product life cycle. It is a prerequisite for improving cross-functional collaboration and exploiting synergies between organizational domains. PLM integrates people, data, processes and business systems and serves as a product information backbone for an organization by providing the right information at the right time and in the right context. The concept is based on coordinated methods, processes and organizational structures and usually uses IT systems for recording and managing the data. In the literature,[2,3,4] PLM does have a few overlaps with the concept of Data Lifecycle Management (DLM), such as data acquisition and capture; data backup and recovery; data management and maintenance; and data retention and secure destruction.

Figure 2 highlights the components of BLDM. Types of data, data sources and data management approaches are important factors for identification of technologies for BLDM. Type of data and data sources need to be identified with respect to the life cycle phase, while data management methods may include a wide range of data acquisition, processing and dissemination methodologies from within an IT infrastructure. For example, cross-industry standard process for data mining (CRISP-DM), an open standard process model, is applicable in lithium-ion battery cell production.[5] Similarly, data analytics for BMS architecture during battery usage was proposed in Ref. [6], in order to drive data and battery life estimation.

In the early 2000s, the main enabling technologies for life cycle management were sensors, indicators, RFIDs and wireless communications, used to facilitate movement of data from real space to virtual space.[7] With the advent of Industry 4.0, there have been major developments from a technological perspective to set up intelligent manufacturing in factories. With time, the mutual integration and optimization

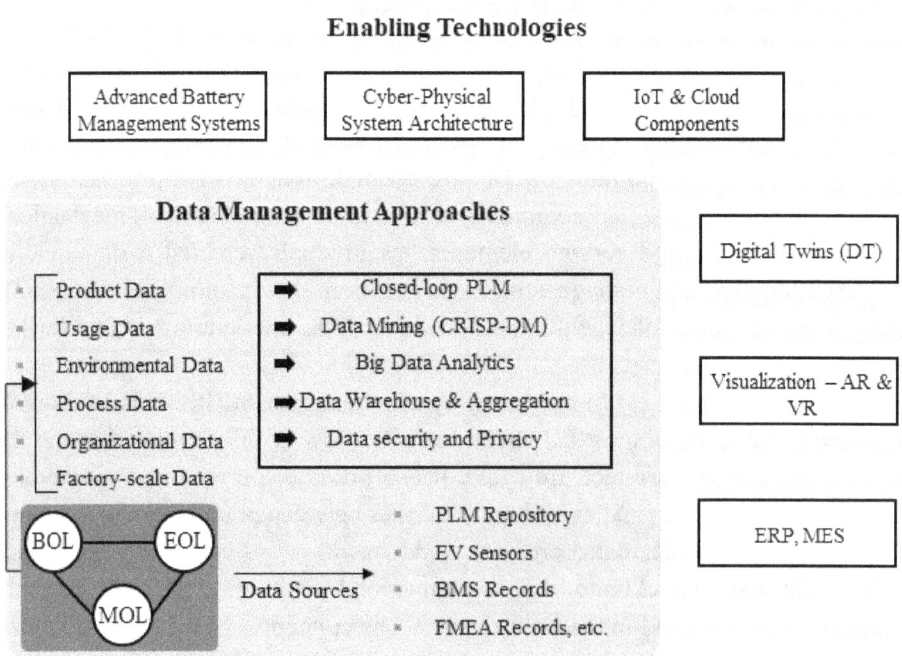

Fig. 2. Enabling technologies for battery life cycle data management at end of life.

of enabling technologies will become more simple, convenient and accurate.[8] The enabling technologies identified in Fig. 2 are further elaborated as follows.

2.1. *CPS architecture*

The term Cyber-Physical Systems "reflects a vision of a technology that deeply connects our physical world with our information world" and can interact with humans through many new methods.[9] It is seen as the core concept of Industry 4.0 for smart factories and a key assistant for future technologies. With recent developments that have resulted in higher availability and affordability of sensors, data acquisition systems and computer networks, the competitive nature of today's industry forces more factories toward implementing high-tech methodologies.[10] Nonetheless, CPS research is still in its infancy. Innovative approaches to abstraction and architectures that enable seamless integration of control, communication and computation must be developed for rapid design and deployment of CPS.

Standardized abstractions and architectures are urgently needed to fully support integration and interoperability and spur similar innovations in cyber-physical systems. One of the newly introduced concepts of CPS is the virtual battery. Due to the rise in demand for the electric and hybrid vehicle and the storage of power plant, batteries have become more and more critical. Battery power levels are impacted by many environmental factors like temperature, humidity, charging level and discharge rate. We need a battery model which evaluates the battery health and failure prediction with the given information, through simulation under different conditions. However, there are many challenges with the CPS system as well. To make the correct decision, it is essential to send the correct data at the right time for the right reason. In addition the security of the transmitted data and the reliability and safety of the CPS need to be ensured.[11]

In order to implement Circular Economy (CE) to enhance the sustainability of firms, Thang *et al.* propose framework technologies including CPS to overcome barriers of implementation and present a use case for that.[12] Many researches have pointed out that data-driven Industry 4.0 can be used to solve the problems of a CE. Cyber-physical systems standards and specifications as well as CPS modeling and modeling integration are named as some of the crucial challenging factors for achieving CE.[13]

2.2. *IoT and cloud components*

Cloud computing is strongly converging with the Internet of Things (IoT), offering novel techniques for data management and visualization on the cloud.

However, system designers and operations managers face numerous challenges to realize IoT cloud systems in practice, mainly due to the complexity involved with provisioning large-scale IoT cloud systems and diversity of their requirements in terms of IoT resources consumption, customization of IoT capabilities and runtime governance.[14] The results of IoT failures can be severe; therefore, the study and research of security issues in the IoT is of extreme significance. The main objective of IoT security is to preserve privacy, confidentiality and ensure the security of the users, infrastructure, data and devices of the IoT, and guarantee the availability of the services offered by an IoT ecosystem.[15]

A novel cloud-based battery condition monitoring platform is proposed which utilizes IoT devices and cloud components. The IoT components including data acquisition and wireless communication components can be implemented in battery modules, which allows a module to communicate with others and the cloud. The cloud components need to include a cloud storage, analytics tool and visualization.[16]

With reference to CE, Rajput *et al.* points out that adopting IoT can be used to retrieve real-time information, which further helps in improving the data collection, sharing the resource consumption and reducing materials wastage. It further enhances the manufacturing systems' accuracy and precision.[13] Furthermore, it is pointed out that through IoT, it is possible to foster CE through the connection of people and things by mobile devices, which leads to economic opportunities for individuals and businesses in multiple domains.[12]

2.3. *Digital twins*

An important prerequisite for smart manufacturing is cyber-physical integration, which is increasingly being embraced by manufacturers. The DT is another concept associated with cyber-physical integration. A DT creates high-fidelity virtual models of physical objects in virtual space in order to simulate their behaviors in the real world and provide feedback.[17]

A DT reflects a bi-directional dynamic mapping process; it breaks the barriers in the product life cycle and provides a complete digital footprint of products.[18] As a result, DTs enable companies to predict and detect physical issues sooner and more accurately, optimize manufacturing processes and produce better products. At present, advances in new IT are enabling the growth of DTs. Since DTs open up a new way to synchronize physical activities with the virtual world, they have become a popular research topic. Regarding battery systems, many parts of the life cycle are not yet under accurate supervision. A digital twin architecture for automotive battery systems is a possibility, on which digital services for various stakeholders along the manufacturing and product life cycle of a battery can be established.[19]

With regard to CE, DTs are described as an important tool by allowing a real-time simulation of processes, material flows and energy consumptions. However, it is pointed out that even though DT could be easily adaptable to CE practices, only a few papers have dealt with this topic so far. One of the described ideas of adopting a DT in CE practice is focused on the virtual optimization of disassembly (e.g. monitoring materials and energy consumptions) or the recycling and remanufacturing processes (e.g. storing knowledge about components and materials embedded into products).[20]

2.4. *Visualization — AR and VR*

Since the origins of Augmented Reality (AR), industry has always been one of its prominent application domains. The recent advances in both portable and wearable AR devices and the new challenges introduced by Industry 4.0 further enlarge the applicability of AR to improve productiveness and to enhance the user experience. The effectiveness of AR does not lie in the visualization process itself: it is how data are visualized that makes AR a very powerful technology.[21]

Virtual Reality (VR) application in Industry 4.0 allows companies to decrease design and production costs, maintain product quality and reduce the time needed to go from product concept to production. The scope of Industry 4.0 is to overlay a simulation on a real-time production line that can investigate poorly understood phenomena and aid the rectification of bottlenecks.[22] A virtual battery helps to collect health, reliability, operational readiness information in real time in visualization form, which helps the designer to address the flaws and design issues and improve life expectancy.[23]

Focusing on CE AR/VR could represent a crucial element for improving disassembly and remanufacturing processes. Furthermore, VR technologies can support CE practices by enabling virtual testing of waste from electrical and electronic equipment (WEEE) in a disassembly plant configuration through a set of dedicated simulation tools. An AR-based CE solution proposed by Demestichas *et al.* enables workers to detect and visualize buried utilities for excavator, thus preventing utility damage (such as water/wastewater pipes, conduits, cables, etc.) and their subsequent economic and environmental impacts.[24]

2.5. *ERP, MES*

Each company needs to improve the processes within the production system in order to enable better monitoring and control, as well as to provide all real-time information that is important for the decision-making process, in order to improve all processes and simultaneously control all elements of the process. Some of the systems that provide process monitoring are Enterprise Resources Planning (ERP) and Manufacturing Execution Systems (MES), but they are often not sufficiently integrated.[25] Information objects are determined by MES, ERP, QMS (Quality Management Software) and PDM (Product Data Management) systems and electronic document management systems. The list of design and production procedures is determined by the automated processes of a particular enterprise, which determines the variability of the digital passport.[26]

The integration of CE principles in ERP systems can help develop a procurement system that involves competitive strategy for remanufacturing and recycling material. Supplier development strategies and managing vendor inventories for outsourcing decisions are described as key for the excellence of remanufacturing operations in CE.[27] Furthermore, it is pointed out that data handling systems support the goals of Circular Economy, as they integrate information produced by heterogeneous data collection systems such as IoT, ERP and CRM systems.[28]

References

1. Terzi S, Bouras A, Dutta D, Garetti M, Kiritsis, D. Product lifecycle management — From its history to its new role. Int J Product Lifecycle Management. 2010;4:360–89.
2. Gharaibeh A, Salahuddin MA, Hussini SJ, Khreishah A, Khalil I, Guizani M, Al-Fuqaha A, Smart cities: A survey on data management, security, and enabling technologies. IEEE Communications Surveys & Tutorials. 2017;19(4):2456–501.
3. Higgins S. The lifecycle of data management. Manag Res Data. 2012:17–45.
4. Reid R, Fraser-King G, Schwaderer WD. Data lifecycles: managing data for strategic advantage. John Wiley & Sons; 2007;11.
5. Schnell J *et al.* Data mining in lithium-ion battery cell production. J Power Sources. 2019;413: 360–6.
6. Zhou K, Chao Fu, Shanlin Yang. Big data driven smart energy management: From big data to big insights. Renew Sust Energy Rev. 2016;56:215–25.
7. Grieves MW. Product lifecycle management: The new paradigm for enterprises. Int J Prod Dev. 2005;2(1/2):71.
8. Wan J, Cai H, Zhou K, Industrie 4.0: Enabling technologies, Proceedings of 2015 International Conference on Intelligent Computing and Internet of Things, Harbin, China, 2015, pp. 135–140.
9. Lee EA, Seshia SA. Introduction to embedded systems: A cyber-physical systems approach. 2nd ed. Cambridge, MA: MIT Press; 2017.
10. Lee J, Bagheri B, Kao H-A. A Cyber-Physical Systems architecture for Industry 4.0-based manufacturing systems. Manuf Lett. 2015;3:18–23.
11. Singh M, Gupta H. Significance of Virtual Battery in Modern Industrial Cyber-Physical Systems. In: Simulation for Cyber-Physical Systems Engineering 2020. pp. 305–319. Springer, Cham.
12. Pham TT, Kuo T-C, Tseng M-L, Tan RR, Tan K, Ika DS, *et al.* Industry 4.0 to accelerate the circular economy: a case study of electric scooter sharing. Sustainability. 2019;11(23):6661.
13. Rajput S, Singh SP. Connecting circular economy and Industry 4.0. Int J Inf Manage. 2019;49:98–113.
14. Nastic S, Sehic S, Le D-H, Truong H-L, Dustdar S. Provisioning software-defined IoT cloud systems. In: 2014 International Conference on Future Internet of Things and Cloud. IEEE; FiCloud 2014, Barcelona, Spain, 2014. pp. 288–295.

15. Mohamad Noor MB, Hassan WH. Current research on Internet of Things (IoT) security: a survey. Comput Netw. 2019;148:283–94.
16. Adhikaree A, Kim T, Vagdoda J, Ochoa A, Hernandez PJ, Lee Y. Cloud-based battery condition monitoring platform for large-scale lithium-ion battery energy storage systems using Internet-of-Things (IoT). In: ECCE: 2017 IEEE Energy Conversion Congress and Exposition: 1–5 October 2017. New York: IEEE; 2017. pp. 1004–1009.
17. Tao F, Qi Q, Wang L, Nee A. Digital twins and Cyber–Physical Systems toward smart manuf Industry 4.0: correlation and comparison. Engineering. 2019;5(4):653–61.
18. Tao F, Cheng J, Qi Q, Zhang M, Zhang H, Sui F. Digital twin-driven product design, manufacturing and service with big data. Int J Adv Manuf Technol. 2018; 94(9–12): 3563–3576.
19. Merkle L, Segura AS, Torben Grummel J, Lienkamp M. Architecture of a digital twin for enabling digital services for battery systems. In: 2019 IEEE International Conference on Industrial Cyber Physical Systems (ICPS 2019): Howards Plaza Hotel Taipei, Taiwan, 2019 May 06–09. Piscataway, NJ: IEEE; 2019. pp. 155–60.
20. Rocca R, Rosa P, Sassanelli C, Fumagalli L, Terzi S. Integrating virtual reality and digital twin in circular economy practices: a laboratory application case. Sustainability. 2020;12(6):2286.
21. Pace F de, Manuri F, Sanna A. Augmented reality in Industry 4.0. Am J Comput Sci Inf Technol. 2018;06(01):17.
22. Liagkou V, Salmas D, Stylios C. Realizing virtual reality learning environment for Industry 4.0. Procedia CIRP. 2019;79:712–17.
23. Singh M, Gupta H. Significance of virtual battery in modern industrial Cyber-Physical Systems. In: Risco Martín JL, Mittal S, Ören T, editors. Simulation for Cyber-Physical Systems engineering: a cloud-based context. 1st ed. Cham: Springer International Publishing; Springer; 2020. pp. 305–19. (Simulation Foundations, Methods and Applications).
24. Demestichas K, Daskalakis E. Information and communication technology solutions for the circular economy. Sustainability. 2020;12(18):7272.
25. Berić D. The implementation of ERP and MES systems as a support to industrial management systems; 2018 [cited 2021 Mar 1]. Available from: http://ijiemjournal.org/images/journal/volume9/IJIEM-3_018.pdf
26. Julia D, Yuriy G. A generalization of approaches to creating a digital passport supporting the stages of the electronic product life cycle and the features of the formation of design decisions based on it Industry 4.0. 2020;5(3):101–104. Available from: https://stumejournals.com/journals/i4/2020/3/101.full.pdf
27. Procurement 4.0 and its implications on business process performance in a circular economy [cited 2021 Mar 10]. Available from: https://www.sciencedirect.com/science/article/abs/pii/S0921344919304082?via%3Dihub
28. Pagoropoulos A, Pigosso DC, McAloone TC. The emergent role of digital technologies in the circular economy: a review. Procedia CIRP. 2017;64:19–24.
29. Wilts, Henning, and Holger Berg. The digital circular economy: can the digital transformation pave the way for resource-efficient materials cycles? Vol. 2017. Wuppertal Institut für Klima, Umwelt, Energie, 2018.

https://doi.org/10.1142/9789811245626_0018

Chapter 17

Smart Measuring — Measurement Assessment for Returned Batteries

Friedrich-Wilhelm Speckmann
Fraunhofer-Institute for Manufacturing Engineering and Automation
Nobelstr. 12, 70563 Stuttgart, Germany
friedrich-wilhelm.speckmann@ipa.fraunhofer.de

Battery manufacturing is a steadily growing industry and it requires a variety of resources. Therefore, battery recycling gains momentum and will become a major part of the sustainable market. Next to the direct recycling of returned battery cells, there is the possibility of a second-life application. Here, returned battery cells that have not reached their end-of-life limits are employed into applications with lower requirements such as stationary battery storage systems. To assess which cells are still useable, there are different battery cell characterization methods. This chapter presents suitable cell assessment processes and highlights their multiple uncertainties.

1. Life Cycle Optimized Usage of Returned Batteries

The lifespan of batteries varies enormously due to different cell chemistries and the field of application. Modern lithium-ion batteries (LIBs) are employed in various mobile devices and, nowadays, more and more in the automotive sector. The traction batteries of electric vehicles (EVs) are especially stressed by the fast charging/discharging cycles as well as harsh temperature environments. Therefore, their useful lifetime is limited to three to five years. After this period, the battery capacity will decrease by around 25%, and the electric vehicle can achieve less distance without recharging. These batteries have to be switched with new ones, and the older batteries are often burned instead of being recycled or reused in second-life applications (e.g. stationary battery storage systems). Nevertheless, that is likely to change going forward for multiple reasons. First and foremost, regulation is forcing change. For instance, in Europe, the European

Directive on End-of-Life Vehicles forces automobile producers to reuse or recycle up to 85% of vehicle content.[1] Additionally, the Battery Directive puts the responsibility on the automobile manufacturer to collect and recycle batteries. Furthermore, the price of lithium and other rare metals has been increasing over the last years, and thus, it can be economically viable to recover these materials.

2. Assessment of Returned Batteries — Smart Measuring

The globally increased amount of EVs led to a strong interest in recycling of LIBs, reclaiming not only cell-specific metals like lithium, cobalt, manganese and nickel but also valued materials like copper, aluminum and graphite.

An additional trend in recent years is refurbishing and reusing retired EV batteries in areas with fewer requirements, such as stationary storage systems. This field of application is called second life. Here, the returned LIBs can serve as stationary storage for peak load management, EV charging stations or to enhance renewable energy grid interaction.[2–4] An automotive battery pack reaches the end of its first-life application when it loses 20% of its initial capacity or doubles the internal resistance.[5] Currently, the pack is not suitable as a traction battery anymore because the capacity loss affects acceleration, range and regeneration. Nowadays, these cells are commonly transferred to recycling processes. There are multiple existing recycling procedures comprising physical and chemical treatments.

On the one hand, the main physical processes are mechanical separation, thermal treatment or dissolution. On the other hand, the most critical chemical processes consist of acid leaching, bioleaching, solvent extraction, chemical precipitation and electrochemical techniques. Furthermore, it is essential to note that these processes are not limited to one, and frequent overlap between the technologies occurs.[6] The recovered raw materials have to be tested on their quality and fulfill the required specifications for new slurries. However, it is most likely that the new slurries will consist of approximately 90% new materials and about 10% recycled components. Figure 1 displays the closed value chain of battery recycling and manufacturing.

Not all returned battery cells in a module have already reached their end of useful life. Therefore, it can be interesting to open returned battery modules and recover the single battery cells instead of directly shredding the entire module. In this case, the recovered cells have to be assessed in order to decide on their further purpose. Are the cells in a state that has not reached the end-of-life criteria? Can

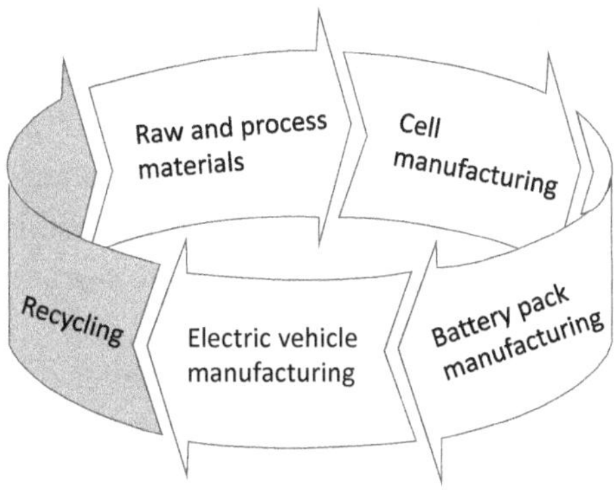

Fig. 1. Proposed battery life cycle value chain loop, starting with the cell production and ending with the recovery of raw materials for the addition in new cells.

the cells still be used for mobile or rather second-life applications such as stationary battery storage, or are they only suited for recycling? Because of the varying battery cell states and the potential for further use cases, it is necessary to test and assess every one of them.

3. Overview of Battery Cell Assessment Methods

There are multiple approaches to characterizing the current state of health (SOH) of a battery cell described in the literature with various benefits and disadvantages. A key parameter to be measured for rechargeable batteries is the state of charge (SOC). This common metric is used in multiple battery characterization methods and is connected to the SOH. Therefore, more extensive investigations have been conducted on the SOC to assume that the battery is in a healthy state. Due to the hidden coupling between these two values, both values' combined analysis is more complex. In such a case, the observable values are rare, and information can only be gathered on terminal voltage and current parameters. Hence, joint estimations on the battery pack become even more difficult in comparison to single cells.[7] Nevertheless, there are multiple approaches to SOC and SOH estimation on a cell level.

For example, open-circuit voltage (OCV) measurements take advantage of the linear relation between the SOC and the OCV for batteries such as lead–acid technologies:

$$\mathrm{OCV}(t) = a_1 \cdot \mathrm{SOC}(t) + a_0 \tag{1}$$

where SOC(t) is the respective charge value at time t, and a_0 and a_1 are the coefficients. However, although, different cell chemistries show nonlinear behaviors such as Li-ion batteries, which have to be determined in offline testing. Here, the SoC estimation is only correctly determinable if the battery is disconnected from the load over a more extended period. Even if this method is not easily applicable for onboard SOC estimation, there are some more accurate battery assessment approaches during operation.[8]

Another standard characterization method is frequency-domain identification, which is based on electrochemical impedance spectroscopy (EIS). This method can decouple phenomena such as electrolyte composition, charge transfer and double-layer capacity, as well as diffusion.[9,10] In contrast to these advantages, the traditional EIS method is not able to run online. Conventionally, most EIS-based methods only use the internal cell resistance value and sometimes add the charge transfer resistance[11] to the input parameters. However, even with these adaptions, the system cannot run online and lacks estimation accuracy. The combination of the EIS results and a neural network (NN) can strongly reduce the characterization time and increase the estimation accuracy. However, the quality depends on the NN configuration, for which parameters are chosen as trainings and input data. A smart way to accelerate the remaining capacity calculation is the usage impedance values in the entry vector, which relate to specific frequencies. Furthermore, the applied weightings of the inputs have to use cleverly guessed parameters. Therefore, the iterative optimization requires less computational time.

An experimental study showed a very fast cell capacity estimation with a low prediction error of below 0.5% after 1,500 charge/discharge cycles.[20] The employed NN was set up with 3–4 artificial neurons in the hidden layer and was trained with only three high-drain Samsung 15L (1,500 mAh) cells. In addition, the calculated resulting cell capacity values can be used to further extrapolate the capacity fade and thus enable the best suited use of the returned battery cell depending on the prediction.

An alternative approach to estimating the SOH is the battery model-based parameter identification for onboard performance assessment. Regarding the operational data, this technique adopts the optimal state algorithms such as the *Kalman* or least square filtering to approximate capacity and internal resistance.[12–14] Other researchers focus on the kinetic of constant-current/constant-voltage charge as a means for calendric aging investigations.[15] All in all, it has to be kept in mind that every battery cell characterization method depends on multiple entry data and, thus, always shows a slight error due to numerous factors. Here, it can be interesting to know the maximum deviation of the calculated results.

4. Uncertainty Management

Uncertainty management is an essential topic for reliable estimation of battery performance. Four main types of uncertainties are classified in measurement uncertainty, algorithm uncertainty, environmental uncertainty, the combined model and model parameter uncertainty. First, measurement uncertainties consist of current and voltage measurement errors and have been included in most researches. For example, the sensor noise value is commonly integrated into the *Kalman* filter. Second, algorithm uncertainties describe the numerical accuracy of the mathematical estimations. Multiple studies focus on the varying fitting quality of algorithms. Li *et al.* compared the SOC estimation based on a *Luenberger* observer, sigma point *Kalman* filter and sequential *Monte Carlo* calculations.[16,17] The central parameter in environmental uncertainties is temperature detection, and it strongly impacts the battery SOC and SOH.[18] In the field of battery modeling, two significant uncertainties have to be regarded for a high prediction quality.

On the one hand, the model parameter uncertainty includes the realization of the physical uncertainty of, e.g. equivalent circuit models or electrochemical models. Here, the manufacturing inaccuracies are the main reason for deviating battery cells.[19] On the other hand, the model uncertainty describes the difference between a battery model attempt and the actual physical system. It is important to note that no battery model will be identical to the actual cell, and thus, there will always be an error. However, in most cases, the error margin is small enough to not represent significant model deviations from the investigated system. Until now, these kinds of uncertainties are only barely considered in battery prognostics.

5. Summary

For successful reuse of returned batteries, it is crucial to characterize each cell's current SOH because there are major deviations in the battery cell states. The main focus lies on fast and accurate SOC and SOH estimation of the measured cells. There are multiple battery cell assessment methods that are described in this chapter. Mainly, impedance spectroscopy-based measurements in combination with artificial intelligence procedures are most promising. They enable a precise cell state estimation and incorporate, for example, neural networks to function with a minimum of measurement data.

Furthermore, it is essential to assess the various uncertainties of the employed characterization methods. These potential errors range from simple measurement inaccuracies of current and voltage values to complex modeling uncertainties. These represent the difference between the physical system and the modeling approach and will always be present. Therefore, the knowledge of all uncertainties describes the quality of the battery cell state estimation method.

References

1. Office P. CL2000L0053EN0140020.0001.3bi_cp 1.1.
2. Keeli A, Sharma RK. Optimal use of second life battery for peak load management and improving the life of the battery. In: 2012 IEEE International Electric Vehicle Conference. IEEE; 2012. pp. 1–6. Greenville, United States.
3. Hamidi A, Weber L, Nasiri A. EV charging station integrating renewable energy and second-life battery. In: 2013 International Conference on Renewable Energy Research and Applications (ICRERA). IEEE; 2013. pp. 1217–21. Madrid, Spain.
4. Koch-Ciobotaru C, Saez-de-Ibarra A, Martinez-Laserna E, Stroe D-I, Swierczynski M, Rodriguez P. Second life battery energy storage system for enhancing renewable energy grid integration. In: 2015 IEEE Energy Conversion Congress and Exposition (ECCE). IEEE; 2015. pp. 78–84. Montreal, Canada.
5. Saxena S, Le Floch C, MacDonald J, Moura S. Quantifying EV battery end-of-life through analysis of travel needs with vehicle powertrain models. J Power Sources. 2015;282:265–76.
6. Ordoñez J, Gago EJ, Girard A. Processes and technologies for the recycling and recovery of spent lithium-ion batteries. Renewable Sustainable Energy Rev. 2016;60:195–205.
7. Li J, Zhou S, Han Y, editors. Advances in battery manufacturing, service, and management systems. Hoboken: New Jersey; John Wiley & Sons Inc; 2017. (IEEE Press series on systems science and engineering).
8. Weng C, Sun J, Peng H. A unified open-circuit-voltage model of lithium-ion batteries for state-of-charge estimation and state-of-health monitoring. J Power Sources. 2014;258:228–37.
9. Hang T, Mukoyama D, Nara H, Takami N, Momma T, Osaka T. Electrochemical impedance spectroscopy analysis for lithium-ion battery using Li4Ti5O12 anode. J Power Sources. 2013;222:442–7.
10. Barré A, Deguilhem B, Grolleau S, Gérard M, Suard F, Riu D. A review on lithium-ion battery ageing mechanisms and estimations for automotive applications. J Power Sources. 2013;241:680–9.
11. Annual Conference of the Prognostics and Health Management Societ, editor. An adaptive recurrent neural network for remaining useful life prediction of lithium-ion batteries; 2010.
12. Wang S, Verbrugge M, Wang JS, Liu P. Multi-parameter battery state estimator based on the adaptive and direct solution of the governing differential equations. J Power Sources. 2011;196(20):8735–41.
13. Remmlinger J, Buchholz M, Soczka-Guth T, Dietmayer K. On-board state-of-health monitoring of lithium-ion batteries using linear parameter-varying models. J Power Sources. 2013;239:689–95.
14. Jannesari H, Emami MD, Ziegler C. Effect of electrolyte transport properties and variations in the morphological parameters on the variation of side reaction rate across the anode electrode and the aging of lithium ion batteries. J Power Sources. 2011;196(22):9654–64.
15. Eddahech A, Briat O, Vinassa J-M. Determination of lithium-ion battery state-of-health based on constant-voltage charge phase. J Power Sources. 2014;258:218–27.

16. Li J, Klee Barillas J, Guenther C, Danzer MA. A comparative study of state of charge estimation algorithms for LiFePO4 batteries used in electric vehicles. J Power Sources. 2013;230:244–50.
17. Li J, Klee Barillas J, Guenther C, Danzer MA. Sequential Monte Carlo filter for state estimation of LiFePO 4 batteries based on an online updated model. J Power Sources. 2014;247:156–62.
18. Chen Z, Mi CC, Fu Y, Xu J, Gong X. Online battery state of health estimation based on Genetic Algorithm for electric and hybrid vehicle applications. J Power Sources. 2013;240:184–92.
19. Xing Y, He W, Pecht M, Tsui KL. State of charge estimation of lithium-ion batteries using the open-circuit voltage at various ambient temperatures. Appl Energy. 2014;113:106–15.
20. Speckmann, F.W, Stroebel, M, Birke, K.P. Accelerated state-of-health estimation for battery recycling, using neural networks. In: 2020 3rd International Conference on Power and Energy Applications (ICPEA); 2020. pp. 10–1. Hong Kong, China, published by IEEE Explore.

https://doi.org/10.1142/9789811245626_0019

Chapter 18

Disassembly — A Requirement for Efficient Circularity of Battery Systems

Sabri Baazouzi*, Max Weeber† and Kai Peter Birke‡
Fraunhofer IPA, Nobelstraße 12, Stuttgart, Germany, 70569
**sabri.baazouzi@ipa.fraunhofer.de*
†max.weeber@ipa.fraunhofer.de
‡kai.peter.birke@ipa.fraunhofer.de

Due to the enormous consumption of resources during the production of electric vehicles, establishing a circular economy is crucial for the sustainability of future e-mobility solutions. Different circular economy strategies are required both to extend the service life of individual vehicle components such as the battery and to ensure high-quality recycling processes that provide pure materials for manufacturing. In this context, product disassembly is of particular significance for implementing closed-loop solutions. In this chapter, we show the importance of battery disassembly. We then outline disassembly scenarios by identifying disassembly modes and provide an overview of disassembly sequence planning. Finally, we give a general picture of disassembly factories and propose a flexible and automated disassembly net for battery systems up to the cell level.

1. Introduction

The disassembly of battery systems is an essential step for implementing circular economy solutions. It is the link between upstream operations such as collection and downstream processes used to treat battery packs and components such as reassembly and recycling. Whatever the selected end-of-life strategy is, the battery has to be disassembled at the end of its useful life. First, the disassembly is necessary to access modules and cells suitable for follow-up applications such as stationary energy storage and, second, to achieve high-quality and cost-effective recycling by increasing the purity of recycled materials.[1] Disassembly is also important to minimize the quantity of recycling materials that require hydrometallurgical or pyrometallurgical processing. This in consequence further reduces the resource demand for additional resources such as water and chemical agents. Disassembly can be performed at different system levels, as illustrated in Fig. 1. The first level aims to remove the modules. For this purpose, Tan *et al.*[2] identified five main steps: The first step involves testing and discharging, followed by the

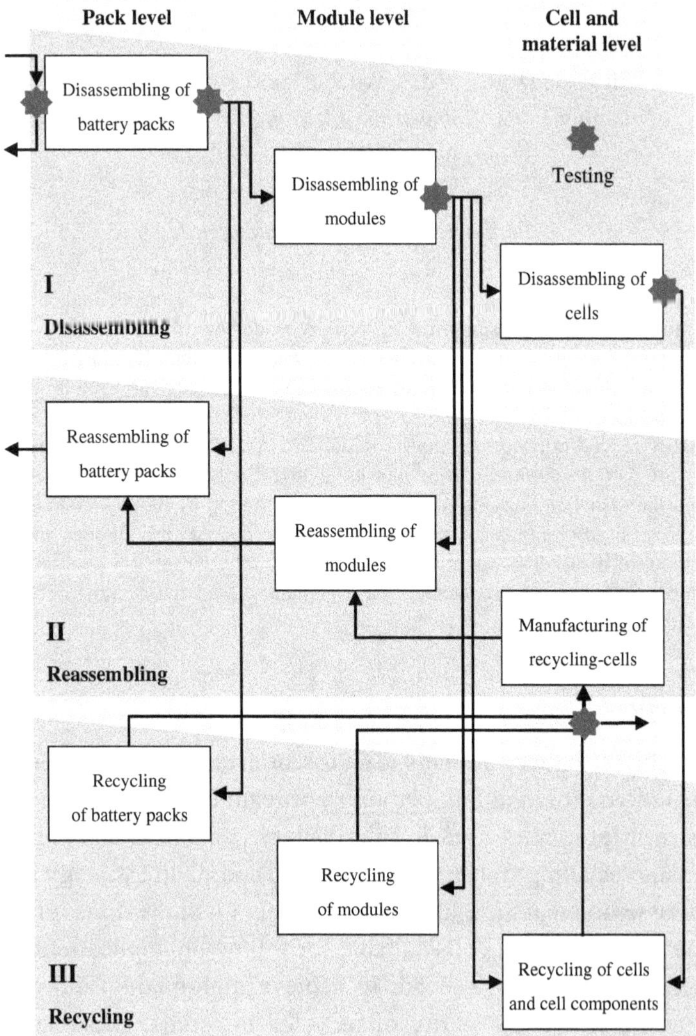

Fig. 1. Significance of the disassembly of batteries at their end-of-life phase.

removal of the battery casing. After that, the external components can be disassembled (BMS, power electronics and heat/cooling system). The fourth step consists of cutting and removing wires, cables and connectors. At last, the modules can be taken out after disassembling the stack holders. The obtained modules can be subsequently disassembled to the cell level if they are not suitable for reuse or if further disassembling is not profitable or coupled with safety risks. Thereby, six components can be extracted from the modules: Battery cells, cell contacting, cell fixation, housing, thermal management and the BMS.[3] Furthermore, the disassembly can go one step further by opening the cells. In this way, active materials for the manufacturing of recycling cells can be recovered, and the recycling processes' efficiency can be increased by ensuring a higher quality of the recovered raw materials.[1]

However, the disassembly of electric vehicle batteries is a sophisticated and challenging technical task. Wegner *et al.*[4] identified three main challenges: the variety of system designs, the treatment of flexible components and the hazards resulting from high voltage and chemicals. Also, Jin *et al.*[5] mentioned the uncertainties regarding the quantity, quality and timing of returns. Tan *et al.*[2] classified the parameters which significantly affect the disassembly of battery systems into three categories: design, safety and costs. Additional challenges are establishing uniform standards for fire protection and occupational health and safety as well as the professional qualification of workers employed in disassembly factories.[6]

All in all, disassembly is very decisive for sustainable e-mobility and associated with multiple challenges in order to create the necessary framework conditions. In the following subsections, modes are focused on to derive potential disassembly scenarios for battery systems. Furthermore, the steps of disassembly sequence planning (DSP) as a pre-process for disassembly are explained, which can be used to improve design for circularity and to support the planning of disassembly processes. Finally, disassembly factories are briefly discussed.

2. Disassembly Scenarios

2.1. *Disassembly modes*

Disassembly modes can be classified in terms of the disassembly depth (complete, incomplete), the number of components disassembled simultaneously (sequential/parallel), the disassembly techniques (destructive, semidestructive, non-destructive) and the automation level (manual, semi-automated, automated), as illustrated in Fig. 2. Thereby, the process depth and the automation level are significant factors that influence the economic performance of disassembly processes.[7–9]

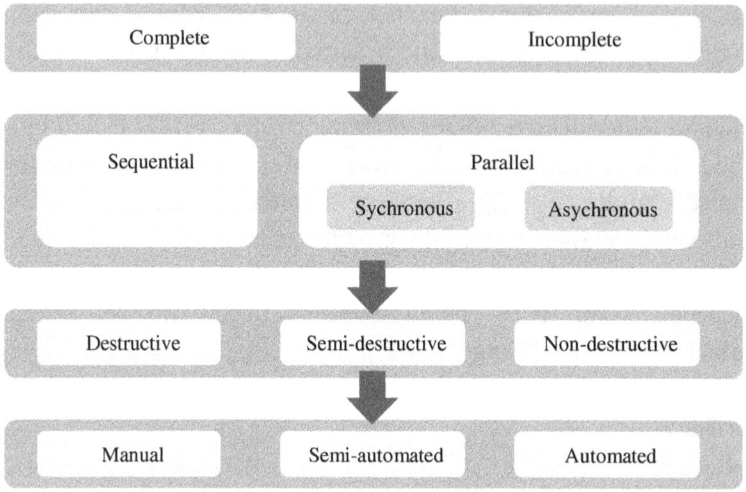

Fig. 2. An overview of the disassembly modes.

2.1.1. *Disassembly depth*

Depending on the disassembly objectives, such as reducing disassembly times, increasing revenues or reducing environmental impact, different disassembly levels can be realized. While complete disassembly involves removing all parts, incomplete disassembly is limited to the extraction of previously selected components during the disassembly sequence planning. This can be done in two ways: high-value strategy through the selective disassembly of valuable parts or high-impact strategy, which involves eliminating hazardous and environmentally harmful items.[10] Complete disassembly has been discussed much more often in the literature.[7] However, the focus of current research is on incomplete disassembly as an option that allows to quickly access specific components and to optimize the economic and environmental performance of the disassembly process.[11] Figure 3 shows the point with the maximum economic profit depending on the disassembly depth. This point represents the optimal disassembly strategy[12] and is commonly called the "stopping point" after which further disassembly reduces possible profits.[11]

2.1.2. *Parallel disassembly*

Depending on the number of manipulators used, disassembly can be classified into two types: (1) Sequential disassembly (SD) and (2) Parallel disassembly (PD). In SD, a single manipulator is used, which allows the product sub-assemblies and parts to be removed only one by one. Alternatively, several parts can be disassembled at the same time when PD is employed with the combination of multiple

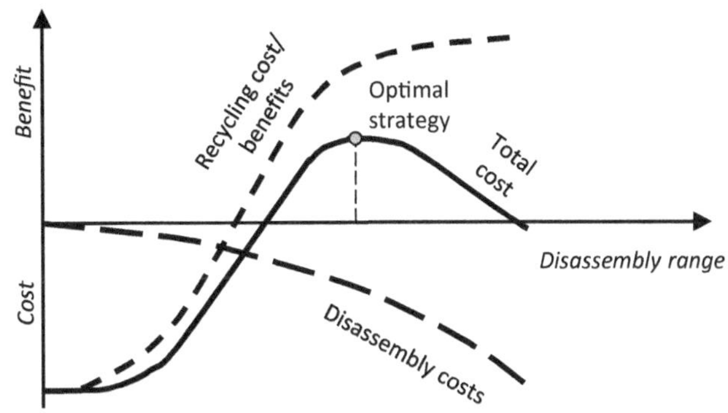

Fig. 3. Optimal disassembly strategy according to Ref. [12].

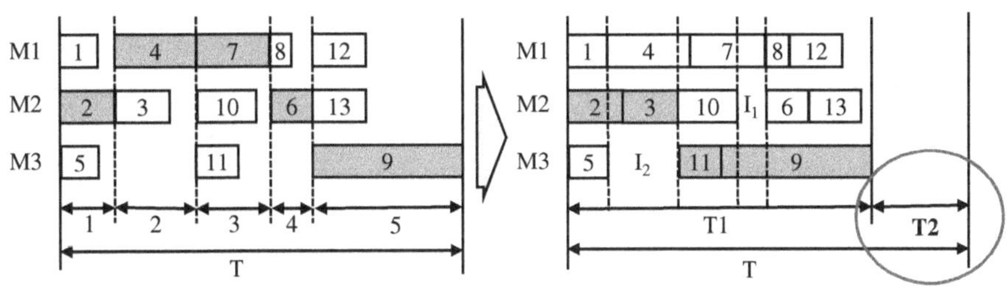

Fig. 4. Time difference between aPD and sPD when using three manipulators according to Ref. [9].

manipulators. If the manipulators start the assigned disassembly tasks simultaneously, it is called synchronous parallel disassembly (sPD). However, this leads to idle times if the disassembly tasks take different lead times. Therefore, asynchronous parallel disassembly (aPD) is outlined in[9] as a further optimization approach to reduce the total product disassembly time. Thereby, the manipulators perform the next disassembly task immediately after finishing the current one. However, this is possible only if the disassembly sequence precedence constraints are not violated, which means that the aPD cannot eliminate all idle times. Figure 4 shows possible disassembly plans of a product consisting of 13 components. Three manipulators are used here. On the left, the total process time T is illustrated when aPD is used. It can be observed that through the parallelization of disassembly tasks, the disassembly can be significantly accelerated compared to sequential disassembly. Furthermore, sPD can further optimize the processing time (Fig. 4 right). In this case, the disassembly is reduced by T2.[9]

Electric vehicle batteries (EVBs) are large and complex products. Therefore, linear sequential disassembly can possibly lead to high disassembly times. Besides, several steps of a disassembly task can be performed simultaneously, such as removing the screws of the battery cover or taking out the modules and the cells. Therefore, parallel disassembly using multiple manipulators can be recommended in this disassembly case. However, this is associated with higher investment costs and additional efforts for the DSP.

2.1.3. *Disassembly techniques*

The disassembly can be divided into non-destructive, semi-destructive and destructive disassembly based on the used separation operations. Non-destructive disassembly is essential for implementing alternative circular economy strategies such as remanufacturing and repurposing, which require undamaged components to be reused. However, this disassembly type is the most expensive and requires a high degree of flexibility, and is, therefore, generally unprofitable.[8] Semi-destructive disassembly can also meet the requirements for establishing closed-loop strategies that extend the service life of product components since only connecting parts, and possibly the housing, are being destroyed while the main components remain intact. In this way, it is possible to handle uncertainties during the disassembly process, such as corroded joints. In destructive disassembly, destroying separation methods like cutting are mainly applied, which are fast, efficient and economically feasible.[8] Moreover, destructive disassembly is the most common method when end-of-life products are being recycled.[13]

2.1.4. *Automation level*

The automation level can be used to classify disassembly processes. Thereby, manual, semi-automated and automated disassembly can be distinguished. Currently, disassembly operations are carried out manually. This leads to high costs, which are additionally increased if manufacturers ignore Design for Disassembly (DfD) guidelines, thus making disassembly economically infeasible. Automation is regarded as a promising approach to improve the profitability of disassembly. Moreover, automation has the potential to reduce safety risks during disassembly. In semi-automated disassembly solutions, robots and human operators perform the disassembly task in collaboration. Robots perform hazardous operations, such as removing dangerous subassemblies, or monotonous, repetitive tasks, e.g. unscrewing bolts. Human workers perform complicated disassembly tasks that are difficult to automate. Fully automated disassembly is associated with many challenges because of the wide range of uncertainties with regard to the variety or conditions of the EoL products. An automated disassembly system typically consists of three subsystems/operating modules: sensorial system, intelligent

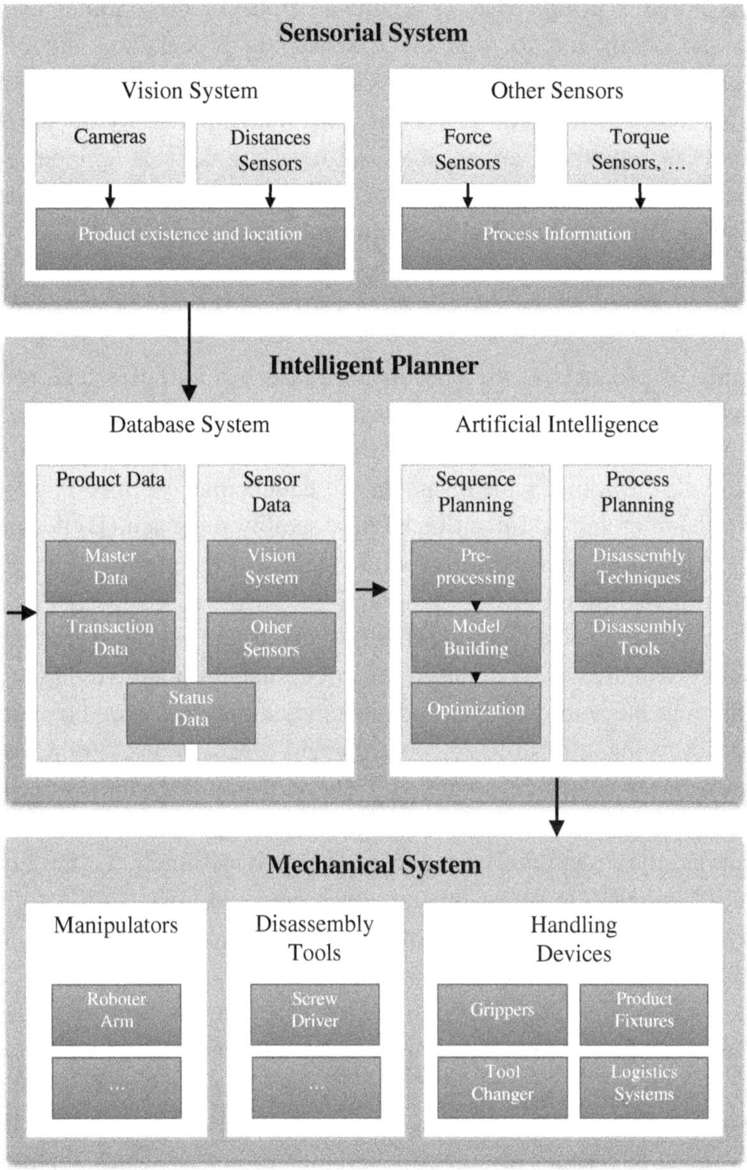

Fig. 5. Subsystems of automated disassembly systems.

planner and mechanical system (Fig. 5). The sensorial system consists of a vision system to detect the position of the products and their components and other sensors necessary for the definition of the process parameters. The intelligent planner is essential to enable product-specific decisions based on history and sensor data, which are stored in a database. It is able to find out the optimal disassembly sequence and depth and to plan the disassembly process autonomously. The mechanical system is in charge of the physical interactions with the EoL products. Its elements can be grouped into three categories: manipulators (robot arm), disassembly tools (screwdriver, cutter, etc.) and handling devices (grippers, logistics system, etc.). It is recommended to have a modular design for automated disassembly stations in order to allow flexibility for future adaptions.[8]

2.2. *Disassembly planning*

Disassembly of products at the end-of-life phase is a complex task resulting in high costs. Therefore, planning the disassembly process is essential to increase disassembly efficiency and reduce the associated costs.[7] When considering the entire life cycle of products, the disassembly sequencing is applied in two applications: (A) planning and optimization of disassembly lines and (B) design for circularity.[14] Moreover, disassembly sequencing is necessary to improve the efficiency of recycling processes.[15]

Disassembly sequence planning (DSP) is a well-known optimization problem, which has been increasingly focused on in circular economy research in recent years. Although the same product is disassembled and the same components are removed, factors that affect the outcome, such as process time, vary depending on which disassembly sequence is applied.[16] Thereby, in a large number of literature sources, the selection of the appropriate EoL strategy is not taken into account when planning the disassembly process. It is mostly assumed that the EoL option is previously known. However, this approach does not always reflect the reality, especially for large and complex products, as the condition of several components is subject to stochastic uncertainties and can often only be determined during disassembly. This is why the DSP optimization problem has been extended by the parallel consideration of the EoL choices in recently published literature.[10,11]

Zhou *et al.* break down the DSP process into three main steps (Fig. 6). First, the disassembly modes are specified (see Section 2.1). Subsequently, the disassembly is modeled.[7] Here, pre-processing is necessary, in which the precedence relationships in the product are identified. This can be done, for example, by using CAD data or by performing manual experiments. The most common disassembly modeling methods are undirected graph, disassembly tree, directed graph, AND/OR graph and Petri nets.[17] These methods can result in matrix representations that allow a relatively simple numerical implementation. The third step is the calculation of the optimal disassembly sequence. This requires an objective such as reducing costs and minimizing the environmental impact, which must be first

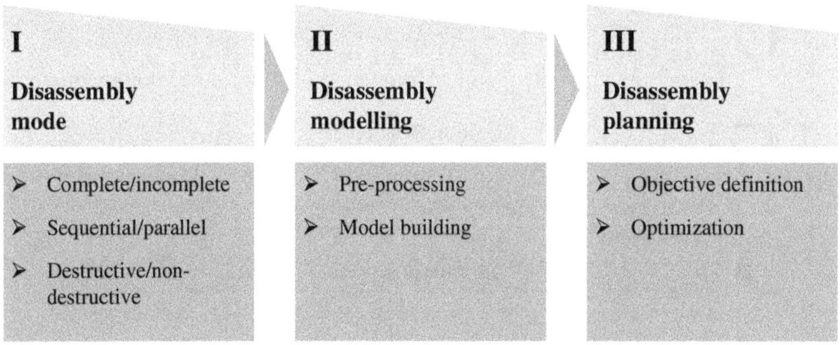

Fig. 6. Steps of disassembly sequence planning according to Ref. [7].

defined and then optimized. For optimization, mostly nature-inspired heuristic algorithms are used because DSP is an NP (nondeterministic polynomial time)-complete problem,[7] which means that linear programming and rule-based methods are time and computation-demanding. Especially genetic algorithms are a simple way to provide the best solution by operating in three steps: selection, crossover and mutation.[8] A more efficient algorithm for more than one objective is the multi-objective ant bee colony algorithm which provides Pareto solutions in less than half of the genetic algorithm's CPU time.[10] Obviously, most of the planning methods are static. Due to the unexpected uncertainties in the disassembly process, the adaptive planner is a dynamic method which trains with data from the real disassembly based on past data. Therefore, machine learning techniques are necessary to generate disassembly sequences.[8]

2.3. *Disassembly scenarios for battery systems*

Disassembly scenarios for electric vehicle batteries depend on the battery design parameters such as variety and complexity and the return quantities. This results in different disassembly scenarios over the next years. While the battery design determines the technological feasibility of effective disassembly solutions through modular product design, the reduction of variant diversity and the standardization of connection techniques, the economic efficiency is driven by the number of batteries returned. The mentioned two factors show a rapid development over time. Thereby, the guidelines of Design for Disassembly (DfD) are becoming increasingly important during the product development phase, and the number of electric cars is rising very rapidly worldwide, which will result in a high number of EoL batteries in the next few years. Disassembly, in general, consists of pre-, main and post-processes.[18] The pre-processes are preparatory steps for the disassembly

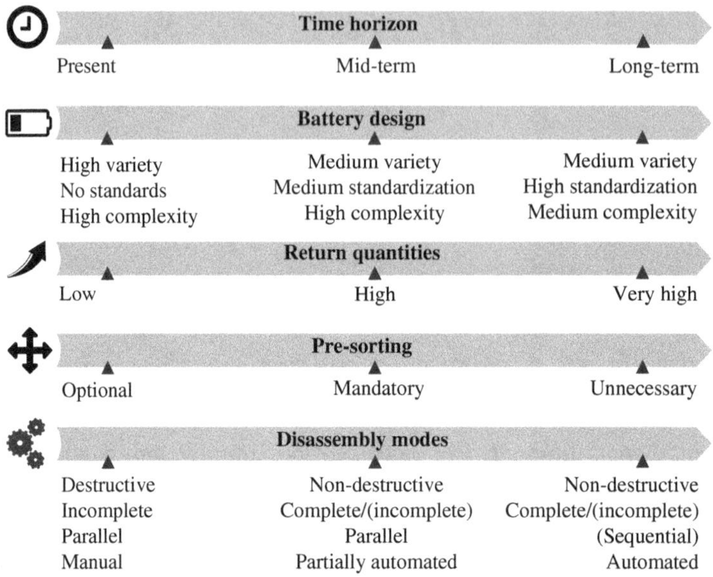

Fig. 7. Battery disassembly scenarios.

operations and include, for example, sorting, cleaning, testing and disassembly sequence planning. During the main process, parts and sub-assemblies are removed using multiple functions, such as clamping, handling, detection, unscrewing and cutting. Post-processes are, for instance, testing, cleaning and packaging. In this section, disassembly refers to the main disassembly steps. Hermmann *et al.*[19] defined three disassembly scenarios using a morphological box. In the following, these scenarios are described based on the mentioned reference. In doing so, some properties of the scenarios will be updated. The general conditions and main characteristics of the different scenarios are illustrated in Fig. 7. At present, battery design is characterized by great variety, lack of standardization and high complexity. In addition, the return volumes are low. That is why disassembly is performed entirely manually without needing any special disassembly equipment in a single disassembly place. This means that there are no transport operations within the disassembly net before the disassembly is completely finished. The transport to and from the disassembly stations is done by means of forklifts. Destructive separation methods are more time-efficient and cost-effective because nearly all batteries are recycled. Furthermore, destructive disassembly is often needed due to the complex product design. Pre-sorting is currently optional and can reduce the disassembly times. The mid-term can be considered as a transition phase. Both standardized and non-standardized batteries will reach the end-of-life stage and need to be processed. That is why pre-sorting is mandatory. Disassembly will be performed semi-automatically in human–robot collaboration. Intralogistics will change. This involves the use of workpiece carriers and roll conveyors within the disassembly net. The use of forklifts to transport batteries to disassembly stations will still be maintained. In the long term, it is assumed that all batteries will be built according to standardized design guidelines. Therefore, the disassembly can be fully automated. Automated disassembly systems are performance-oriented and can handle high volumes of all standard battery systems, which makes pre-sorting unnecessary. However, post-sorting will be required to ensure high qualitative recycling. Non-destructive disassembly will be economically feasible and will be needed primarily to enable alternative circular economy strategies at the component level, such as reuse. Material transport in disassembly factories will be carried out the same way as in the previous transition scenario with the possibility of replacing forklifts with automated guided vehicles (AGVs).

3. Disassembly Factories for Battery Systems

The following sections provide an overview of disassembly factories. First, classification criteria are outlined. Then, disassembly networks are discussed in a general way. We also present a flexible disassembly layout for battery systems up to the cell level.

3.1. *Classification*

Disassembly factories are required elements for the end-of-lie phase of products, and they specify the material flows in a circular economy.[20] Thereby, they can perform two main functions: the service function by offering services such as disposal tasks, and the producer function, such as providing disassembled components suitable for further use.[18] Three types of disassembly factories can be identified depending on the relationships between manufacturers and disassembly companies. If the product manufacturer carries out the disassembly itself, this is known as manufacturer disassembly. Moreover, manufacturers can commission specialized factories to disassemble their products at the end-of-life stage, which is called cooperative disassembly. The third disassembly model does not involve explicit cooperation. In this case, the disassembly is performed anonymously.

3.2. *Networks*

When planning processing networks for retired battery systems, the optimized choice of factory locations is of high importance. In this respect, the focus should be on minimizing transport costs,[15] as they represent 40–50% of the total costs when the batteries are recycled.[21] This is also an important issue to consider when setting up and placing disassembly factories, which define the main material flow in a circular economy. Therefore, network optimization is mandatory when planning the infrastructure for battery disassembly. This issue is included in the research project DeMoBat, which is funded by the state Baden-Württemberg in Germany and mainly targets automated battery disassembly solutions. Different scenarios can be taken into consideration where the service areas of the disassembly factories are configured differently. Consequently, factories will be dimensioned according to the chosen network strategy. For example, we can look at the possible scenarios for the EoL batteries in Baden-Württemberg. Five scenarios are possible: local, regional, national, European and international. In the local scenario, the batteries will be collected and disassembled where they are generated. For instance, the batteries from the city of Stuttgart will be disassembled in a factory located in the Stuttgart area. This will result in many small disassembly sites throughout the state, resulting in decentralized disassembly infrastructures with very low collection transport costs. In the regional scenario, few larger disassembly factories would be built. Furthermore, the batteries can be processed in a national or even European disassembly network.[6] In this way, they might not be disassembled in Baden-Württemberg at all. The last scenario assumes that the batteries will be treated outside the European Union (EU). This scenario could have economic benefits. Nevertheless, it does not clearly contribute to ensuring the resilience of supply chains of critical battery raw materials in the EU.

3.3. *Layouts*

Considering the challenges for battery disassembly factories in the future, we propose the following layout for disassembly nets in which batteries are disassembled in an automated way down to the cell level (Fig. 8). In particular, we consider the processes of testing, discharging and disassembling both at the system and module level. Immediately before and after the disassembly net, precisely dimensioned warehouses or buffers — 1 and 10 — are placed to enable fluent disassembly. This is done according to the just-in-time principle. In this way, the safety risks resulting from battery storage in the disassembly plant can be minimized. The batteries are brought to the disassembly net by forklifts or AGVs. Thereby, a certain set of the transport fleet orbits the disassembly net and picks up collected sub-assemblies. The material flow within the disassembly net is carried out by means of differently dimensioned roll conveyors, which can transport entire battery systems, modules or cells. The disassembly process starts when the

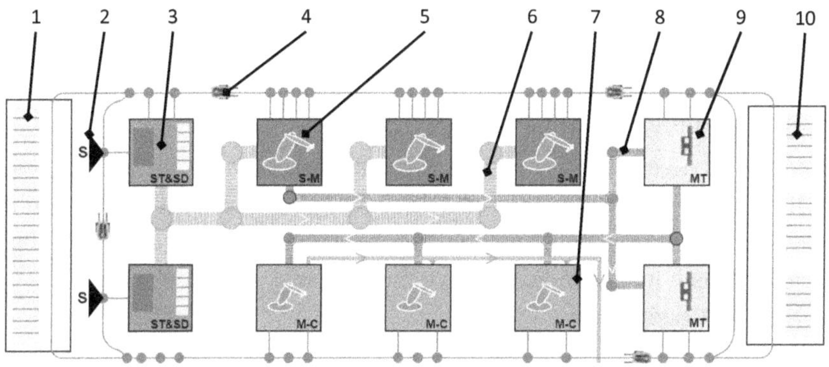

Fig. 8. Disassembly net configuration in disassembly factory for battery systems.

Notes: 1. Input warehouse; 2. Disassembly net entry; 3. Station: System testing and discharging; 4. AGV; 5. Station: Disassembling Systems to module level; 6. Roll conveyor for battery systems; 7. Station: Cell extraction from modules; 8. Roll conveyor for battery modules; 9. Station: Module testing and discharging; 10. Output warehouse.

batteries are transported to the station — 3 — where they will be first tested. If they are suitable for reuse and only need repair operations or if they pose any safety risks for further disassembly, they will not be forwarded and will be rejected immediately. For these two cases, two collection points are designed at this station. After testing, the next route for the battery can be determined, and a discharge option can be selected depending on the optimal end-of-life strategy, which can be specified using multiobjective optimization techniques implemented in the intelligent disassembly planner, see Fig. 5. Afterward, the battery is transported to one of the disassembly stations — 5 — arranged in a matrix form. These stations can be supplied independently and can be configured differently in order to reduce investment costs. For example, the first station must handle only deeply discharged batteries, while the second one can operate independently of the state of charge. However, both of them can only disassemble the most common and widespread batteries. This can reduce the flexibility degree, which is associated with high costs. The high system flexibility can then be offered by the third station. Thereby, all stations are closed cells and are equipped with air cleaners to prevent possible hazards caused by battery outgassing. The stations are also equipped with other safety devices such as firefighting systems. During disassembly, disassembled parts will be pre-sorted locally during the disassembly process and then placed in specially designed containers or workpiece carriers to the warehouse — 10. Apart from modules that will be maintained in the disassembly net for further treatment, the rest of the disassembly products from a battery can be classified into four categories: (a) mechanical scrap (screws, module holders, etc.), (b) electrical components (BMS, cables, etc.), (c) housing bottoms and

(d) housing covers. That is why we have defined four collection points at the disassembly stations to the module level. Next, the removed modules are taken to one of the stations — 9 — where they are tested and discharged if they need to be recycled. Reusable or hazardous modules are collected at this station by the circulating transport fleet. The remaining modules are brought to stations — 7 — where they are further disassembled. The battery cells leave the disassembly net by means of a roll conveyor, while other accumulated mechanic and electrical products such as housing and cables are moved to the output warehouse — 10 — by the transport fleet.

4. Summary

Disassembly is a crucial technology to implement circular economy solutions. In this chapter, we discussed the high importance of disassembly for battery systems. We then described the disassembly modes and presented disassembly scenarios for batteries. Moreover, we gave an overview of disassembly factories and proposed a layout for a flexible and automated disassembly net, which allows the implementation of diverse circular economy strategies for battery systems.

Today, disassembly represents a bottleneck process in this context. Therefore, implementing disassembly networks for battery systems is of particular significance for a sustainable future e-mobility. However, disassembly factories will face several challenges to ensure their competitiveness in the long term. Handling the following challenges will be essential: the wide range of battery variants, the increasing return volumes, the uncertainties regarding quality, quantity and timing of returns, the short life cycle of cell technologies and the diversity of circular economy strategies.

References

1. Wegener K, Chen WH, Dietrich F, Dröder K, Kara S. Robot assisted disassembly for the recycling of electric vehicle batteries. Procedia CIRP. 2015;29:716–21.
2. Tan WJ, Chin CMM, Garg A, Gao L. A hybrid disassembly framework for disassembly of electric vehicle batteries. Int J Energy Res. 2021;45(5), S. 8073–82. DOI: 10.1002/er.6364.
3. Gerlitz E, Greifenstein M, Hofmann J, Fleischer J. Analysis of the variety of lithium-ion battery modules and the challenges for an agile automated disassembly system. Procedia CIRP. 2021;96:175–80.
4. Wegener K, Andrew S, Raatz A, Dröder K, Herrmann C. Disassembly of electric vehicle batteries using the example of the Audi Q5 hybrid system. Procedia CIRP. 2014;23:155–60.
5. Jin X. A review on end-of-life battery management: challenges, modeling, and solution methods. In: Li Z, *et al.* (Hg.) 2016 — advances in battery manufacturing. pp. 79–98. Hoboken, NJ, USA John Wiley & Sons, Inc. DOI:10.1002/9781119060741.

6. acatech/Circular Economy Initiative Deutschland/SYSTEMIQ. Resource-efficient battery life cycles: driving electric mobility with the circular economy; 2020 Oct 06. Available from: https://www.acatech.de/publikation/ressourcenschonende-batteriekreislaeufe/
7. Zhou Z, Liu J, Pham DT, Xu W, Ramirez FJ, Ji C, *et al.* Disassembly sequence planning: recent developments and future trends. Proceedings of the Institution of Mechanical Engineers. Part B J Eng Manuf. 2019;233(5):1450–71.
8. Vongbunyong S, Chen WH. Disassembly automation. Cham: Springer International Publishing; 2015.
9. Ren Y, Zhang C, Zhao F, Xiao H, Tian G. An asynchronous parallel disassembly planning based on genetic algorithm. Eur J Oper Res. 2018;269(2):647–60.
10. Ren Y, Jin H, Zhao F, Qu T, Meng L, Zhang C, *et al.* A multiobjective disassembly planning for value recovery and energy conservation from end-of-life products. IEEE Trans Automat Sci Eng. 2020:18:1–13. DOI: 10.1109/TASE.2020.2987391.
11. Alfaro-Algaba M, Ramirez FJ. Techno-economic and environmental disassembly planning of lithium-ion electric vehicle battery packs for remanufacturing. Resour Conserv Recycl. 2020;154:104461.
12. Feldmann K, Trautner S, Meedt O. Innovative disassembly strategies based on flexible partial destructive tools. *Ann Rev Control.* 1999;23:159–64. DOI: 10.1016/S1367-5788(99)90079-2.
13. Chung C, Peng Q. Evolutionary sequence planning for selective disassembly in de-manufacturing. Int J Comput Integr Manuf. 2006;19(3):278–86.
14. Lambert AJD. Disassembly sequencing: a survey. Int J Prod Res. 2003;41(16):3721–59.
15. Yun L, Linh D, Shui L, Peng X, Garg A, Le MLP, *et al.* Metallurgical and mechanical methods for recycling of lithium-ion battery pack for electric vehicles. Resour Conserv Recycl. 2018;136:198–208.
16. Ke Q, Zhang P, Zhang L, Song S. Electric vehicle battery disassembly sequence planning based on frame-subgroup structure combined with genetic algorithm. Front Mech Eng. 2020;6. DOI: https://doi.org/10.3389/fmech.2020.576642, https://www.frontiersin.org/articles/10.3389/fmech.2020.576642/full
17. Ren Y, Tian G, Zhao F, Yu D, Zhang C. Selective cooperative disassembly planning based on multiobjective discrete artificial bee colony algorithm. Eng Appl Artif Intell. 2017;64:415–31.
18. Huber A, Marx-Gómez J. Ansatz zur Ermittlung der Wiedereinsetzbarkeiten von Demontageerzeugnissen im Rahmen reaktiver Demontageplanung; [cited 2021 Mar 4]. Available from: http://enviroinfo.isep.at/UI%20200/HuberA%20MarxG%F3mezJ.ath.pdf
19. Herrmann C, Raatz A, Andrew S, Schmitt J. Scenario-based development of disassembly systems for automotive lithium ion battery systems. AMR. 2014;907:391–401. DOI: https://doi.org/10.4028/www.scientific.net/AMR.907.391
20. Ciupek M. Beitrag zur simulationsgestützten Planung von Demontagefabriken für Elektro- und Elektronikaltgeräte. Stuttgart: Fraunhofer-IRB-Verl.; 2006. (Berichte aus dem Produktionstechnischen Zentrum Berlin).
21. Beaudet A, Larouche F, Amouzegar K, Bouchard P, Zaghib K. Key challenges and opportunities for recycling electric vehicle battery materials. Sustainability. 2020;12(14):5837.

https://doi.org/10.1142/9789811245626_0020

Chapter 19

Battery Recycling — Material Recovery Processes

Steffen Kiemel*, Inga Landwehr† and Max Weeber‡
Fraunhofer Institute for Manufacturing Engineering and Automation IPA
Nobelstraße 12, Stuttgart, Germany
**steffen.kiemel@ipa.fraunhofer.de*
†inga.landwehr@ipa.fraunhofer.de
‡max.weeber@ipa.fraunhofer.de

This chapter presents battery recycling including its current state and future challenges. The focus is on introducing the main process steps, including thermal, mechanical, pyro-, hydro- and biometallurgical processing, required to recover the precious materials from a battery system at its end of life. This chapter also provides an extensive summary of globally available recycling capacities clustered according to the different recycling processes and with reference to the respective companies. An excursus on the topic of direct recycling, its prospects and constraints concludes this chapter and gives an outlook on the emerging processing route for recycling Li-Ion batteries.

1. Introduction

As described in Chapter 18, the disassembly of battery systems is necessary for the pre-treatment in subsequent processing steps. Today, this is mostly handled manually. However, with the expected increase of EoL batteries in the near future, concepts for automation of single steps or the entire dismantling process are currently under development. Related research projects are, among others, AutoBattRec and DeMoBat. Yet, there is still a big gap between the TRL of manual and automated pre-treatments. Reversely, automation bears the potential of increasing the turnover of disassembly workshops significantly as soon as a homogenous input of EoL batteries can be ensured, or the automated infrastructure is equipped with effective self-learning algorithms (compare Fig. 1). Also, the exposure of workers to electrical risks is reduced by using robots instead of a human workforce.[1] The dismantling process itself is a sequence of the generalized seven steps visualized in Fig. 1. Thereby, the draining of the coolant is optional as not all battery systems include liquid thermal management systems.

The resulting peripheral components (housing, cables, BMS, etc.) are directed toward conventional recycling routes, e.g. for copper, aluminum, stainless steel and electronic scrap.[2,3] Most battery disassembly workshops operating on an industrial scale end their processes at the module level. However, there are concepts for subsequent processing of the obtained battery components that necessitate further dismantling down to the cell level. The modules and/or cells are either introduced in material recovery processes on-site or are transferred to centralized large-scale processing plants (e.g. Umicore). This reduces the transportation costs, which equal around 40–50% of the costs for recycling Li-ion batteries.[1] The mass which has to be transported is reduced significantly. Furthermore, if only material fractions or the black mass is transferred, the safety risks associated with the transportation of battery systems reduce significantly.

In general, there exist five approaches for the processing of battery modules/cells (compare Section 2). The individual recycling operation of most companies comprises a selection of one or more of these five sub-processes. Industrially applied combinations are described in Section 3. Furthermore, current recycling operations are mentioned and located. Besides, a short introduction to the concept of direct recycling is provided.

2. General Types of Li-Ion Battery Processings

The following subsections describe the five types of processings for Li-ion batteries that aim to recover material fractions and are currently applied at an industrial scale. Fig. 1 illustrates a condensed overview.

2.1. *Thermal processing*

Thermal processing is applied in order to evaporate the electrolyte and incinerate plastic components contained in the modules/cells. This is especially relevant for the removal of the binder between the cathode material and aluminum foil within the cell (mostly PVDF). Depending on the material, which has to be thermally decomposed, the necessary temperatures vary significantly. The distillation or pyrolysis is conducted in retort furnaces (e.g. Accurec), rotary kilns (e.g. Mitsubishi Heavy Industries) or shaft furnaces (e.g. Umicore).[4,5] After thermal treatment, the cathode material can be easily separated by further mechanical processing.[6] The recovery of the evaporated electrolyte is possible by introducing the gaseous

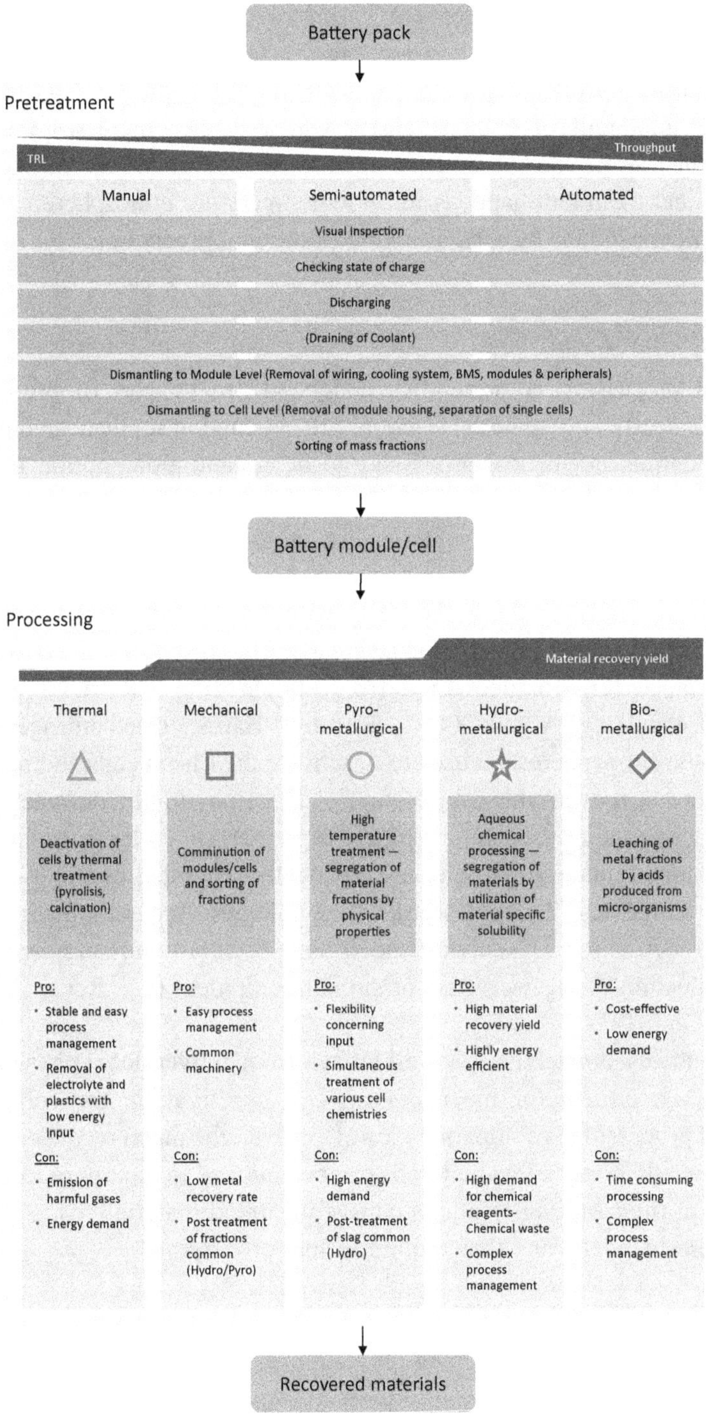

Fig. 1. Pre-treatment and processing of EoL LIB.

fraction in a downstream condenser.[5] The stable and easy process management enables large-scale treatment of Li-ion batteries. However, the decomposition of materials during pyrolysis emits ecologically harmful gases.[6] This necessitates end-of-pipe treatment of exhaust gases, which further increases the resource demand of the process and consequently reduces its ecological benefit. Occasionally, the mechanical treatment happens in a low-temperature atmosphere.[7] Such cryogenic treatment can also be seen as a kind of thermal processing.

2.2. *Mechanical processing*

Mechanical processing is sometimes understood to be part of pre-treatment.[6] However, as nearly pure material fractions are obtained in sophisticated processing routes, we define mechanical processing to be a stand-alone option for material recovery processes. The mechanical processing starts with the comminution of the modules/cells. Commercial operations use large-scale shredders like hammer/impact mills (e.g. Recupyl, Retriev, Accurec). AEA Technology applies a different approach by initially opening the cells with lasers and mechanical scissors. Afterward, the fractions are treated in a shredder.[5] In most cases, the crushing takes place in a protected environment. Liquid (e.g. lithium brine — Retriev Technologies) or inert gaseous (e.g. Argon/CO_2 — Recupyl, Batrec; dried nitrogen — AEA Technologies) atmospheres are used to deactivate the Li-ion cells and reduce reactivity in order to prevent the risks of fire.[7,5,8] Electrohydraulic defragmentation is another way to conduct mechanical processing currently under research.[9,10]

The obtained material particles are physically separated by the use of vibrating sieving tables (e.g. Recupyl, Retriev Technologies, Accurec), magnetic sifting (e.g. Recupyl, Accurec, Düsenfeld GmbH, Sony & Sumitomo) and density sifting by the application of zigzag sifters or similar equipment (e.g. Recupyl, Accurec, Düsenfeld GmbH).[7,5]

Hence, the used machinery is well known from conventional physical separation processes. Except for the necessary cell deactivation, the complexity of mechanical processing is comparably low. However, the metal recovery rate is low due to the fact that only a few materials can be obtained in their pure form. A large proportion of the resulting fractions consist of mixed materials which cannot be separated mechanically and thus require further processing.[6]

2.3. *Pyrometallurgical processing*

Pyrometallurgical processing applies high and ultra-high temperatures to reduce metals in shaft furnaces (e.g. Umicore) or electric arc furnaces (e.g. Accurec).[11,12] The individual melting points of the metals contained in battery modules/cells enable the segregation of material fractions with similar physical properties in the form of alloys.[5] The thermal treatment (compare Section 2.1.) is sometimes seen as part of the pyrometallurgical processing as it is mostly conducted before or even in the same furnace (Umicore uses during its Val'Eas process a three-stage shaft furnace, where the first two stages equal thermal processing).[4] Thereby, the organic compounds act as fuel and reduce the necessary resource input for reaching the higher temperatures of the subsequent stages. Moreover, the graphite and aluminum included in the cells act as reducing agents and further reduce the need for fuel.[4,11] On the contrary, this decreases the amount of recoverable material significantly. For example, aluminum and other metals like manganese and lithium mostly end up in the slag.[4] Although there exist approaches for extracting lithium from the slag,[13] most of it is currently downgraded due to its use as a filler material in the building industry.[11] Furthermore, the melting of the metallic parts requires high temperatures, resulting in high energy demand and consequently causes adverse effects on the environment.[2] Direct emissions even intensify the negative environmental impact during pyrometallurgical processing due to fuel combustion and thermal decomposition of materials.[2] Toxic emissions are treated with end-of-pipe technologies, which additionally increases the demand for energy.[13] Moreover, the obtained alloys need to be further processed. Most commercial operations thus affiliate hydrometallurgical processing to their pyrometallurgical treatment.[4]

Advantages of the pyrometallurgical processing are its process robustness and its high throughput.[6,11] The mature process does not necessarily need an elaborated pre-treatment and allows the treatment of different cell chemistries simultaneously.[4,11]

2.4. *Hydrometallurgical processing*

Hydrometallurgical processing necessitates pre-treatment in terms of mechanical or pyrometallurgical processing.[11] The obtained alloys or granulated material fractions, which mainly involve the cathode materials, are introduced into aqueous chemical processes.[4] Most commercial operations start with the leaching of the introduced material to dissolve the contained metals.[4,5] Sulfuric acid is a common leaching agent used by various companies (e.g. Umicore, Recupyl).[5] Although this enables a high material yield, the use of inorganic acids results in significant environmental hazards. Potential alternatives, which are investigated extensively, are organic acids.[6] After dissolving, further purification and separation are achieved, for example, by solvent extraction (e.g. Nickelhütte Aue, Glencore) and/or chemical precipitation (e.g. Recupyl, Accurec, Düsenfeld, Nickelhütte Aue, Glencore), ion-exchange (e.g. Glencore) and liquid–liquid reactions.[4,14] Depending on the processing route, the chemical agents used as well as the preceding processing applied, varying derivates of materials (Co, Ni, Mn, Cu, Al, Li and Fe) can be obtained.

Some of these materials cannot be recovered by the processing routes described earlier (compare Sections 2.1.–2.3.) or at least not in the same quality and purity.[6] This pertains especially to the valuable cathode materials. Compared to pyrometallurgical processing, the hydrometallurgical treatment is highly energy-efficient.[11,15] This results in lower emissions of greenhouse gases.[4,2,1] Although hydrometallurgical processing ensures the highest metal recovery yield at present, it has to be stated that it cannot stand by itself as pre-processing in terms of mechanical and/or pyrometallurgical treatment of the modules/cells is required.[1]

The disadvantage of this type of processing is the necessity for considerable amounts of chemical agents.[1,6] This results in high costs and large amounts of chemical waste. Partially, these have to be treated as hazardous waste.[4,11] Although performing significantly better than pyrometallurgy in the ecological impact category of global warming potential (IPCC (2013): GHG-emissions), hydrometallurgical processing results in a negative ecological impact concerning marine eutrophication.[3] The process control is rather complex, and the parameters of most processes need to be adjusted depending on the cathode chemistry treated. This increases the necessary effort for sorting and pre-processing of the introduced material fractions.[4]

2.5. *Biometallurgical processing*

Biometallurgical processing is mentioned as a promising prospect for future processing of Li-ion battery recycling. The treatment itself is an alteration of the hydrometallurgical process. The aim is to reduce the environmental impact of material recycling. The acids necessary for leaching the metal fractions are not fed from external sources but are produced by microbial organisms.[6] The produced leaching agents depend on the species of introduced microorganisms (bacteria = inorganic acids, fungi = organic acids).[4] Following the bioleaching, conventional hydrometallurgical processes as described in Section 2.4. are conducted subsequently (e.g. chemical precipitation).[4] While not being applied on an industrial scale, biometallurgical processing has been investigated extensively in recent research. Zheng *et al.* provide a detailed overview of the respective research activities.[6]

Compared to conventional hydrometallurgical processing, bioleaching is more cost-effective and is characterized by a low energy demand.[6] The yield of recovered materials is at a similar level as conventional treatment. However, the metabolization of the microorganisms needs time. Furthermore, the cultivation of the organisms is complex.[6] Contamination of the population needs to be prevented as it either kills the bacteria or fungi strain or impairs its performance.[4] Hence, bioleaching is currently associated with a larger effort than conventional processing.

3. Industrially Applied Processing Routes for Li-Ion Battery Recycling

Figure 2 illustrates the processing routes for recycling Li-ion batteries that are currently applied on an industrial scale. It can be observed that several strategies for recovering materials from EoL batteries are applied. However, all industrial processes are based on a combination of the general procedures described in Section 2. While there are variations in applied process parameters and partially the sequence of the sub-processes, the selection of general procedures is similar for several companies. The process flow diagrams shown in Fig. 2 are abstracted for reasons of clarity and comprehensibly. Specific industrial processing routes of individual companies recover various derivates of the listed elements. Concerning its great potential and significance in current research activities, direct recycling is described in detail in Section 4 as another alteration which allows the direct recovery of active material. Several companies only apply one of the presented processes. However, some of these businesses transfer their obtained alloys, powders or other derivates to subcontractors for further processing (e.g. VW group passes its obtained black powder to business partners for hydrometallurgical treatment).[16,17] Despite our careful review of the process flows from different companies presented in the literature, we make no claim with regard to their completeness. Considering these circumstances and the facts, that high recovery rates can only be achieved by a sensible combination of processes, Fig. 2 neglects the illustration of single process routes. The overall result of the literature review concerning commercially applied processing routes and the respective companies is illustrated in Figs. 3 and 5. Literature sources are stated for information concerning annual recycling capacities, the location of the facilities and the selection of general recycling procedures (compare Table 1). Companies referred to in the literature, but with no active web presence and/or current information concerning their battery recycling activity are neglected (e.g. AEA, Pilagest and many more). Furthermore, some companies were acquired by others during the past few years (e.g. Xstrata = Glencore, Recupyl = TES-AMM, Toxco = Retriev). As shown in Fig. 4, battery recycling is mostly conducted in Europe (especially Germany,

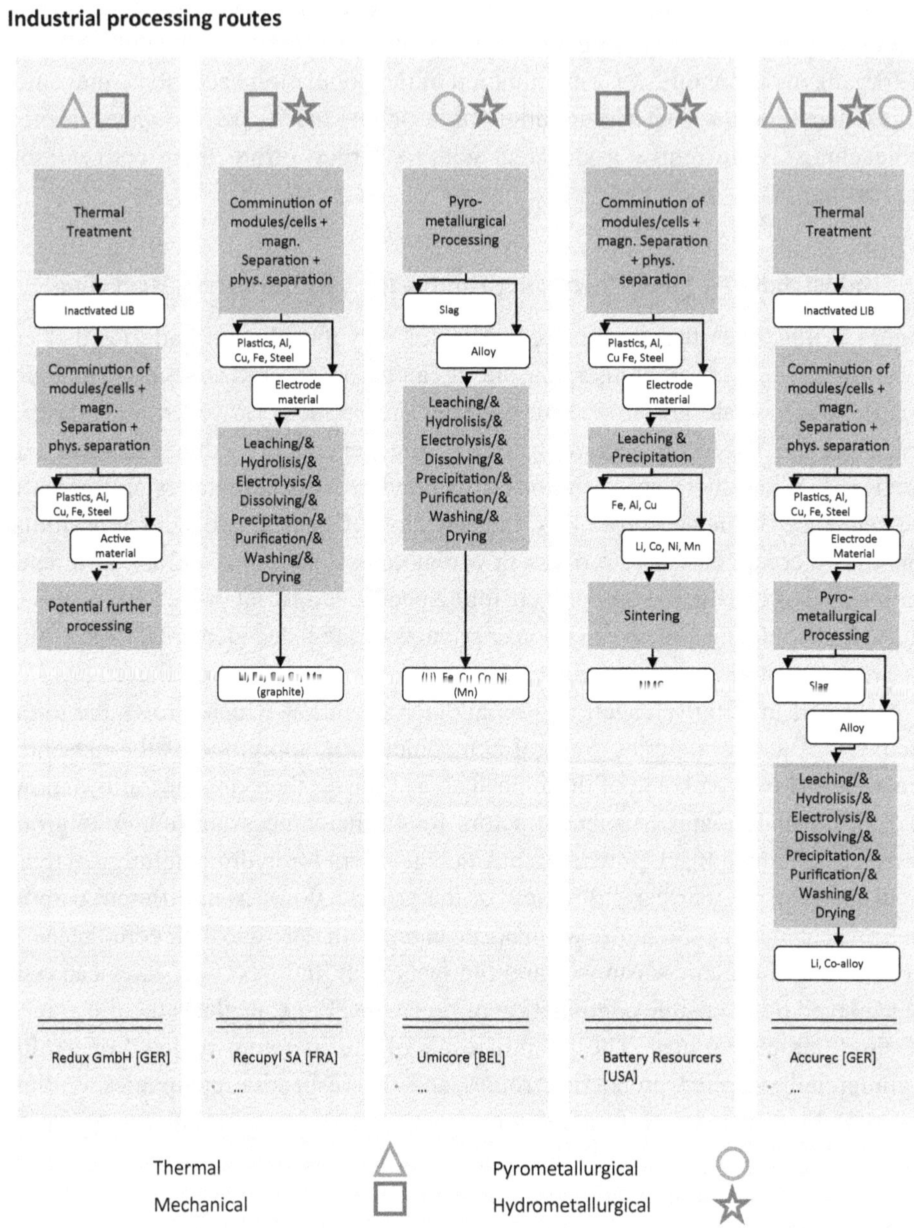

Fig. 2. Industrial processing routes for LIB recycling.

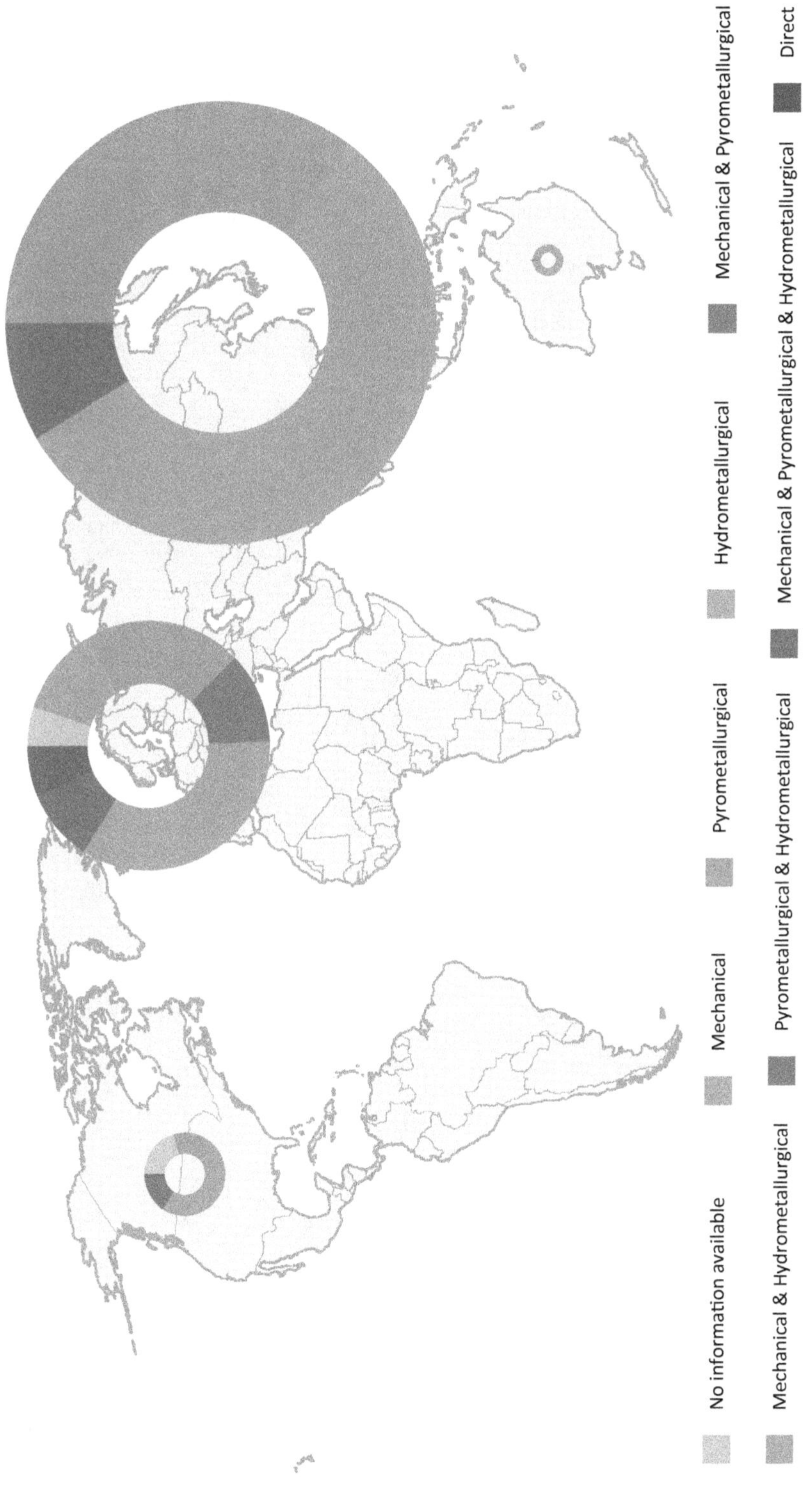

Fig. 3. Shares of LIB recycling capacity.

Table 1. Li-ion battery recycling operations (literature sources).

Company Name	Country	Capacity [t/a]	Processing: Thermal	Mechanical	Hydro	Pyro	Direct	Source
Accurec	Germany	6,000	x	x	x	x		[1, 5, 8, 11, 18, 19]
AkkuSer Oy	Finland	4,000		x	x			[1, 7, 8, 11, 18, 19, 20]
American Manganese	Canada	3,650		x	x			[4, 11, 21–23]
Anhua Taisen Recycling Technology Co. Ltd.	China	20,000		x	x			[11, 22]
BatRec Industrie AG	Switzerland	200	x	x	x			[1, 5, 8, 11, 18]
Battery Resourcers	USA	500		x	x			[4, 7, 11, 24]
Chemetall bzw. Rockwood Lithium GmbH	Germany	5,000						[8, 11]
Clean Earth (AERC Recycling Solutions)	USA					x		[11]
Dowa Eco-System Co., Ltd.	Japan	1,000				x		[1, 8, 18, 25]
Düsenfeld GmbH	Germany	3,000		x	x			[5, 11]
Envirostream Australia	Australia	3,000		x	x			[22]
ERAMET	France	20,000				x		[18, 26]
ERLOS	Germany	1,500					x	[22]
EST Energetics	Germany					x		[27]
Fortum Corporation	Finland	3,000		x	x			[28, 29]
Fuoshan Bangpu Ni/Co High-Tech Co.	China	3,600		x	x			[8, 11, 30, 19]
GEM Co., Ltd.	China	30,000		x	x			[4, 18, 31]
GHTECH	China	1,800		x	x			[4, 11, 32–34]
Glencore (former Xstrata)	Canada, Norway	7,000			x	x		[1, 11, 18, 19]
Huayou Cobalt New Material Co Ltd.	China	60,000				x		[11, 35, 36]
Hunan Brunp Recycling Technology Co.	China	10,000			x	x		[1, 11, 18, 31, 37]
Hydrovolt (from 2021)	Norway	8,000		x				[38]
Inmetco	USA	6,000				x		[4, 7, 8, 11]
JX Nippon Mining and Metals Co.	Japan	5,000			x	x		[8, 11, 18, 39]
Kyburz	Switzerland							[40]
Li-Cycle Corp.	Canada, USA	10,000		x	x			[11, 41, 42]
Lithion Recycling	Canada	200			x			[11, 43]
Neometals Ltd.	Canada	pilot		x	x			[44–46]
Nickelhütte Aue GmbH	Germany				x	x		[1, 11, 47]
Nippon Recycle Center Corp.	Japan					x		[11, 18, 48]

Table 1. (*Continued*)

Company Name	Country	Capacity [t/a]	Processing: Thermal	Mechanical	Hydro	Pyro	Direct	Source
Northvolt AB	Sweden	100		x	x			[49, 50]
OnTo Technology	USA			x	x		x	[1, 4 11]
Primobius (Neometals & SMS Group)	Germany	20,000		x	x			[11, 51]
PROMESA GmbH & Co. KG	Germany	3,200		x				[11, 52]
REDUX GmbH	Germany	10,000	x	x		x		[11, 53, 54]
Retriev Technologies	Canada, USA	4,500		x	x			[1, 5, 11, 18, 19]
REVATECH	Belgium	3,000		x	x			[8, 11]
S.N.A.M.	France	300	x	x	x	x		[1, 8, 11, 18]
SARP Industries (Euro Dieuze Industrie)	France	200			x			[8, 11, 18]
Shenzhen Highpower Technology Co., Ltd.	China				x	x		[4, 11, 55]
Shenzhen Tele Battery Recycle Co., Ltd	China	6,000		x	x			[11, 56]
Sitrasa	Mexico			x				[57]
SMCC (SungEel MCC Americas)	USA	3,000						[4, 58–60]
Soundon New Energy Tech. Co. Ltd.	China							[11, 61]
Sumitomo Sony	Japan	150			x	x		[1, 8, 11, 18, 19, 62]
SungEel HiTech	South Korea	24,000		x	x			[4, 11, 63, 64]
TES-AMM (+Recupyl SAS)	Singapore	14,110		x	x			[4, 4, 5, 11, 65, 66]
Umicore	Belgium	7,000			x	x		[1, 5, 8, 11, 18]
Veolia Environmental STS, L.L.C.	USA					x		[11, 53]
VW Group Components	Germany	1,500		x				[17, 67]

France, Scandinavia and Benelux), the USA, Canada, China and Japan. Figure 3 illustrates the LIB recycling capacity per continent. Scattered facilities for Li-ion battery recycling can be found in South Korea, Singapore, Australia and Mexico. From the identified 50 recycling operations, 30 apply a hydrometallurgical treatment, whereas only 17 conduct pyrometallurgical treatment during their battery

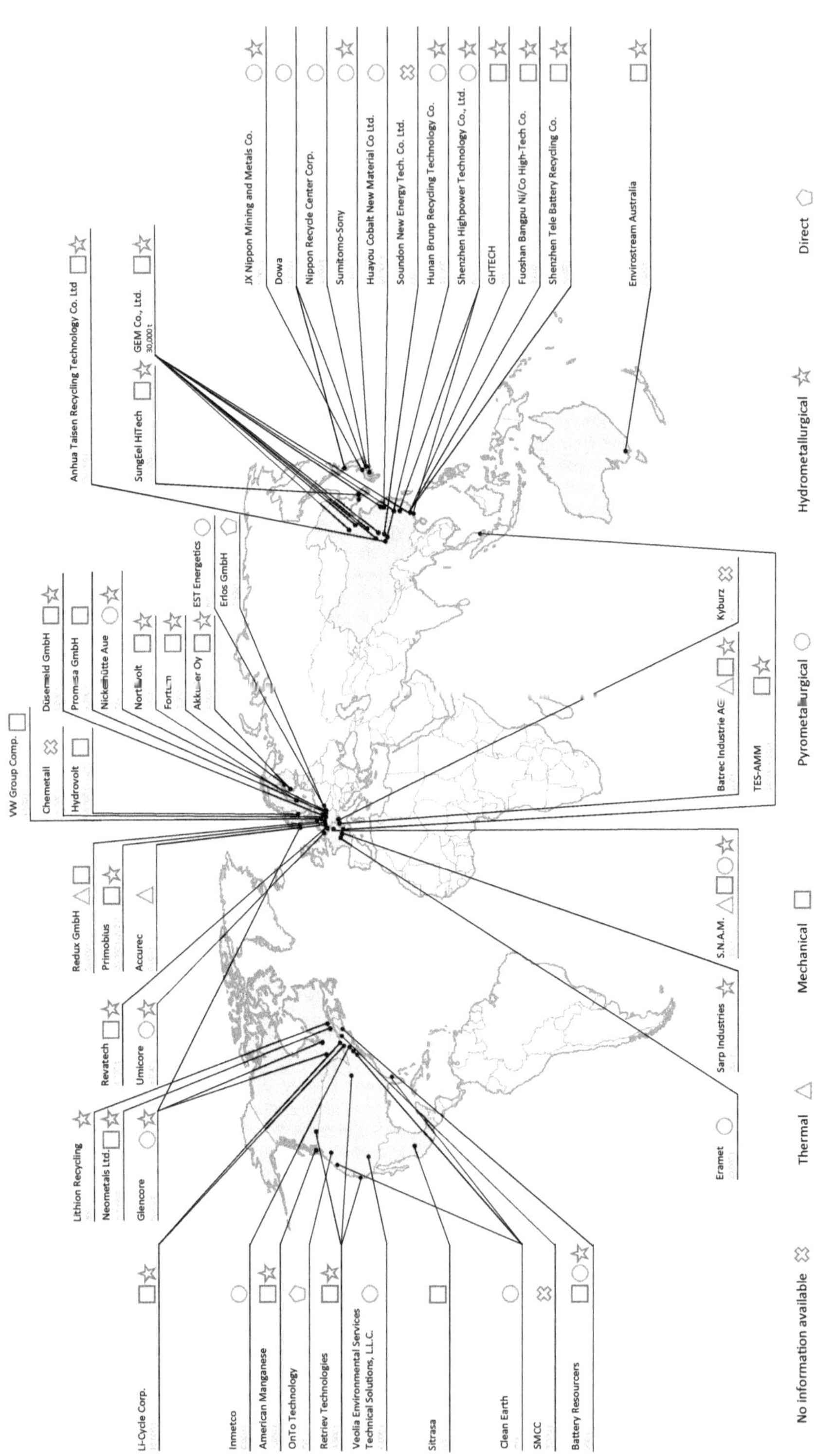

Fig. 4. Li-ion battery recycling operations.

recycling process. While the selection of processing steps of European- and American companies is rather heterogeneous, mechanical in combination with hydrometallurgical treatment is dominant in China. While being applied all around the world, pyrometallurgical processing is especially prevalent in Japan. Currently, direct recycling is solely conducted in Germany and the United States.

4. Excursus: Direct Recycling as an Emerging Processing Route for Recycling Li-Ion Batteries

Direct recycling of battery electrodes describes a relatively new possibility in which the structure and morphology of the active material are preserved.[68–70] This is made possible by mechanically separating the anode, cathode and separator components at an early stage of the process and recycling them separately.[7,69,71] Thus, the active materials do not have to be extracted from a complex solution or alloy by hydrometallurgical or pyrometallurgical methods. Since pyrometallurgical digestion is very energy-intensive, the use of direct recycling can save about 50% of the associated greenhouse gas emissions per kWh of battery capacity.[72]

The quality of the recyclate depends on the cells introduced.[1] For a high recyclate quality, the processing of uniform cell chemistries is required.[73] If the material is manufacturing scrap, for example, from electrode production, it can be returned directly to cell production after recovery. Recovered materials from used battery cells usually show aging characteristics, i.e. cathode materials has a changed lithium stoichiometry compared to fresh material or impurities based on side reactions during cycling. Properties of the original active material can be almost recreated by additional process steps. This treatment is called revitalization, which is accomplished, for instance, by adding of LiOH and subsequent heating steps. Material analyses show that this secondary raw material is structurally indistinguishable from unused material.[69,74] However, laboratory-scale studies show that only a small amount of recyclate can be added without affecting battery performance.[75] The quality should be significantly improved through optimized process control so that cells with 100% secondary raw material could also be used on an industrial scale in the future.

One possible and simple process for direct recycling is shown in Fig. 5. After cell opening, mechanical separation of the anode, cathode and separator is performed. Subsequently, the electrode material is washed off the surface of the current collector with a slightly heated water jet. The current collector is consequently dried and can be recycled. In addition to the active material, the particle-containing mixture also contains a conductive additive (e.g. carbon black) and electrolyte and binder residues. These can be removed in a further washing step.[71,69] Additional purification steps, as mentioned before, improve the material quality.

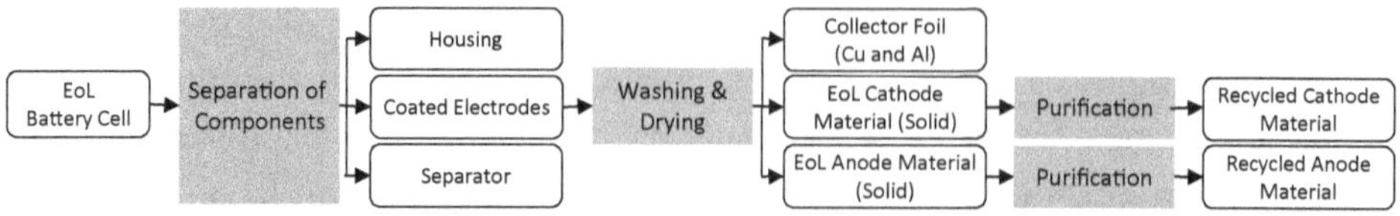

Fig. 5. Flow diagram of a direct recycling process — material recovery by electrode washing.

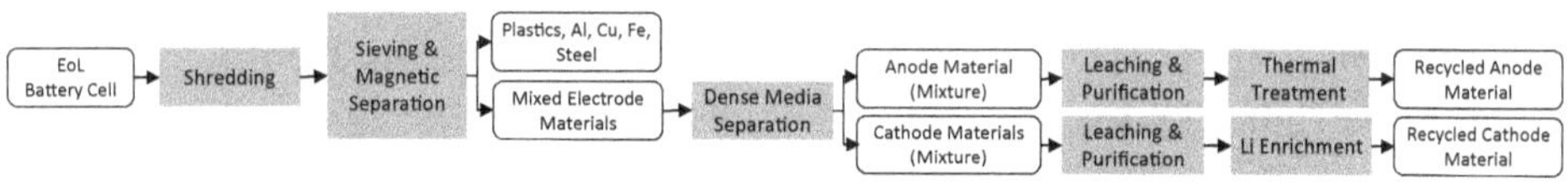

Fig. 6. Flow diagram of a direct recycling process — material recovery after cell shreddering.

During cell design, the recyclability of cells should already have been considered. Through water-based electrode manufacturing and the use of water-soluble binders, this process can be made even more efficient with respect to energy use. Additionally, the quality of the recovered cathode material can be increased.[69]

Another opportunity of direct material recycling is similar to the classic recycling routes (ref. Fig. 6). The cell will be shredded first, followed by removing magnetic components, for example, from the cell housing. Subsequently, anode material (graphite) and cathode material are separated from each other by a dense media separation. Since this process depends not only on density but also on particle size, the cathode material can be found in the anode mixture and vice versa. These impurities, for example, are leached and removed from the relevant fractions without changing the structure of the given active materials. Using a pyrometallurgical step, both anode and cathode materials can be recycled. This process is used by OnTo Technology.[7]

5. Summary

There exist various approaches in recovering materials from end-of-life lithium-ion batteries. While all recycling strategies necessitate an initial dismantling of the battery system, the subsequent extraction steps are a company-specific combination of the four most common processing schemes (mechanical, thermal, pyrometallurgical and hydrometallurgical processing). Relevant clusters of battery recycling facilities can be found in descending order in Asia, Europe and Northern America. While there is a distinct focus on mechanical and hydrometallurgical recycling as well as pyrometallurgical recycling in Asia, companies in Northern America and especially in Europe are more heterogeneous in the choice of their approaches in recycling LIB.

Until now, only a few companies focused on direct recycling, which aims at retrieving active material whose structure and morphology are preserved. While having the potential of reducing the ecological impact of the life cycle of lithium-ion batteries, this approach in recycling is rather new and thus is currently under research.

References

1. Larouche F, Tedjar F, Amouzegar K, Houlachi G, Bouchard P, Demopoulos GP, *et al.* Progress and status of hydrometallurgical and direct recycling of li-ion batteries and beyond. Materials (Basel). 2020;13(3):801.
2. Kiemel S, Koller J, Karabelli D, Singh S, Full J, Weeber M, *et al.* Untersuchung: Kreislaufstrategien für Batteriesysteme in Baden-Württemberg: Endbericht. Fraunhofer Institut für Produktionstechnik und Automatisierung (IPA): Stuttgart; 2020.
3. Buchert M, Jenseit W, Merz C, Schüler D. Verbundprojekt: Entwicklung eines realisierbaren Recyclingkonzepts für die Hochleistungsbatterien zukünftiger Elektrofahrzeuge — LiBRi: Teilprojekt: LCA der Recyclingverfahren. Darmstadt: Öko-Institut e.V.; 2011. Available from: https://www.oeko.de/presse/archiv-pressemeldungen/presse-detailseite/2012/oekobilanzen-fuer-zwei-verfahren-zum-recycling-von-lithium-ionen-batterien-aus-elektrofahrzeugen
4. Chen M, Ma X, Chen B, Arsenault R, Karlson P, Simon N, *et al.* Recycling end-of-life electric vehicle lithium-ion batteries. Joule. 2019;3(11):2622–46.
5. Becker J, Beverungen D, Winter M, Menne S, editors. Umwidmung und Weiterverwendung von Traktionsbatterien. Wiesbaden: Springer Fachmedien Wiesbaden; 2019.
6. Zheng X, Zhu Z, Lin X, Zhang Y, He Y, Cao H, *et al.* A mini-review on metal recycling from spent lithium ion batteries. Engineering. 2018;4(3):361–70.
7. Velázquez-Martínez O, Valio J, Santasalo-Aarnio A, Reuter M, Serna-Guerrero R. A critical review of lithium-ion battery recycling processes from a circular economy perspective. Batteries. 2019;5(4):68. Available from: https://www.mdpi.com/2313-0105/5/4/68
8. Knights B, Saloojee F. Lithium battery recycling — Keeping the future fully charged: green economy research Report No. 1, Green Fund. Midrand; 2015.
9. Berg N, Eisert S. Bessere Verwertungsmöglichkeiten von Verbundmaterialien mittels Schockwellentechnologie. Deutsches Zentrum für Luft- und Raumfahrt (DLR): Freiberg, Germany; 2016. (Ressourceneffizienz vor Ort). Available from: http://docplayer.org/53757986-Bessere-verwertungsmoeglichkeiten-von-verbundmaterialien-mittels-schockwellentechnologie.html
10. ImpulsTec GmbH. Zerkleinerung mittels Schockwellen: Das Verfahren von ImpulsTec setzt sich in der Recyclingindustrie durch; 2020. Available from: https://eu-recycling.com/Archive/28013
11. Werner D, Peuker UA, Mütze T. Recycling chain for spent lithium-ion batteries. Metals. 2020;10(3):316.
12. Georgi-Maschler T, Friedrich B, Weyhe R, Heegn H, Rutz M. Development of a recycling process for li-ion batteries. J Power Sources. 2012;207:173–82.
13. Buchert M, Sutter J. Stand und Perspektiven des Recyclings von Lithium-Ionen-Batterien aus der Elektromobilität: Synthesepapier erstellt im Rahmen des vom Bundesministerium für Umwelt, Naturschutz und nukleare Sicherheit geförderten Verbundvorhabens MERCATOR "Material Effizientes Recycling für die Circular Economy von Automobilspeichern durch Technologie ohne Reststoffe". Darmstadt: Öko-Institut e.V.; 2020. Available from: https://www.oeko.de/publikationen/p-details/stand-und-perspektiven-des-recyclings-von-lithium-ionen-batterien-aus-der-elektromobilitaet

14. Brückner L, Frank J, Elwert T. Industrial recycling of lithium-ion batteries — A critical review of metallurgical process routes. Metals. 2020;10(8):1107.
15. Mayyas A, Steward D, Mann M. The case for recycling: overview and challenges in the material supply chain for automotive li-ion batteries. Sustainable Mater Technol. 2019;19:e00087.
16. Schaal S. Volkswagen Group Components nimmt Pilot-Recyclinganlage in Betrieb; 2021. Available from: https://www.electrive.net/2021/01/29/volkswagen-group-components-nimmt-pilot-recyclinganlage-in-betrieb/
17. Volkswagen AG. Lithium to lithium, manganese to manganese; 2020. Available from: https://www.volkswagen-newsroom.com/en/stories/lithium-to-lithium-manganese-to-manganese-4662
18. Lebedeva N, Di Persio F, Boon-Brett L. Lithium ion battery value chain and related opportunities for Europe. Petten; 2016. Available from: https://ec.europa.eu/jrc/en/publication/eur-scientific-and-technical-research-reports/lithium-ion-battery-value-chain-and-related-opportunities-europe
19. Maiyalagan T, Elumalai P. Rechargeable lithium-ion batteries: trends and progress in electric vehicles. Boca Raton: CRC Press; 2020.
20. AkkuSer Oy. AkkuSer Oy — Sustainable battery recycling; 2021. Available from: https://www.akkuser.fi/en/company/
21. Jacoby M. It's time to get serious about recycling lithium-ion batteries. Chem Eng News. 2019;97(28). Available from: https://cen.acs.org/materials/energy-storage/time-serious-recycling-lithium/97/i28
22. Agarwal S, Rosina M. Lithium-ion battery recycling market & technology trends 2020: Market and Technology Report 2020 [Sample]: Yole Development; 2020. Available from: https://www.google.com/url?sa=t&rct=j&q=&esrc=s&source=web&cd=&ved=2ahUKEwiDu-r449ruAhWgwAIHHWWpCtc4ChAWMAJ6BAgGEAI&url=https%3A%2F%2Fwww.i-micronews.com%2F%3Faction%3Ddownload-attachment%26file_id%3D129114%26source_id%3D128950%26key%3De27b34e2933009dcf4914f041f3e615b%26type%3Dsample&usg=AOvVaw1OMN5iX4WpxGxaXKHNmIlG
23. Jung JC-Y, Chow N, Warkentin DD, Chen K, Melashvili M, Meseldzija Z, *et al.* Experimental study on recycling of spent lithium-ion battery cathode materials. J Electrochem Soc. 2020;167(16):160558.
24. Battery Resourcers. Battery resourcers pilot plant online; 2018. Available from: https://www.batteryresourcers.com/news
25. Dowa Eco-System Co., Ltd. Dowa Eco-System increases processing of lithium-ion batteries and enables both safe treating and efficient metal recycling; 2018. Available from: https://www.dowa-eco.co.jp/en/news/archive/20181217.html
26. Recycling Today Staff. Eramet acquiring French metal recycler; 2009. Available from: https://www.recyclingtoday.com/article/umicore_battery_belgium/
27. EST Energetics. Informationen zum Unternehmen; 2019. Available from: https://www.est-energetics.com/de/downloads__51/
28. Fortum Corporation. Fortum to acquire Crisolteq, a recycling specialist of valuable metals in batteries; 2020. Available from: https://www.fortum.com/media/2020/01/fortum-acquire-crisolteq-recycling-specialist-valuable-metals-batteries
29. Fortum Corporation. Fortum expands its EV battery recycling operations with a new mechanical processing plant in Finland; 2021. Available from: https://www.fortum.com/media/2021/01/fortum-expands-its-ev-battery-recycling-operations-new-mechanical-processing-plant-finland
30. Company-List.org. Bangpu Nickel & Cobalt Technique Co., Ltd.: Company Details; 2020. Available from: https://www.company-list.org/bangpu_nickel_cobalt_technique_co_ltd.html

31. Gu F, Guo J, Yao X, Summers PA, Widijatmoko SD, Hall P. An investigation of the current status of recycling spent lithium-ion batteries from consumer electronics in China. J Cleaner Prod. 2017;161:765–80.
32. Guangdong Guanghua Sci-Tech Co., Ltd. About us: brief introduction; 2021. Available from: https://www.ghtech.com/Eabout/about_100000000298674.html
33. Large Custom Lithium Ion Battery Pack. Only 5 companies are listed in the compliance list of power battery recycling market; 2019. Available from: https://www.large.net/news/7pu43n3.html
34. Guangdong Guanghua Sci-Tech Co., Ltd. Comprehensive utilization of traction batteries; 2018. Available from: https://www.ghtech.com/Eapplication/recovery_100000000297454.html
35. Zhejiang Huayou Cobalt. About us; 2021. Available from: http://en.huayou.com/about.html?introId=29
36. Tanaka Precious Metals. How China is cornering the lithium-ion cell recycling market; 2019. Available from: https://tanaka-preciousmetals.com/en/elements/news-cred-20191128-04/
37. Lee A. Blast at Chinese recycler shows battery supply chain risks: Bloomberg L.P.; 2021. Available from: https://www.bloombergquint.com/business/blast-at-chinese-recycler-shows-battery-supply-chain-risks
38. Breivik Ø. Hydro and Northvolt launch joint venture to enable electric vehicle battery recycling in Norway; 2020. Available from: https://www.hydro.com/en/media/news/2020/hydro-and-northvolt-launch-joint-venture-to-enable-electric-vehicle-battery-recycling-in-norway/
39. Suzuki T, Fukao K. Rare metal miners dig into used batteries for lithium and cobalt: Recycled resources to fill shortages in an electric car future: Nikkei Inc.; 2018. Available from: https://asia.nikkei.com/Business/Business-trends/Rare-metal-miners-dig-into-used-batteries-for-lithium-and-cobalt
40. Randall C. Kyburz opens battery recycling facility in Zurich; 2020. Available from: https://www.electrive.com/2020/09/07/kyburz-opens-battery-recycling-facility-in-zurich/
41. Li-Cycle Corp. Making lithium-ion batteries a truly circular and sustainable product: in Progress. & Milestones Reached.; 2021. Available from: https://li-cycle.com/about/
42. Fernandes L. Li-cycle announces commercial lithium-ion battery recycling plant now operational in Rochester, New York: New spoke is the largest capacity lithium-ion battery recycling facility in North America: Li-Cycle Corp.; 2020. Available from: https://li-cycle.com/news/li-cycle-announces-commercial-lithium-ion-battery-recycling-plant-now-operational-in-rochester-new-york/
43. Lithion recycling. Lithion recycling's lithium-ion battery recycling plant development; 2019. Available from: https://www.lithionrecycling.com/lithium-ion-battery-recycling-plant/
44. NS Energy. Neometals starts lithium battery recycling pilot plant; 2019. Available from: https://www.nsenergybusiness.com/news/neometals-starts-lithium-battery-recycling-pilot-plant/
45. Maisch M. Neometals commissions li-ion battery recycling pilot plant; 2019. Available from: https://www.pv-magazine-australia.com/2019/02/14/neometals-commissions-li-ion-battery-recycling-pilot-plant/
46. Neometals Ltd. Batterie-Recycling; n.d. Available from: https://www.neometals.com.au/de/unsere-projekte/core-projects/recycling/
47. Nickelhütte Aue GmbH. Wir verwerten Lithium/NiMh-Batterien aus der Automobilindustrie; 2021. Available from: https://nickelhuette-aue.de/de/services/batterierecycling
48. Nippon Recycle Center Corp. Business overview: Recycling of rechargeable batteries; 2021. Available from: https://www.recycle21.co.jp/recycle-e/service/pro.html
49. Schaal S. "Revolt": Northvolt steigt in Batterie-Recycling ein; 2019. Available from: https://www.electrive.net/2019/12/13/revolt-northvolt-steigt-in-batterie-recycling-ein/
50. Northvolt AB. Closing the loop on batteries; 2021. Available from: https://northvolt.com/loop

51. Sagermann T. Primobius unterzeichnet MoU mit InoBat über den Betrieb einer kommerziellen Recyclinganlage für Lithium-Ionen-Batterien in Osteuropa: SMS Group; 2020. Available from: https://www.sms-group.com/de/presse-medien/pressemitteilungen/presseinformationen/primobius-unterzeichnet-mou-mit-inobat-ueber-den-betrieb-einer-kommerziellen-recyclinganlage-fuer-lithium-ionen-batterien-in-osteuropa-1544
52. Promesa GmbH & Co. KG. Promesa Unternehmenspräsentation: Sichere Verwertung von Li-Batteriesystemen; n.d. Available from: https://vvg.eu/unternehmen/downloads
53. Bernardes A, Espinosa D, Tenório J. Recycling of batteries: A review of current processes and technologies. J Power Sources. 2004;130(1–2):291–8.
54. Redux GmbH. Geschichte; 2017. Available from: https://www.redux-recycling.com/de/unternehmen/geschichte/
55. Shenzhen Highpower Technology Co., Ltd. Recycling & re-utilization: recycling project; 2021. Available from: https://www.highpowertech.com/products-recycling
56. Tele Battery Recycle Co., Ltd. Products and services; 2021. Available from: http://www.telerecycle.com/
57. Sistemas de Tratamiento Ambiental S.A. de C.V. Treatment and recycling of batteries and fluorescent lamps; n.d. Available from: http://www.sitrasa.com/SITRASA/Treatment_and_recycling_of_batteries_and_fluorescent_lamps.html
58. SMCC Recycling. Metallica Commodities Corp. and SungEel HiTech Announce Joint Venture; 2018. Available from: https://www.prnewswire.com/news-releases/metallica-commodities-corp-and-sungeel-hitech-announce-joint-venture-to-redefine-north-american-metals-recycling-and-energy-markets-300622736.html
59. SMCC Recycling. SungEel MCC Americas; 2021. Available from: https://tracxn.com/d/companies/smccrecycling.com
60. Recycling Technology Worldwide. SungEel MCC Americas announces lithium-ion battery recycling plant location in New York State; 2018. Available from: https://www.recovery-worldwide.com/en/news/sungeel-mcc-americas-announces-lithium-ion-battery-recycling-plant-location-in-new-york-state_3274245.html
61. Soundon New Energy Technology Co. Ltd. Recycle secondary utilization; n.d. Available from: http://en.e-soundon.com/index/Ecology/Recycling
62. Sumitomo Metal Mining Co., Ltd. Utilizing unused resources: achieving Japan's first "battery to battery" recycling; n.d. Available from: https://www.smm.co.jp/en/sustainability/activity_highlights/article_15/
63. South China Morning Post. SungEel HiTech leads the way in eco-friendly recycling of lithium-ion batteries; 2020. Available from: https://www.scmp.com/country-reports/country-reports/topics/south-korea-business-report-2020/article/3079850/sungeel
64. DGAP. EcoGraf Limited: Agreement with leading lithium-ion battery recycler located in South Korea; 2020. Available from: https://www.dgap.de/dgap/News/corporate/ecograf-limited-agreement-with-leading-lithiumion-battery-recycler-located-south-korea/?newsID=1403111
65. TES-AMM. TES to acquire assets of Recupyl SAS: Acquisition of Recupyl's assets accelerates TES's move into the battery processing market in Europe; 2018. Available from: https://www.tes-amm.com/press-release/tes-to-acquire-assets-of-recupyl-sas-2
66. TES-AMM. Commercial battery recycling helps your business and the planet; 2021. Available from: https://www.tes-amm.com/battery-recycling
67. Schaal S. Volkswagen Group Components nimmt Pilot-Recyclinganlage in Betrieb; 2021. Available from: https://www.electrive.net/2021/01/29/volkswagen-group-components-nimmt-pilot-recyclinganlage-in-betrieb/
68. Li Z. A cost-effective lithium-ion battery direct recycling process. Meet Abstr. 2018.

69. Li J, Lu Y, Yang T, Ge D, Wood DL, Li Z. Water-based electrode manufacturing and direct recycling of lithium-ion battery electrodes-a green and sustainable manufacturing system. iScience. 2020 [cited 2021 Jan 26];23(5):101081.
70. Sloop SE, Trevey JE, Gaines L, Lerner MM, Xu W. Advances in direct recycling of lithium-ion electrode materials. ECS Trans. 2018 [cited 2021 Feb 24];85(13):397–403.
71. Schmidt M, Göckerlitz M, Paesold RD, Göbel F, Wienold H, Göckeritz M, *et al.* Inventors; Weck Poller holding GMBH [DE]; WKS Technik GMBH [DE]. Recycling Method for Lithium Ion Batteries. EP3563446 (B1). 2019 Nov 6.
72. Vekic N. Lithium-Ionen-Batterien Für Die Elektromobilität: Status, Zukunftsperspektiven, Recycling. Thinktank Industrielle Ressourcenstrategie; 2020 März [cited 2021 Feb 23].
73. Thompson DL, Hartley JM, Lambert SM, Shiref M, Harper GDJ, Kendrick E, *et al.* The importance of design in lithium ion battery recycling — A critical review. Green Chem. 2020 [cited 2021 Mar 12];22(22):7585–603.
74. Sloop SE, Crandon L, Allen M, Lerner MM, Zhang H, Sirisaksoontorn W, *et al.* Cathode healing methods for recycling of lithium-ion batteries. Sustainable Mater Technol. 2019 [cited 2021 Jan 27];22:e00113.
75. Shi Y, Zhang M, Meng YS, Chen Z. Ambient-pressure relithiation of degraded Li x Ni 0.5 Co 0.2 Mn 0.3 O 2 ($0 < x < 1$) via eutectic solutions for direct regeneration of lithium-ion battery cathodes. Adv Energy Mater. 2019 [cited 2021 Feb 24];9(20):1900454.

https://doi.org/10.1142/9789811245626_0021

Chapter 20

Business Models for Second-Life Battery Applications

Sonja Rosenberg*, Sandra Huster†, Simon Glöser-Chahoud‡ and Frank Schultmann¶

Karlsruhe Institute of Technology, Institute for Industrial Production, Hertzstraße 16 76131 Karlsruhe, Germany

**sonja.rosenberg@kit.edu*

†sandra.huster@kit.edu

‡simon.gloeser-chahoud@kit.edu

¶frank.schultmann@kit.edu

Business Models (BMs) can be seen as a conceptual plan for a business to achieve economic profitability. The demand and supply side in combination with the business's internal value creation has to be investigated when describing a BM. For Second Use Batteries (2uBs), potential BMs must be developed to succeed in the market. In connection with 2uBs, we distinguish between BMs that envisage reuse in vehicles (remanufacturing) and those that use batteries for non-automotive purposes (repurposing).

Remanufacturing BMs is likely to involve the vehicle producer since they have an overview and partial control over the demand and supply side through their brand workshops. Remanufactured batteries could serve as cost-efficient spare parts for private customers or second-hand car dealers who want to upgrade their cars.

Repurposed batteries can be combined to industrial or large-scale storage systems. Industrial battery systems might, for example, be used as electric vehicle (EV) fast-charging stations or for providing power in case of power outages. Large-scale battery storage serves as stationary short-term energy storage. Possible applications are, for instance, peak shaving or power arbitrage. Vehicles' Original Equipment Manufacturers (OEMs) and energy suppliers likely collaborate in BMs focusing on repurposing.

1. Introduction

This chapter takes a closer look at possible BMs that can be established for reintroducing reassembled batteries into a second use phase. Currently, 2uBs are barely found on the market. Nevertheless, with the expected growth in EV registration, the supply of used electric vehicle batteries (EVBs) will increase within the next decade. Attention to potential BMs is needed, as today's design and production decisions might influence the economic and ecological success of 2uB BMs.

We start this section with a short introduction to the meaning and usage of the term business model in Section 2 to understand our reasoning within the further explanation. Business models that operate with Second Use Batteries that serve the same purpose as in their first life, in other words, as a component of EVs, are presented in Section 3. Thereafter, we address repurposing business models in Section 4. The BMs are discussed within their sections; therefore, we close the chapter with a short conclusion and summary.

2. Background and Usage of Business Models

Osterwalder *et al.*[1] see a business model as a "conceptional tool" with a set of dimensions that allow conveying the logic of doing business in the case of an individual company. Various authors describe fundamental elements of BMs.[2–7] A rough classification is made into factors belonging to a business's supply side, the demand side and the internal value creation.[3,7] At the same time, the financial or capital perspective needs to be included.[8] According to Amit and Zott,[9] the description of BMs must be adjusted to the purpose of use. Meanwhile, Morris *et al.*[10] discuss that in a strategic perspective, which is most important for businesses not existing yet, BMs should be focused on critical components such as information about stakeholder interactions and alliances and the value creation to describe business positioning in a market. According to Geissdorfer *et al.*,[11] circular business models include circular economy thinking, such as "cycling, extending and/or dematerializing material and energy loops to reduce [...] waste [...]". These circular economy elements are presented in detail in Chapter 15.

We take a conceptual, strategic view of businesses working and trading with 2uBs. We are presenting and discussing BM that can be enrolled soon if sufficient used EV batteries exist.

Therefore, we look at a business's supply and demand side and discuss which products and services with 2uBs can match both sides. Furthermore, we glance at the revenue generation schemes. However, we omit the internal processes and resources of disassembling and remanufacturing or repurposing, as they are presented in Chapters 18 and 19.

When one considers that the described BMs do not exist today, we look further at external factors, such as barriers for our proposed BM, and discuss how they may be overcome.

3. Business Model with Remanufacturing of Batteries

Business Models that focus on reusing remanufactured batteries in automotive applications presume a large EV market, preferably consisting of a first- and second-hand market. These two markets together allow potential business opportunities in the aftermarket or the used car trade business.

As Germany has a large first- and second-hand vehicle market compared with other European countries,[12] it seems to be an appropriate location for starting a potential business for 2uBs. Thus, our further argumentation builds on the German market. German premium OEMs have been actively involved in the remanufacturing of conventional vehicle components such as engines and have provided them to the aftermarket.[13] Moreover, some OEMs announced circular-economy-related strategies for the decades to come.[14–17] As the remanufacturing business can be a cornerstone of the circular economy (cf. Chapter 19), OEMs or their 1st-tier suppliers seem likely to take a central position in a circular battery BM. One possible outline of such a BM is depicted in Fig. 1, omitting other parts of the OEM business, such as production and selling of new vehicles.

As shown in Fig. 1, the OEMs engage in collaboration with various partners. They can use existing supplier relations for procuring necessary material for remanufacturing, while recovered waste material during remanufacturing can be forwarded to recycling plants. Of tremendous significance are the partners that supply and demand the remanufacturing service. Here, the existing brand network of dealerships and workshops (often belonging together) is used. Workshops are the center of introducing remanufactured spare parts into the aftermarket. Battery spare parts can either be single battery modules or complete battery systems if they have failed. A replacement of a complete battery system might arise as part of guarantee claims.

Exchanges because of a battery guarantee claim should only account for a small percentile of the overall sold electric vehicles. However, the OEM is responsible for providing a replacement and might want to use remanufactured batteries if these are cheaper than new ones. Table 1 shows that OEMs currently offer an eight-year battery guarantee or a high km equivalent. The total number of guarantee claims will rise together with the number of EVs.

In conventional repairs, the customer may choose if a remanufactured or brand-new part is installed. In Germany, surveys show that customers often decide based on the price comparison and that many drivers are generally open-minded to using remanufactured spare parts.[24] Using brand workshops for repairs is common, especially as customers do not like to change their workshops.[25] Comprehensive advice from the employees of the workshops allows one to

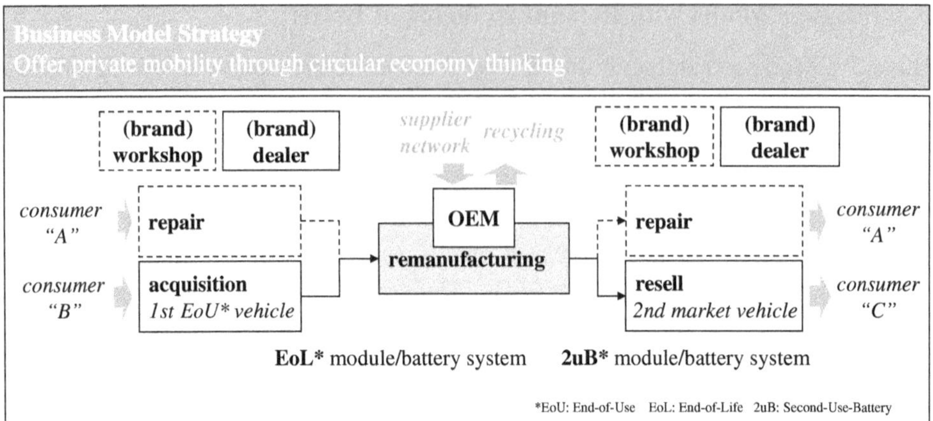

Supply
- EoL batteries from workshops or dealerships (cooperations)
- Supplier/recycling network for additional material supplies and recovery

Demand
- Aftermarket business; B2C: Private consumer requests/pays product
- Upside potential for second-hand market; B2B: dealership pays for remanufactured product

Collaboration
- Brand dealership and workshops
- Supplier network and recycling for remanufacturing itself
- Consumers have to see an added value by remanufactured product (either battery itself or in combination with vehicle)

Finance
- OEM invests in industrial remanufacturing factory (including disassembling and reassembling)
- Logistic and transport costs have to be covered (usage of existing structure with dealers)
- Prices/Cost for remanufactured batteries depend partly on quality of returned 1st EoL battery
- Private consumer pays in workshops for remanufactured product
- Dealership pays for remanufactured product and charges it to buyer of 2nd market vehicle

Fig. 1. BM with remanufacturing.

explain the possibility of using remanufactured battery spare parts to the customers.

In contrast to the aftersales business through workshops, demand for remanufactured batteries might also exist when reselling used EVs. Considering that about 30% of revenue by the German car trade is accounted for by sales of used cars by brand dealers, while new vehicle sales made up 39% in 2019, the significance of a well-functioning second-hand vehicle market is apparent.[26] Approximately seven million used vehicles were registered in Germany in 2020, which is about twice as much as new vehicles.[27] In 2020, the share of electric

Table 1. Battery guarantee conditions from various OEMs in Germany.

Vehicle/OEM	Guarantee Conditions	Guaranteed SOH	Source
Audi e-tron	160000 or 8 years	70%	[18]
BMW i3	Since 08/2019: 160000 km or 8 years	70%	[19]
Mercedes EQC	160000 km or 8 years	70%	[20]
Opel	160000 km or 8 years	70%	[21]
Tesla Model S	240000 km or 8 years	70%	[22]
VW	160000 km or 8 years	70%	[23]

vehicles of the used car market was less than 1 %, showing that this is still a niche market.[28] Nevertheless, we expect the second-hand market of EVs to grow exponentially with increased EV sales.[29,30] A closer look at the properties of used vehicle sales can demonstrate a potential market for remanufactured OEM batteries.

Second-hand vehicles offered by brand dealerships are typically sold at higher prices than in private sales, as such vehicles are often younger and of premium quality.[24,31] Tolksdorf[28] argues that the second-hand EV sales price depends on the battery quality represented by the State-of-Health (SOH) (cf. Chapter 17). Erich[32] predicts similar or even slightly higher prices for second-hand EVs than for conventional vehicles. Because the battery system accounts for the largest part of the overall vehicle price in new electric vehicles, its quality will significantly influence the vehicles' selling price.[33]

Currently, there is nearly no knowledge about the most influencing factors on customer preference when buying an EV.[12] Nevertheless, looking at conventional used vehicle buyers, reliability and purchasing price[31,34] are considered very important for second-hand car buyers. Meanwhile, range anxiety has been pointed out as a drawback for potential EV buyers.[35,36]

These arguments show that (brand) dealers will only achieve high selling prices for used EVs if the battery is of the best quality. Integrating an updated second-life battery in the form of a remanufactured battery might increase the battery's reliability and thus the electric vehicle's value. Especially for vehicles whose batteries are close to the ending of the battery guarantee or whose battery guarantee expired, remanufactured batteries can be the right choice. The remanufactured battery should come with a new battery guarantee.

It is worthwhile to mention that in some cases, second-hand vehicles of brand dealers are still of such a young age that the battery does not need to be remanufactured. In Germany, only 15% of vehicles sold by authorized dealers are more than six years old.[31,34] Despite this, EVs are less maintenance-intensive[33] and have fewer functional components that are subject to wear and tear than conventional vehicles, and older EVs with a remanufactured battery might offer more attractive resale options than comparable aged conventional vehicles.

Dealers might increase their selling profit if a remanufactured battery costs less than the vehicle's value increase by an updated battery. One must keep in mind that a dealer would trade in vehicles' original batteries with remanufactured ones. Using all existing information, including available digital knowledge about the original battery (cf. Chapter 16), helps dealers to deduce for which vehicles such an upside process is valuable. A possible cost structure for remanufacturing can be matched to the SOH condition of the 1st-life battery. Of course, fix costs, e.g. for the logistical processes, have to be covered. Branding partners and OEMs are already working together, and essential business structure and logistic workflow standards (spare part logistics) exist, which can be used for remanufactured batteries as well.[37] Thus, these costs can be lower than for completely independent partners.

One of the biggest challenges for remanufactures is securing the overall core supply, in other words, having enough but not excessive products to remanufacture to achieve economic processes.[38] An OEM with an extensive dealership network and a concept of "changing old components with new remanufactured ones" will have a good fitting between core acquisition and sold remanufactured products.

As the OEM is also responsible for the design and production of batteries, they have profound knowledge about the remanufacturing possibilities and can actively influence it.[39] A design for remanufacturing (and repair) might have a more profound position in the future design processes than in the past if it fits the overall circular economy business strategy.

Challenges and threats for such a BM can be battery specific or more routed in the general remanufacturing business. As there is only a little information about real battery degradation, it is difficult to predict how many and when batteries for remanufacturing will be available. A second significant barrier for introducing remanufactured batteries into the market can be the development of prices of new batteries, as the price of remanufactured batteries depends on the price of new batteries at the time of remanufacturing.[40] It is generally expected that prices for batteries will further decrease, but less rapidly than in the last years.[41] Thus, to be economical, remanufacturing costs have to be even lower in the future.

The continuously developing battery technology can be a further challenge for spare part supply for remanufacturing. As the supply of new spare parts is secured for a limited time after end-of-production due to supplier obligations, it might stop before the demand for remanufactured batteries ends. Therefore, reconfiguration of new battery modules and systems by using parts of several returned batteries is unavoidable. Standardization of battery modules among several vehicle types seems to be the best way to increase supply reliability by increasing the number of reconfiguration possibilities.

Additionally, remanufacturing is always accompanied by some market challenges and risks, such as competition or legal barriers.[39] The European Committee announced a proposal for a battery regulation in December 2020 covering some existing grey areas for battery remanufacturing.[42] Within the proposal, legal requirements are made for reintroducing remanufacturing batteries into the market, e.g. quality testing must be performed.[42,59] Such regulations can protect the market against remanufactured products of low quality, as they increasingly appear on the market for conventional spare parts.[39] Due to the overall OEM benefits such as a collaboration network, lower logistical costs and possible (automated) industrial processes, independent remanufacturers have higher risks for remanufacturing. They will most likely target a different customer group.

4. Business Model with Repurposed Batteries

BMs for repurposed batteries have been gaining more attention in the recent past than remanufacturing batteries for reuse in vehicle applications.[40,43–45] The overall idea is to use returned battery systems or their components in less demanding applications once they cannot be used in the vehicle anymore to lengthen the usage by cascading. Table 2 includes a collection and brief description of battery storage system (BSS) projects involving the repurposing of EVBs in Germany. According to Figgener *et al.*,[46] BSS can be distinguished into battery home storage systems (BHSSs), industrial storage systems (BISSs) and large-scale storage systems (BLSSs). BLSSs have a battery capacity of over 1 MWh or battery power of over 1 MW. In comparison, BISSs are categorized based on their capacity being between 30 kWh and 1 MWh. A BHSS has a capacity smaller than 30 kWh.[47]

As the battery storage market is intensively growing in combination with the growing amount of renewable energy, reusing second-life batteries is a promising market.[41] BHSSs are almost exclusively used in combination with renewable energy production such as photovoltaic and heat pumps in private households. Typical benefits of a BHSS are time shifts between energy generation and self-consumption or the option to feed electricity to the grid when needed.[59]

BISSs are applied in commercial or industrial applications and, for instance, used for increasing the share of renewable energy within production processes to achieve a lower CO_2 footprint for products. The CO_2 footprint of products is gaining importance due to the European Green Deal[60] and the overall higher sustainability awareness of consumers.[61] BLSSs are used to overcome curtailments by

Table 2. Collection of battery storage projects with repurposed batteries in Germany.

Type	Description	Actors	Source
BLSS	• 13 MWh capacity/13 MW power, using repurposed (and new) EVBs • 1000 EVBs from smart EQ fortwo • Used for stabilization of primary balancing power • Storage can be used as "living storage" to store new EVBs as spare parts	Daimler; The Mobility House AG, GETEC, REMONDIS SE	[48,49]
BLSS	• 2600 battery modules from 100 EV (test fleet of Active E and BMW i3) • 2 MW power and 2.8 MWh capacity • Tested and employed as energy storage of PV energy and for performance buffer for fast-charging station	BMW; Bosch; Vattenfall	[50,51]
BISS	• Repurposing as a mobile fast-charging station with storage capacity up to 360 kWh and 100 kW power • Can charge up to 4 vehicles at the same time (has DC- and AC connections) • Usable without power grid connection, e.g. at public parking lots or for events • Built with repurposed and remanufactured battery modules • End of 2020: 9 charging stations were installed; series production about to start	VW, E.ON	[52,53]
BLSS	• Energy storage of 1.9 MWh capacity with 20 EVBs from test vehicles • Used for grid stability, peak shaving and storing PV energy	Audi, BELECTRIC	[54]
BISS	• Replacement of lead–acid batteries of forklift trucks • 36 modules of an Audi e-tron are disassembled, and 24 are reassembled • The 24 modules have the same size and weight as the replaced lead–acid battery	Audi	[55,56]
BLSS	• Started in 2017: 500 used and new EVBs of BMW i3 are used as a stationary BSS • Modular layout; 700 EVBs could be integrated; expansion possible • Located on the BMW production site for storing energy from the company-owned wind turbines • Lowering CO2-print of products • Attached to the public electricity network, energy might also be sold to the market	BMW, Energy2Market GmbH	[57]

Table 2. (*Continued*)

Type	Description	Actors	Source
BISS/ BLSS	• Started June 2020: Planning of a reference 2uB storage • Reference storage is tested with different configurations and usage profiles as technical and process model • Power and capacity still unknown • Usage for energy company's PV-/wind parks planned (peak shaving/lowering curtailments) • Focus on long-term cooperation • Potential offering of BSS for other industrial companies in the future	Audi, EnBW	[58]

wind and solar energy, one of the biggest issues of renewable energy, and to stabilize the power distribution system.[62]

A potential BM for repurposing batteries, again from an OEM's perspective, is presented Fig. 2. While the OEM is in the center of the business concept and performs the repurposing, close collaboration, especially to energy suppliers, extends over the complete repurposing and second use processes. The collaboration of OEM and energy suppliers has been instrumental in the success in the existing 2uB projects depicted in Table 2. The repurposing activities are supplied with End-of-Use battery systems or modules collected from EV dismantlers and workshops.

The business's value proposition is the creation and potential operation of BISSs or BLSSs with repurposed battery components. Both product types have in common that they can be understood as individually designed applications. This is of importance for 2uBs, as repurposed modules have unique characteristics (e.g. varying SOH).

These individual qualities are one reason to exclude BHSSs in the BM as private customers are used to buying products of a precise and accurate product quality.[63] Additionally, BHSSs are expected to have shorter second usage phases than BISS applications,[64] and higher end-of-life collection costs compared to other BSS types arise due to their decentralized distribution in various households.

One potential application for industrial battery storage is its use as or with EV fast-charging stations, where it is used to buffer energy to avoid peak loads if several vehicles are fast charged simultaneously.[65] The market potential for EV charging infrastructure will grow in the upcoming years as the EV diffusion rate

Business Model Strategy

Cascading and maximum lifetime utilization of batteries

Involved Stakeholders along BM processes

(brand) workshop · *supplier network* · *recycler* · **Company "X"** · *recycler*

EoL vehicle dismantler* · **OEM** · **energy supplier** · **energy supplier** · **OEM**

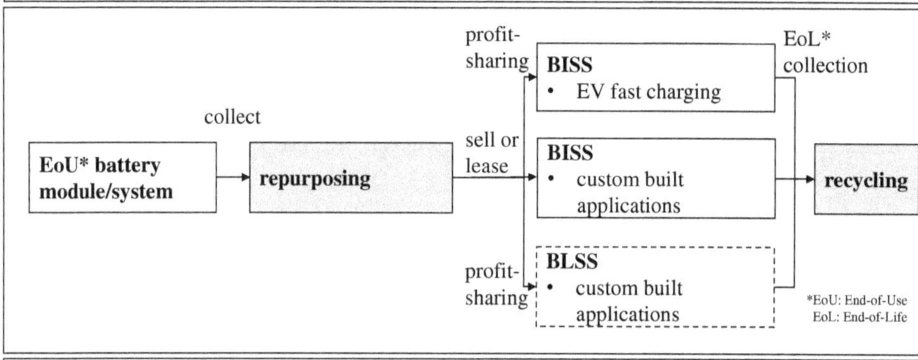

Supply

- Supply with EoU battery systems or modules through the EoL vehicle dismantler or workshops

Demand

- EV fast charging stations for extended product service system, operated by energy supplier
- Businesses that need battery storages for industrial applications
- BLSS operated by energy supplier for grid support and national energy supply

Collaboration

- OEM is BM responsible and can be in close collaboration with energy supplier (Joint-Venture possible)
- BISS are build to custom order, intensive exchange with companies necessary
- Battery collection through specialized 3 PL provider or collection network
- Supplier network for repurposing material (e.g., inverter)
- Recycling companies

Finance

Cost:

- Collection cost of EoU batteries was priced in the new-EV price by OEM
- Cost for repurposing (disassembly, reconfiguration, additional materials)
- Collection at the end of 2nd use is priced in as well as recycling

Revenue:

- EV fast charging stations are leased to energy suppliers or profit-sharing of OEM/energy supplier for received based on charging station turn-over
- BLSS energy supplier are responsible for maintenance and storage, profit-sharing of operation with OEM
- OEM sells/leases BISS to business customers, while energy supplier can offer maintenance and service of storage

Fig. 2. BM with battery repurposing.

increases. By offering charging infrastructure next to the EV, an OEM introduces further value-added services to its vehicles and can learn about the EV driver's behavior during usage. Energy suppliers can use charging services as a new revenue stream in the changing energy sector and a possibility of building decentralized, flexible short-term energy storage.[66] In the described BM, they are responsible for the energy supply, the maintenance of the repurposed BISSs for EV fast-charging stations and the digital processes for payment. Thus, next to the OEM, the energy supplier is a leading partner in the proposed business model.

BISSs can show great potential for a variety of commercial and industrial applications. Next to their use to store excess renewable energy, they might improve the resilience of companies or public buildings such as schools in case of power outages.[67] Repurposed batteries could also be used as mobile BISSs to ensure cold food supply chains, where they outperform current lead–acid batteries.[68] A further collection of BISS applications can be found in the papers of Fischhaber *et al.*[40] and Heymans *et al.*[69]

The application area for BLSSs is less diversified than for BISSs: Due to their larger sizes, they are always stationary and are used for instant peak shaving of energy demand and supply or for power arbitrage.[70] BLSSs are expected to be managed and maintained through energy suppliers as various legal regulations exist, and further regulation is in development.[71,72]

The unique product value for business customers in this BM is precisely tailored energy storage. Therefore, a comprehensive energy demand analysis should be carried out before a storage system is installed. Rallo *et al.*[64] present a calculation process for finding the best economical 2uB storage in industrial applications. Based on the energy profile of a specific use case for which a second use BISS is constructed, alternative options are calculated. If a storage configuration with a positive ROI within the expected lifetime of the 2uB storage is found, it can be considered economical.

Such an approach is reasonable when constructing BISSs and BLSSs, as several factors influence the economic performance. First and foremost, the power and capacity of the storage must be appropriate for the application. Meanwhile, the operating costs increase with the system size, and capital costs must not be neglected.[89] Due to their expandable and modular system setup, the construction of a BLSS should be based on the same guidelines.[73]

BISSs, built to a customer's requirement, can either be sold or leased. The buying or leasing company might also have a contract with an energy supplier for service and maintenance.

For BLSSs, a profit-sharing approach between the OEM, which repurposes the batteries, and an energy supplier, who operates and maintains the 2uB system, seems to be a promising manner of remuneration. Nevertheless, finding an adequate sharing of the profit will be a challenge, as the different costs of the partners must also be considered. Profit sharing can be used for the BISSs for EV fast charging, although different profit shares might be applicable.

As pointed out in some cases, using repurposed batteries for BSSs is not always economically viable today,[69,74–76] which means that the BM would not be profitable today. Furthermore, since currently there are too few available used batteries, it would also not be possible to carry out the repurposing as a day-to-day operational business today.

Today's cost and revenue calculations for repurposing in the future are insecure due to various factors, such as the quality of available batteries for repurposing, the future prices of new batteries, but also the dynamics and price volatilities of the energy market, where they are used for storage.[40,77–78]

OEM can influence the economic attractiveness of future repurposing by applying a standardized battery design. Standardization of modules will lead to lower repurposing costs because battery disassembling and reassembling costs decrease, while reconfiguration gets easier. Furthermore, the cell chemistry itself can influence which 2uB storage applications are promising.[79]

It is conceivable that an OEM plans for the secondary use of batteries when launching new EVs and corresponding batteries. Such a planning concept can include a leasing concept of 2nd-life-optimized EVBs for selected customer groups that hold larger vehicle fleets such as car-sharing providers. If the OEM has real-time SOH data during battery use, the optimal replacement time can be forecasted to maximize the overall economic and ecological best battery life.[80,81]

As the overall political and social developments tend to foster the circular economy, special attention should also be drawn to the ecological footprint of BSSs, which influences the overall environmental performance of the stored (renewable) energy. Standardized guidelines of calculating the ecological footprint for 2uB storage do not exist today, but are already addressed in research.[82–85] Offering 2uB storage with lower CO_2 footprints than comparable BSSs from new batteries can be a decisive selling argument.

At least, missing legal frameworks and approved standards for repurposing can be a major drawback for OEMs or their stakeholders to invest and enhance in repurposing activities in Europe.[72,86] Such guidelines are needed for the testing and quality standards of the repurposed products and the usage of 2uB storage itself. The first standard covering the evaluation of repurposing batteries has been approved by the American National Standard and the Canadian National Standard in 2018.[87] This standard sets a guideline for various processes, e.g. grading of batteries and possible reconfiguration of cells and modules if defined criteria are met. Comparable European or worldwide standards are still missing.[88]

5. Summary and Conclusion

Within the last sections, we explained how potential BMs for repurposing and remanufacturing of used EVBs can be designed from an OEM's perspective. We focused on discussing the most important factors such as supply and demand and added value for customers. Furthermore, we addressed potential drawbacks and barriers for business activities.

The two proposed BMs share the common elements of using existing collaboration structures to secure the supply for remanufacturing and repurposing. In particular, the repurposing BM enlarges on OEM's current business activities as the created products are not used within vehicles. Therefore, finding new collaboration partners, most of all energy suppliers, appears to be an indispensable success factor for such a BM.

While we displayed the two circular economy activities of remanufacturing and repurposing in separate BMs, a combination especially for an OEM can be employed. An OEM can benefit from additional flexibility when performing both remanufacturing and repurposing. This will make it even easier for remanufacturing to meet the entire demand for remanufactured batteries, while fewer spare parts for remanufacturing will have to be newly procured from suppliers. Since the two recovery concepts also include many identical processes such as disassembling, the company might have additional benefits, such as common economies of scale, thereby reducing costs.

Nevertheless, both BMs face similar barriers, such as insecurity about future legal frameworks. These barriers can hold back needed investment into 2uB-friendly design, today. Furthermore, the expected use phase of EVBs exceeds the usual planning time horizon for strategic business decisions. It seems reasonable that OEMs might invest more in circular economy strategies if they can expect benefits before products are returned. A benefit does not necessarily need to be monetary, but could also be a kind of reimbursement when deciding on the ecological footprint or degree of circularity of products. This shows that society and politics can set incentives and influence the attractiveness of circular second use battery business models.

References

1. Osterwalder A, Pigneur Y, Tucci CL. Clarifying business models: Origins, present, and future of the concept. CAIS: Communications of the Association for Information Systems, 2005;16: 1–25.

2. Zott C, Amit R, Massa L. The business model: Recent developments and future research. J Manage. 2011;37(4):1019–42.
3. Bieger T, Reinhold S. Das wertbasierte Geschäftsmodell — Ein aktualisierter Strukturierungsansatz. In: Bieger T, Knyphausen-Aufseß D zu, Krys C, editors. Innovative Geschäftsmodelle. Berlin, Heidelberg: Springer Berlin Heidelberg; 2011. pp. 13–70.
4. Teece DJ. Business models, business strategy and innovation. Long Range Plann. 2010;43(2–3):172–94.
5. Al-Debei M, Avison D. Developing a unified framework of the business model concept. Eur J Inf Syst. 2010;19:359–76.
6. Gassmann O, Frankenberger K, Csik M. Geschäftsmodelle entwickeln: 55 innovative Konzepte mit dem St. Galler Business Model Navigator. 2., überarbeitete und erweiterte Auflage. München: Hanser; 2017. Available from: https://www.researchgate.net/publication/318212328_Geschaftsmodelle_entwickeln_55_innovative_Konzepte_mit_dem_St_Galler_Business_Model_Navigator
7. Morris M, Schindehutte M, Allen J. The entrepreneur's business model: Toward a unified perspective. J Bus Res. 2005;58(6):726–35.
8. Petrovic O, Kittl C, Teksten RD. Developing Business Models for Ebusiness (October 31, 2001). Available at SSRN: https://ssrn.com/abstract=1658505 or http://dx.doi.org/10.2139/ssrn.1658505 Titel anhand dieser DOI in Citavi-Projekt übernehmen
9. Amit R, Zott C. Value creation in E-business. Strat Mgmt J. 2001;22(6–7):493–520.
10. Morris M, Schindehutte M, Richardson J, Allen J. Is the business model a useful strategic concept? Conceptual theoretical and empirical insights. J Small Bus Strategy. 2006;17(1):27–50. Available from: https://libjournals.mtsu.edu/index.php/jsbs/article/view/62
11. Geissdoerfer M, Pieroni MP, Pigosso DC, Soufani K. Circular business models: A review. J Cleaner Prod. 2020;277:123741.
12. Bienias K, Kowalska-Pyzalska A, Ramsey D. What do people think about electric vehicles? An initial study of the opinions of car purchasers in Poland. Energy Rep. 2020;6:267–73.
13. Casper R, Sundin E. Addressing Today's challenges in automotive remanufacturing. J Remanuf. 2018;8(3):93–102.
14. Audi AG. Circular economy: Value creation and Produktion; 2019. Available from: https://www.audi.com/en/company/sustainability/core-topics/value-creation-and-production/circular-management.html
15. Daimler AG. Special "Lebenszyklus Batterie"; [cited 2021 Feb 10]. Available from: https://www.daimler.com/nachhaltigkeit/lebenszyklus-batterie/
16. emove306°. Interview mit Markus Schäfer, Daimler AG: "Das VISION AVTR unterstreicht den Mut und Pioniergeist unserer Marke"; [cited 2021 Feb 18]. Available from: https://www.emove360.com/de/interview-mit-markus-schaefer-daimler-ag-das-vision-avtr-unterstreicht-den-mut-und-pioniergeist-unserer-marke/
17. Volkswagen. Volkswagen with new corporate mission statement environment "goTOzero"; [cited 2021 Feb 18]. Available from: https://www.volkswagen-newsroom.com/en/press-releases/volkswagen-with-new-corporate-mission-statement-environment-gotozero-5180
18. Audi AG. Audi Garantie: Neuwagengarantie der AUDI AG; [cited 2021 Feb 24]. Available from: https://www.audi.de/dam/nemo/customer-area/warranty-guarantees/pdf/audi_garantie.pdf
19. BMW AG. BMW i Battery Certificate für die Hochvolt- Batterie — Leistungen und Bedingungen; 2019. Available from: https://www.bmw.de/content/dam/bmw/marketDE/bmw_de/topics/offers-and-services/warranty/pdf/Zertifikat%20BMWi_BEV_Ab%2001.08.2019_final.docx%20V2.pdf

20. Jordan M. Erstmals mit dem EQC bestellbar: After-Sales Service Pakete; 2019 [cited 2021 Feb 24]. Available from: https://blog.mercedes-benz-passion.com/2019/05/erstmals-mit-dem-eqc-bestellbar-after-sales-service-pakete/
21. Opel. Batterien für Elektroautos; 2021 [cited 2021 Feb 24]. Available from: https://www.opel.de/e-mobilitaet/fakten-vorteile/batterien.html
22. Tesla. Neuwagengarantie; 2021 [cited 2021 Feb 24]. Available from: https://www.tesla.com/de_DE/support/vehicle-warranty
23. Volkswagen AG. Hochvoltbatterie: Garantie und Pflege; 2021 [cited 2021 Feb 24]. Available from: https://www.volkswagen.de/de/elektrofahrzeuge/id-technologie/batteriegarantie-und-pflege.html
24. Endlein M. DAT report 2017. Ostfildern; 2018.
25. Endlein M. DAT report 2018. Ostfildern; 2019.
26. ZDK. Umsatz des deutschen Kraftfahrzeuggewerbes im Jahr 2019 nach Geschäftsfeldern (in Milliarden Euro); 2020 [cited 2021 Feb 5]. Available from: https://de.statista.com/statistik/daten/studie/13266/umfrage/umsatz-im-kfz-gewerbe-in-deutschland/
27. KBA. Neuzulassungen, Besitzumschreibungen, Außerbetriebsetzungen nach Zulassungsbezirken (FZ 5), Jahr 2019; 2020 [cited 2020 Feb 10]. Available from: https://www.kba.de/DE/Statistik/Produktkatalog/produkte/Fahrzeuge/fz5_nua_uebersicht.html?nn=1146130
28. Tolksdorf B. So entwickeln sich die Restwerte von Elektroautos; 2020 [cited 2021 Feb 5]. Available from: https://www.mobile.de/magazin/artikel/restwert-entwicklung-elektroautos-41404
29. Deloitte. Elektromobilität in Deutschland: Marktentwicklung bis 2030 und Handlungsempfehlungen; 2020 [cited 2021 Feb 5]. Available from: https://www2.deloitte.com/de/de/pages/consumer-industrial-products/articles/elektromobilitaet-in-deutschland.html
30. Nationale Plattform Zukunft der Mobilität. Wege zur Erreichung der Klimaziele 2030 im Verkehrssektor: Arbeitsgruppe 1 Klimaschutz im Verkehr. Berlin; 2019 [cited 2021 Feb 19]. Available from: https://www.plattform-zukunft-mobilitaet.de/wp-content/uploads/2020/03/NPM-AG-1-Wege-zur-Erreichung-der-Klimaziele-2030-im-Verkehrssektor.pdf
31. Endlein M. DAT report 2020. Ostfildern; 2020.
32. Erich M. Future residual values of battery electric vehicles benefit from increased range; 2019 [cited 2020 Feb 5]. Available from: https://www.readkong.com/page/future-residual-values-of-battery-electric-vehicles-benefit-4002175
33. Letmathe P, Suares M. A consumer-oriented total cost of ownership model for different vehicle types in Germany. Transp Res Part D: Transp Environ. 2017;57:314–35.
34. Endlein M. DAT report 2016. Ostfildern; 2017 [cited 2021 Feb 19]. Available from: https://www.dat.de/fileadmin/media/download/DAT-Report/DAT-Report-2016.pdf
35. Linzenich A, Arning K, Bongartz D, Mitsos A, Ziefle M. What fuels the adoption of alternative fuels? Examining preferences of German car drivers for fuel innovations. Appl Energy. 2019;249:222–36.
36. Letmathe P, Suares M. Understanding the impact that potential driving bans on conventional vehicles and the total cost of ownership have on electric vehicle choice in Germany. Sustainable Futures. 2020;2:100018.
37. Volkswagen Original Teile Logistik GmbH & Co. KG. Wettbewerbsüberlegene Logistik liefert alles aus einer Hand; 2016 [cited 2021 Feb 18]. Available from: https://www.volkswagen-otlg.de/wettbewerbsueberlegene-logistik
38. Kamper A, Triebs J, Hollah A, Lienemann C. Remanufacturing of electric vehicles: challenges in production planning and control. Procedia Manuf. 2019;33:280–7.

39. Parker D, Riley K, Robinson S, Symington H, Tewson J, Jansson K, *et al.* Remanufacturing market study: for horizon 2020, grant agreement No 645984, November 2015; 2015 [cited 2021 Feb 19]. Available from: https://www.remanufacturing.eu/assets/pdfs/remanufacturing-market-study.pdf
40. Fischhaber S, Regett A, Schuster S, Hesse H. Second-Life-Konzepte für Lithium-Ionen-Batterien aus Elektrofahrzeugen; 2016.
41. Lebedeva N, Tarvydas D, Tsiropoulos I. Li-ion batteries for mobility and stationary storage applications: scenarios for costs and market growth. Luxembourg: Publications Office of the European Union; 2018. (EUR, Scientific and technical research series; vol 29440).
42. Proposal for a Regulation of the European Parliament and the Council concerning batteries and waste batteries, repealing Directive 2006/66/EC and amending Regulation (EU) No 2019/1020; 2020 [cited 2021 Feb 19]. Available from: https://eur-lex.europa.eu/legal-content/EN/TXT/HTML/?uri=CELEX:52020PC0798&from=EN
43. Bowler M. Battery second use: A framework for evaluating the combination of two value chains; 2014 [cited 2021 Feb 6]. Available from: https://tigerprints.clemson.edu/cgi/view content.cgi?article=2379&context=all_dissertations
44. Martinez-Laserna E, Gandiaga I, Sarasketa-Zabala E, Badeda J, Stroe D-I, Swierczynski M, *et al.* Battery second life: Hype, hope or reality? A critical review of the state of the art. Renewable Sustainable Energy Rev. 2018;93:701–18.
45. Plenter F, Menne S, Hindersmann C, Bräuer S, Voscort J, Mittmann R. Szenarien und Geschäftsmodelle für die Vermarktung umgewidmeter Traktionsbatterien. In: Becker J, Beverungen D, Winter M, Menne S, editors. Umwidmung und Weiterverwendung von Traktionsbatterien. Wiesbaden: Springer Fachmedien Wiesbaden; 2019. pp. 179–258.
46. Figgener J, Stenzel P, Kairies K-P, Linßen J, Haberschusz D, Wessels O, *et al.* The development of stationary battery storage systems in Germany — A market review. J Energy Storage. 2020;29:101153.
47. Figgener J, Stenzel P, Kairies K-P, Linßen J, Haberschusz D, Wessels O, *et al.* The development of stationary battery storage systems in Germany — Status 2020. J Energy Storage. 2021;33:101982.
48. The Moblity House. Ein zweites Leben für Fahrzeugbatterien: Speicherprojekte von The Mobility House; 2018 [cited 2021 Feb 18]. Available from: https://www.mobilityhouse.com/de_de/magazin/Unternehmen/second-life-speicherprojekte.html
49. Daimler AG. Weltweit größter 2nd-Use-Batteriespeicher geht ans Netz; 2018 [cited 2021 Feb 18]. Available from: https://media.daimler.com/marsMediaSite/de/instance/ko/Weltweit-groesster-2nd-Use-Batteriespeicher-geht-ans-Netz.xhtml?oid=13634457
50. Vattenfall GmbH. Second Life Batterien: Potential für Energiespeicher; 2018 [cited 2021 Feb 18]. Available from: https://group.vattenfall.com/de/newsroom/blog/2018/september/second-life-batterien-energiespeicher
51. BMW Group. BMW i Batterien werden als "Second Life Batteries" flexible Speicher für erneuerbare Energien und sichern die Stabilität des Stromnetzes. Hamburg; 2014 [cited 2021 Feb 18]. Available from: https://www.press.bmwgroup.com/deutschland/article/detail/T0193200DE/bmw-i-batterien-werden-als-"second-life-batteries"-flexible-speicher-fuer-erneuerbare-energien-und-sichern-die-stabilitaet-des-stromnetzes?language=de
52. Werwitzke C. VW und E.ON: Gemeinsamer Rollout mobiler Schnellladesäulen; 2020 [cited 2021 Feb 18]. Available from: https://www.electrive.net/2020/02/12/vw-und-e-on-gemeinsamer-rollout-mobiler-schnellladesaeulen/
53. Lambert F. VW deploys first electric car charging stations with giant integrated batteries; 2020. Available from: https://www.electrive.net/2020/02/12/vw-und-e-on-gemeinsamer-rollout-mobiler-schnellladesaeulen/

54. BELECTRIC. BELECTRIC builds battery storage facility for Audi. Berlin; 2019 [cited 2021 Feb 18]. Available from: https://belectric.com/belectric-builds-battery-storage-facility-for-audi/
55. Reintjes M. Audi batteries find second life in forklift trucks; 2019 [cited 2020 Nov 11]. Available from: https://recyclinginternational.com/batteries/audi-battery-recycling/18993/
56. Green Car Congress. Audi testing second-life EV batteries in factory vehicles; 2019 [cited 2020 Oct 20]. Available from: https://www.greencarcongress.com/2019/03/20190307-audi2ndlife.html
57. Energy2market GmbH. Auf dem Weg zur energiewirtschaftlichen Optimierung eines Automobilproduktionsstandortes: Wie die e2m mit der Speicherfarm der BMW Group am Standort Leipzig Flexibilitätspotenziale hebt; 2018 [cited 2021 Feb 18]. Available from: https://www.e2m.energy/de/newsbeitrag-anzeigen/auf-dem-weg-zur-energiewirtschaftlichen-optimierung-eines-automobil-produktionsstandortes.html
58. Still B, Busse J. Audi und EnBW kooperieren bei Batteriespeichern: Gemeinsame Pressemitteilung; 2020 [cited 2021 Feb 26]. Available from: https://www.enbw.com/media/presse/images/pressemitteilungen/2020/20200623__gempm_audi_enbw_2ndlife.pdf
59. EUROBAT. Battery energy storage in the EU: services and technologies; 2020 [cited 2021 Feb 17]. Available from: https://www.eurobat.org/images/members/Battery_Energy_Storage_in_the_EU_report.pdf
60. European Commission. The European Green Deal. Brussels; 2019 COM(2019) 640 final [cited 2021 Feb 17]. Available from: https://www.eea.europa.eu/policy-documents/com-2019-640-final
61. Groening C, Inman J., William Jr. The role of carbon emissions in consumer purchase decisions. IJEPDM. 2015;1(4):261.
62. Chen Q, Zhao T. Heat recovery and storage installation in large-scale battery systems for effective integration of renewable energy sources into power systems. Appl Therm Eng. 2017;122:194–203.
63. Rust RT, Inman JJ, Jia J, Zahorik A. What you don't know about customer-perceived quality: the role of customer expectation distributions. Marketing Sci. 1999;18(1):77–92.
64. Rallo H, Canals Casals L, La Torre D de, Reinhardt R, Marchante C, Amante B. Lithium-ion battery 2nd life used as a stationary energy storage system: ageing and economic analysis in two real cases. J Cleaner Prod. 2020;272:122584.
65. Hussain A, Bui V-H, Baek J-W, Kim H-M. Stationary energy storage system for fast EV charging stations: simultaneous sizing of battery and converter. Energies. 2019;12(23):4516.
66. Sbordone D, Bertini I, Di Pietra B, Falvo MC, Genovese A, Martirano L. EV fast charging stations and energy storage technologies: a real implementation in the smart micro grid paradigm. Electr Power Syst Res. 2015;120:96–108.
67. Laws ND, Anderson K, DiOrio NA, Li X, McLaren J. Impacts of valuing resilience on cost-optimal PV and storage systems for commercial buildings. Renewable Energy. 2018;127:896–909.
68. Lange B, Preisemann C, Geiss M, Lambrecht A. Promoting food security and safety via cold chains: technology options, cooling needs and energy requirements. Eschborn; 2016 [cited 2021 Feb 19]. Available from: https://www.giz.de/de/downloads/giz_2016_Food_Security_Cold_Chains.pdf
69. Heymans C, Walker SB, Young SB, Fowler M. Economic analysis of second use electric vehicle batteries for residential energy storage and load-levelling. Energy Policy. 2014;71:22–30.
70. Wankmüller F, Thimmapuram PR, Gallagher KG, Botterud A. Impact of battery degradation on energy arbitrage revenue of grid-level energy storage. J Energy Storage. 2017;10:56–66.
71. Klausen M. Market opportunities and regulatory framework conditions for stationary battery storage systems in Germany. Energy Procedia. 2017;135:272–82.
72. DeRousseau M, Gully B, Taylor C, Apelian D, Wang Y. Repurposing used electric car batteries: a review of options. JOM. 2017;69(9):1575–82.

73. Hossain E, Murtaugh D, Mody J, Faruque HMR, Haque Sunny MS, Mohammad N. A comprehensive review on second-life batteries: current state, manufacturing considerations, applications, impacts, barriers & potential solutions, business strategies, and policies. IEEE Access. 2019;7:73215–52.
74. Faessler B, Kepplinger P, Petrasch J. Decentralized price-driven grid balancing via repurposed electric vehicle batteries. Energy. 2017;118:446–55.
75. Huang N, Wang W, Cai G, Qi J, Jiang Y. Economic analysis of household photovoltaic and reused-battery energy storage systems based on solar-load deep scenario generation under multi-tariff policies of China. J Energy Storage. 2021;33:102081.
76. Mirzaei Omrani M, Jannesari H. Economic and environmental assessment of reusing electric vehicle lithium-ion batteries for load leveling in the residential, industrial and photovoltaic power plants sectors. Renewable Sustainable Energy Rev. 2019;116:109413.
77. White C, Thompson B, Swan LG. Repurposed electric vehicle battery performance in second-life electricity grid frequency regulation service. J Energy Storage. 2020;28:101278.
78. Wu W, Lin B, Xie C, Elliott RJ, Radcliffe J. Does energy storage provide a profitable second life for electric vehicle batteries? Energy Econ. 2020;92:105010.
79. White C, Thompson B, Swan LG. Comparative performance study of electric vehicle batteries repurposed for electricity grid energy arbitrage. Appl Energy. 2021;288:116637.
80. Wang T, Jiang Y, Kang L, Liu Y. Determination of retirement points by using a multi-objective optimization to compromise the first and second life of electric vehicle batteries. J Cleaner Prod. 2020;275:123128.
81. You G, Park S, Oh D. Real-time state-of-health estimation for electric vehicle batteries: a data-driven approach. Appl Energy. 2016;176:92–103.
82. Bobba S, Cusenza MA, Di Persio F, Eynard U, Mathieux F, Messagie M, *et al.* Sustainability Assessment of Second Life Application of Automotive Batteries (SASLAB): JRC exploratory research (2016-2017): final technical report, August 2018. Luxembourg: Publications Office of the European Union; 2018. (EUR, Scientific and technical research series; vol 29321) [cited 2021 Feb 25]. Available from: https://publications.jrc.ec.europa.eu/repository/bitstream/JRC112543/saslab_final_report_2018_2018-08-28.pdf
83. Bobba S, Mathieux F, Ardente F, Blengini GA, Cusenza MA, Podias A, *et al.* Life Cycle Assessment of repurposed electric vehicle batteries: an adapted method based on modelling energy flows. J Energy Storage. 2018;19:213–25.
84. Schulz M, Bey N, Niero M, Hauschild M. Circular economy considerations in choices of LCA methodology: how to handle EV battery repurposing? Procedia CIRP. 2020;90:182–6.
85. Yang J, Gu F, Guo J. Environmental feasibility of secondary use of electric vehicle lithium-ion batteries in communication base stations. Resour Conserv Recycl. 2020;156:104713.
86. Catton J, Walker SB, McInnis P, Fowler M, Fraser R, Young SB, *et al.* Comparative safety risk and the use of repurposed EV batteries for stationary energy storage. In: 2017 IEEE International Conference on Smart Energy Grid Engineering (SEGE). IEEE; 2017. pp. 200–9. Oshawa, ON, Canada.
87. ANSI/CAN/UL. Standard for evaluation for repurposing batteries. 1st Ed; 2018 Oct 25.
88. Di Persio F, Ruiz V. Standards for the performance and durability assessment of electric vehicle batteries: possible performance criteria for an Ecodesign Regulation. Luxembourg: Publications Office of the European Union; 2018. (EUR, Scientific and technical research series; vol 29371).
89. Mongird K, Viswanathan V, Balducci P, Alam J, Fotedar V, Koritarov V, *et al.* An evaluation of energy storage cost and performance characteristics. Energies. 2020;13:1–51.

Part VI

Outlook

https://doi.org/10.1142/9789811245626_0022

Chapter 21

Beyond Lithium Ion: All Solid-State Batteries

Duygu Karabelli* and Kai Peter Birke†
Fraunhofer IPA, Nobelstraße 12, Stuttgart, Germany
**duygukarabelli@gmail.com*
†kai.peter.birke@ipa.fraunhofer.de

Electrolytes are key components in electrochemical storage systems, which provide the ion transport mechanism between the cathode and anode of a cell. As the battery technologies are in continuous development, there has been a growing demand for more efficient, reliable and environmentally friendly materials. Solid-state lithium-ion batteries (SSLIBs) are considered as next-generation energy storage systems and solid electrolytes are the key components for these systems. Compared to liquid electrolytes, SEs are thermally stable (safer), less toxic and provide a more compact (lighter) battery design. However, the main issue is the ionic conductivity, especially at low temperatures. This chapter presents the most efficient SEs and compares them in terms of their performances and costs. The challenges associated with the current manufacturing of large-scale SSLIBs and their commercialization potentials are described.

1. Top-Down: Why Solid-State Batteries?

Since Sony presented and marketed the first Li-ion battery (LIB) technology in 1991, LIBs have been widely developed and are now present in all portable devices requiring a rechargeable battery (telephone, laptop, etc.). Moreover, the lower manufacturing cost of LIBs makes them also a market-leading technology today for their applications in electromobility (e-mobility). However, as e-mobility (especially Electric Vehicle, EV) is an increasing market and becoming more and more attractive for millions of customers, there is a need for higher energy density cells with increased charge–discharge and thermal performances.

Additionally, the greater its capacity, the better the driving range; however, the recharging time is proportionally longer.

Existing conventional LIB technology is reaching its performance limits, since there will be no compromise on lifetime or safety. The most recent "advanced" Li-ion systems with silicon anode are not expected to exceed the energy densities of 800 Wh L^{-1} or 300 Wh kg^{-1} on a cell scale.[1] It is possible to use Li metal instead of graphite as a negative electrode for achieving higher energy densities. Li metal has a specific capacity (3,860 mAh g^{-1}) about ten times higher than that of graphite[2] and that could reduce the charging time of the battery significantly.

The initial Lithium batteries were already based on "Li-metal" technology, using metallic lithium as the negative electrode, theoretically achieving the highest energy densities.[3] However, the use of lithium in metallic form coupled with an organic liquid electrolyte caused dendrite formations, and conventional porous separators do not prevent the cathode from dendrites which eventually led to an internal short circuit and thus to a thermal runaway. The serious safety problems of this system did not lead to growth during its years on the market. The replacement of a liquid electrolyte with a solid electrolyte, which also acts as a separator, would create an inert, solid system that could solve the problems mentioned above.

For many experts, the so-called "All Solid-State Battery" (ASSB) is currently the greatest hope for equipping electric cars in the medium term with cheaper, lighter, safer and longer-lasting batteries. Its special feature is that it uses a solid electrolyte instead of the liquid and flammable electrolytes. The solid electrolyte offers the prospect of gains in energy density and reduced dependence on critical materials such as cobalt.

In recent years, many companies and start-ups like Toyota, QuantumScape and Solid Power announced their projects and partnerships on solid-state systems. In 2011, the Bolloré group (Blue Solutions) commercialized an EV (Bluecar) running with a solid-state battery[4,5] within the project frame of Autolib based on the work of Hovington et al.[6] This ASSB consisted of a Li-metal anode, LFP cathode and PEO/LiTFSI SE. This battery pack delivered an energy density of 100 Wh kg^{-1} at the pack level and could run more than 1000 cycles.

The reason why this technology has not yet reached the market is that researchers still have to solve a major problem: the low ionic conductivity of the solid electrolyte materials at ambient temperatures. To enable rapid charging, it is essential that lithium ions can pass through the electrolyte quickly without extra sophisticated measures.

Briefly, for a successful ASSB, the solid electrolyte must meet several key criteria such as (i) high ionic conductivity, (ii) wide electrochemical stable window and chemical stability, (iii) simple management of the interfaces between the components of the cell and (iv) affordable cost.[7] There have been many studies to find the best-fitting and -suiting solid electrolyte in order to make ASSBs competitive with today's Li-ion technology.

2. Solid-State Electrolytes as Powders: The Challenge of Contacting

One of the most challenging technical difficulties is the interface between the solid electrodes and the solid electrolyte. When the electrolyte is liquid, it can be absorbed by finely structured electrodes like a sponge. Nevertheless, two adjacent solids cannot be joined together so easily without gaps.

Unlike liquid electrolytes, solid electrolytes are too viscous to adapt to the morphological changes of the surface. Therefore, there may be a decrease in the electrode/electrolyte contact area and as a result an increase in the contact resistance. Since a Li-metal anode has a smoother and denser surface compared to the porous anode, this contact resistance between cathode and solid electrolyte is more challenging.[8]

The volume change of the cathode during the charge–discharge process is another issue which must be taken into account while designing the solid-state cell. According to the study of Cheng *et al.*,[9] the $LiFePO_4$ (LFP) cathode expands up to 6.6%. As the solid electrolyte is directly neighboring the cathode, it has to show enough flexibility in order to adjust its form with the volume change of electrodes during charging and discharging. Among solid electrolytes, polymer electrolytes are much more flexible with lower Young's Modulus (E'_{PEO} = 70MPa) and show better elasticity compared to the LGPS electrolyte (E'_{LGPS} = 38 GPa).[10,11] However, presenting low Young's modulus (E′) is also not enough. The material must show good strength at the same time in order to resist dendrite growths. It has been reported that, for a dendrite-free deposition, the shear modulus (G′) of an SE should be at least twice that of Lithium metal (G^{Li} = 3.4 GPa, G^{LGPS} = 14.35, G^{PEO} = 26.2 MPa).[8,10,11]

A combination of different solid electrolytes might be necessary to achieve good full cell functionality.[12] Even with the use of a small volume fraction of a liquid electrolyte, it is possible to reduce contact impedances (see Section 5). Accordingly, there has been increasing work in recent years on composite electrolytes, e.g. (gel) polymer-oxide or (gel) polymer-thio-phosphate composites, and/or bilayer inorganic solid electrolytes, e.g. LGPS//Li_2S-P_2S_5 or LGPS//Li_2S-P_2O_5.[13,14]

3. The Five Main Paths of Manufacturing Solid-State Cells

Considering potential ideas of how to manufacture solid-state cells, the guide and advice could be to use the "best electrical contacts" and a preparation sequence which can satisfy mass production capability. In this regard, the following five ideas may come up[15]:

(a) **Powder filling and compressing in half shells:** This is a little bit similar to manufacturing coin cells. Profiting from the fact that many solid Li-ion conductors such as sulphides, phosphides and iodides exhibit an inherent ductility, one could expect that this behavior could be used in a favorable way to compress ion conducting membranes and electrodes containing such ductile solid ion conductors. In an ideal case, no additional binder is necessary or the binder content can be at least distinctly and favorably lowered. Since metallic half coin housings are very similar to well-known coin cells that enable stiffness and mechanical integrity, this simple approach could develop into an interesting production approach which, however, may be limited in size, since large coin cells without circle shapes are very challenging in manufacturing.

(b) **Gluing:** Since solid ion conductors as well as active materials in electrodes consist of powders, one could, mainly following the basic idea of the earlier approach, create some gluing electrolytes, e.g. by incorporating inert binders with gluing properties such as Oppanol (BASF). The huge advantage of this approach could be that no additional pressure is required to maintain the contacts. Also, electrolyte solvents in a liquid electrolyte exhibiting such gluing properties could be an option. The liquid electrolyte acts as glue for the particles and is thus immobilized. The composite could then be considered as a quasi-solid. Polystyrol and limonene could act as compounds of such a gluing liquid electrolyte.

(c) **Laminating:** The earlier mentioned glue could also be specialized in a way that it reacts upon lamination (heat + pressure). Again, it could be some inert binder or plasticizer, which of course has to be electrochemically stable, or in case of a plasticizer, that does not perform this task, must be subsequently extracted. The unique pores after such an extraction could be filled by a liquid or gel which is immobilized afterward such that the result is a quasi-solid again.

(d) **Pressing:** This option describes option (a) but without any stabilizing half shells. It may be compared to pressed graphite foils which obtain their characteristics only from the unique mechanical properties of graphite. They look like perfect foils, but without any binder. Again, one could imagine using ductile properties of solid ion conductors to achieve similar results.

(e) **Sintering:** An alternative solution could be sintering. High-temperature fuel cell membranes (ZrO_2) are sintered to be thin, flexible and gas tight. But, in battery cells, there is no easy contact with gases and steam. It needs to be a well-connected solid–solid border. However, sintering electrodes and solid electrolyte together can lead to undesired reactions due to the required high sintering temperatures.

Having a top-down "eagle-eye" view of all these proposals can raise the ultimate question: which of these options can compete with actual Li-ion GWh factories, in terms of cost, mass production capability and saving energy densities, without compromising on losing valuable energy density by thick electrolytes, extremely heavy housings and very high solid electrolyte contents in the electrodes? But, this is not the only, though obvious, battle. Li-ion technology has a huge drawback over aqueous systems, i.e. the ion conductivity. The solution in case of Li-ion technology is easy — compensate conductivity by reduced thickness. Just look at typical electrode and separator thicknesses. This is in the regime of only μm (e.g. negative electrode 70–100 μm, Separator 20 μm, positive electrode up to 150 μm.) But now, solid Li-ion electrolytes have worse ionic conductivity properties than Li-ion cells. This means the best idea would be a copy and adaption of the present Li-ion production process to use solid ion conductor-based Li-metal cells to compensate as much as possible for worse ion conductivities. Following the above discussion, options (b–d) could come nearest to these overall demands. And, as a preliminary result of this discussion, some options that are further developed are already present to push Li-metal batteries with solid ion conductors as competitors for conventional Li-ion technology.

4. Hybrid Approaches

Hybrid approaches are an already well-known form of Li-ion batteries employing liquid organic electrolytes, though many people may not be aware of this. Here, an insulator, more precisely an insulating polymer with voids filled by the liquid organic electrolyte, is present. These polymers are mostly polyolefin-based separators such as polyethylene, polypropylene or sequences thereof. Also, other compounds such as Teflon, Nylon, polyamides (Aramid, Kevlar) and glass fibers are under investigation,[16,17] and all of them have a similar insulating behavior. The ionic conductivity is provided by the liquid electrolyte in the pores of the polymer. The porosity is adjusted in a way that the best filling effect of the liquid organic electrolyte is combined with the necessary insulating properties vs.

electron transport, shorts and avoidance of dendrite formation. Nowadays, "hybrid approaches" have a renaissance and in the case of the presence of solid ion conductors, this means the combination of solid ion conductor and liquid electrolyte. In principle, the above-mentioned separator materials are replaced by the solid electrolyte, which has a huge advantage of bringing up ion conductivity instead of total insulation.

Following this important argument, it may be considered as an inherent disadvantage of the earlier mentioned separator materials (e.g. polyolefin) that they are complete insulators, meaning that where the material is physically present, there is a complete blockage. On the contrary, electrodes of modern Li-ion cells have a structure where grains of active material, such as graphite in the negative electrode and Lithium-Nickel-Manganese-Cobalt-Oxide in the positive electrode, with typical grain sizes in the regime of 10–30 μm, are connected by a binder. In the case of the positive electrode, additionally highly conducting carbon modifications are added. The residual porosity, even after a calendaring (densification) step, is still in the regime of typically 20–30%, and is employed afterward to host the liquid electrolyte. Since the absolute grain size is small, the relatively poor diffusion coefficients for Lithium are enough to preserve the required kinetics of such batteries. So, why not transfer this concept to the electrolyte/separator membrane? Instead of the complete blocking plastic regimes, solid ion conducting particles are present and the voids take up liquid electrolyte. This yields a much more homogeneous distribution of ion conducting areas.

But, this approach includes many challenges. It is not known how the liquid electrolyte acts positively in the case of metallic lithium as the negative electrode material. Organic electrolyte parts stay flammable. The solid electrolyte particles are often highly moisture and carbon dioxide sensitive.[12] And, contrary to the polyolefin, no connected homogeneous structure consisting of one material is present. The particles of the solid electrolyte are only glued — attached together — by the binder, but the resulting structure is usually highly sensitive to pinholes and metallic dendrite growth in between the particles. Much smaller particle sizes of the solid electrolyte ($<1\,\mu$m) are required, because the membranes have to become, in an ideal case, as thin as polyline membranes. The required small particles, however, enhance the reactive behavior with the environment (moisture). Small particles are a challenge in fabrication. To compensate for the more deficient structure over polyolefin, the electrolyte membrane still unusually becomes much thicker (30–100 μm) in order not to fail in a mass production process. The competitors are cheap polyolefin separators that are easy to fabricate, with a range between 8 and 25 μm. Many promising solid electrolytes are actually much more expensive than polyolefin — and probably will stay much more expensive, even if delivered in tons — and contribute negatively to the energy density balance due to the enhanced thickness and often their inherent weight. However, the hybrid approach can still make some sense, because ions are not conducted in channels, but can be much more homogeneously distributed over the area. This avoids high current densities and can suppress dendrite formation. Also, Solid Electrolyte Interphases (SEIs) between the electrolyte/separator and electrode area can be build up more smoothly. The absolutely reduced contact area of the liquid electrolyte part with a metallic lithium electrode can be favorable. And, in case the polymeric binder immobilizes or jellifies the liquid electrolyte, the resulting "solid" composite can become an interesting alternative. The obvious highly important question of "How solid should solid be?" will be answered in the following subsection.

5. How Solid is Solid?

"Solid", "solid electrolytes", and "solid-state batteries" are all not meant to be light fare. These approaches are also not meant to solve all sustainably problems such as safety and lifetime, providing perfect "green batteries". On the contrary, one should interrogate these definitions regarding their rationale. What is really required in practical for advanced batteries? It makes indeed sense to think about solid electrolytes. In case of Li-ion or Li-metal-based cells, solid electrolytes should and must only accomplish a situation that, within the operating regime of the battery, only Li ions are sufficiently mobile (besides the electrons in the electrodes and the outer circuit of course). If Li ions are the only species which are mobile, side reactions are, as a consequence, sufficiently suppressed, and the battery becomes at least kinetically stable within elapsed periods of time, if defined operating conditions such as temperature limits are not harshly transgressed. However, in this case, "solid" means somehow "immobilized", as far as other species besides Li ions and electrons are considered in the battery cell. This can, however, be also easily accomplished by polymer, gel or equivalent approaches. It does not consequently demand ceramic solid electrolytes. On the contrary, such an approach, at least to partly consider jellified regimes in the solid composite electrolyte, introduce Li-ion conducting polymers and even look at solid–liquid hybrid approaches. These approaches can have tremendous benefits for improved Li-ion conductivity. Another simple but highly important issue is the stability window. If sufficient solidification is present, it can highly favor beneficial Solid Electrolyte Interphases (SEIs) between the negative electrode and the solid electrolyte (separator) as well as between the positive electrode and the solid electrolyte. If Li ions are just immobile to guarantee a good SEI (which is, for example, not re-dissolved and well conductive for Li ions), it is a perfect definition of "How solid must solid be". Last but not least, there are two important mechanical issues raised: potential Li-dendrite formation and mass production capability. The solid electrolyte must be as solid to successfully suppress metal dendrite formation. This still seems to be one of the unsolved challenges of employing Li-metal and not having a thin film approach such as with sputtered LiPON,[18] which is not feasible for commercial mass production of cells with sufficient capacity and cost corridor. Furthermore, the electrolyte must show a consistency to unify dendrite suppression and sufficient thinness without pinholes (which is an important production prerequisite) in order not to compromise volumetric and gravimetric energy densities. To conclude, in principle, no solid-state battery is really needed but a Li-metal one is needed, because this, and not any solid electrolyte, shifts energy densities. But, SSBs seem actually be the only and best approach for the energy density push by Li metals.

6. Manufacturing as the Highest Hurdle of Solid-State Cells

Li-ion cells have one inherent miracle. One can manufacture them easily in terms of GWh with accepted cost structures, and this is not self-evident at all! If it becomes feasible to transfer the manufacturing steps of Li ion cells to all solid-state variants, it would be simple to issue a certificate for a great "all solid-state battery future". However, an indispensable prerequisite, and unfortunately not the only one, is to avoid any handling of metallic lithium in production. There is no way to put this into a roll-to-roll production and to control atmospheric influences (such as water, carbon dioxide, nitrogen and sulphuric compounds). Apart from this, no mass manufacturing techniques for Li foils with thicknesses in the regime between 10 and 30 μm are yet known, but these thicknesses of Li foil are required. And, at this stage, the accompanied costs for manufacturing and handling (e.g. heavy-duty dry rooms) of such Li foils are not discussed at all. Therefore, Lithium should be deposited upon the first charge (and not before) out of the positive electrode, and the negative copper (or nickel) current collector should, e.g. be prepared by a special seed layer to accomplish this easily. In that case, however, Li-metal solid-state cells can even show huge advantages in saving the expensive graphite negative electrode which is present in current Li-ion cells. The second issue is to maintain pressure on the cells before formation. If volume changes upon formation lead to any contact losses in the solid cells, this cannot be repaired any more afterward, and this happens once the pressure is released. The results are capacity loss and impedance increase. So, it is not only important to maintain pressure after formation but also to do so earlier and never interrupt this pressure! This can develop to a tremendous demand, because if cells are kept under pressure before and during formation, how can one accomplish the transfer into a battery without losing this pressure? In a battery, of course, the cells can get pressure again (this is already done and practiced in many commercialized large batteries and modules). In terms of Li-metal solid state, it means solving these two hurdles successfully. There is some solution for the first one (*in-situ* formation of metallic Lithium), but not yet for the second one. To summarize, there is actually no GWh factory solution available for Solid-state cells based on metallic Lithium, and this can develop to a final game changer in the direction for keeping Li ions. Many people seem to not yet be aware of this tremendous hurdle!

7. Solid Electrolyte Cost as Game Changer for Promising Candidates

Although inorganic solid electrolytes show excellent ionic conductivities and can compete with organic liquid electrolytes, they are much more expensive compared to SPEs. Currently, their high cost prevents ASSBs from being a viable option for cost-effective manufacturing. It must be noted that the prices for sulfide-based electrolytes decrease if significant production capacities are set up in the future. However, in the case of LGPS, the high cost of germanium still hinders the possibility of its large-scale production. Li_2S, P_2S_5 and GeS_2 are the starting materials and the synthesis of LGPS solid electrolyte takes place between 500 and 600°C in an inert atmosphere.[19] As germanium is an expensive element (1300 US\$ kg^{-1}),[20] the substitution of the critical element Ge has been achieved by isovalent elements such as Si or Sn (starting materials SiS_2 and SnS_2), which are much more affordable while only moderately affecting the conductivity performances.[21]

Schnell *et al.* built a calculation model to compare manufacturing costs of complete ASSBs and compared them with conventional LIB manufacturing.[22] According to their calculations, the manufacturing costs for the sulfide-based ASSBs vary from 132 to 86 \$kW h^{-1}, which is even higher than that of LIB with Si/C anode (~50 \$$kg^{-1}$). However, the higher cell manufacturing costs are mainly due to the use of Li-metal foil in their process. Considering the electrolyte part, even though a cost of a solid electrolyte is higher than that of organic liquid electrolytes, cost savings are still possible due to the elimination of electrolyte filling and formation processes.

8. How Digital Can Become Manufacturing of Solid-State Cells

Answering the above-posed question also required an answer to the following question in advance: How digital can Li-ion cell manufacturing become? Contrary to many other cases, Li-ion cell production distinctly suffers from reproducibility of results in different manufacturing steps. Prominent examples are particle distribution and contacts in the electrodes, porosity and uniform electrolyte penetration as well as penetration times. And, there are similar challenges beforehand. Because active materials of electrodes exhibit grain size distributions, they can show different particle shapes dependent on the production lots. Additionally, many active electrode materials react with the environment, especially with humidity but also sometimes with carbon dioxide. Looking back into electrode structures, local deviations are present in every electrode. However, measurable quantities are statistic such as capacity, impedance and voltage drop after storage. "Statistic" means none of the above-mentioned individual cell deviations are known and recorded in actual cell productions. This raises the following important question: What is an appropriate digitalization approach in a Li-ion battery cell manufacturing line at all? Before digitalization can take place, the reproducibility of results must be ensured, which is not thoroughly possible at the moment in a typical Li-ion cell production. But, two approaches may help. The collection of environmental data is highly important to accomplish at least one important digitalization step, e.g. heating up a machine during production changes tolerances. Keeping temperatures constant is often highly important to avoid mechanical tolerances in the final product. This is only one example. Additionally, it would help to collect as much individual cell data as possible such as weight and to combine them with the environmental data. In an ideal case a digital genetic fingerprint of each cell would be the result. These digital genetic fingerprints of many cells together can be seen as "big production data" and may help to improve machines and their interconnection as well as to optimize the production environment (e.g. dry room conditions). Also, a much better cell backtracking would thus be feasible. And, environmental aspects of the production would at least be better included into and understood for future process optimizations. What is the ultimate outlook? Some but not all of the processes and outputs will stay as random as now in a Li-ion cell production and thus deny digitalization and believing that you can improve random numbers and statistical results with digitalization, the simplification of the processes as well as alternative machine designs must be addressed. Digitalization can here only help in the sense of additional data collection but not solve the entire problem. And now, solid-state batteries can help. Some of the manufacturing processes can become simple. For example, a solid metal Li anode does not suffer from particle size distributions, and environmental influences are thus only present on the metal surface and not inside the whole 3D electrode like in graphite anode. Liquid electrolyte distribution issues vanish since solid electrolytes are present, which can be included before, and do not need not to enter a pore structure afterward as a liquid electrolyte has to do. To conclude, digitalization only will not suppress all deviations caused by manufacturing processes if these processes are not changed themselves. Simplification is the magic word, and this can only come from a change in battery technology itself. In this regard, digitalization has the prospect of being able to assist in the fast ramp-up of high-quality manufacturing of future solid-state battery cells at scale.

9. Next Steps for Solid-State Batteries

Solid electrolytes should not deteriorate to smoke dischargers or smoke grenades in case of advanced batteries. However, unfortunately, they have some inherent weaknesses, and the story may end up like that of ceramic superconductors. What are the indispensable demands for a good battery?

1. Energy density
2. Power density
3. Lifetime expectation (cyclic + calendric)
4. Safety
5. Costs (the overall ruler)

This needs to be, nowadays, supplemented by an LCA (Life cycle assessment) and a "virtual test bench" in advance, if the technology can be produced in terms of GWh factories or not. However, solid electrolytes seem, as per the definition, not to be a part of these indispensable demands for a good battery. So, the outlook ends up with a simple question: Which of these demands can solid electrolytes successfully address, or are there inherent limits which finally obviate their breakthrough in advanced batteries?

References

1. Li C, Zhang H, Otaegui L, Singh G, Armand M, Rodriguez-Martinez LM. Estimation of energy density of Li-S batteries with liquid and solid electrolytes. J Power Sources. 2016;326:1–5. Available from: https://www.sciencedirect.com/science/article/pii/S0378775316308217
2. Liu B, Zhang J-G, Xu W. Advancing lithium metal batteries. Joule. 2018; 2(5):833–45. Available from: http://www.sciencedirect.com/science/article/pii/S2542435118300977
3. Whittingham MS. Electrical energy storage and intercalation chemistry. Science. 1976;192(4244):1126–7.
4. Mauger A, Armand M, Julien CM, Zaghib K. Challenges and issues facing lithium metal for solid-state rechargeable batteries. J Power Sources. 2017;353:333–42. Available from: http://www.sciencedirect.com/science/article/pii/S0378775317304950

5. Zhang B, Zhang Y, Zhang N, Liu J, Cong L, Liu J, *et al.* Synthesis and interface stability of polystyrene-poly(ethylene glycol)-polystyrene triblock copolymer as solid-state electrolyte for lithium-metal batteries. J Power Sources. 2019;428:93–104. Available from: https://www.sciencedirect.com/science/article/abs/pii/S0378775319304367
6. Hovington P, Lagacé M, Guerfi A, Bouchard P, Mauger A, Julien CM, *et al.* New lithium metal polymer solid state battery for an ultrahigh energy: nano C-$LiFePO_4$ versus nano $Li1.2V_3O_8$. Nano Lett. 2015;15(4):2671–8.
7. Wanga L, Zhoua Z, Yan X, Hou F, Wen L, Luo W, Liang J, Xue Dou S. Engineering of lithium-metal anodes towards a safe and stable battery. Energy Storage Mater. 2018;14:22–48.
8. Zhao Q, Stalin S, Zhao C-Z, Archer LA. Designing solid-state electrolytes for safe, energy-dense batteries. Nat Rev Mater. 2020;5(3):229–52. Available from: https://www.nature.com/articles/s41578-019-0165-5
9. Cheng SH-S, He K-Q, Liu Y, Zha J-W, Kamruzzaman M, Ma RL-W, *et al.* Electrochemical performance of all-solid-state lithium batteries using inorganic lithium garnets particulate reinforced PEO/LiClO4 electrolyte. Electrochimica Acta. 2017;253:430–8. Available from: http://www.sciencedirect.com/science/article/pii/S0013468617318224
10. Barai P, Higa K, Srinivasan V. Lithium dendrite growth mechanisms in polymer electrolytes and prevention strategies. Phys Chem Chem Phys. 2017;19(31):20493–505. Available from: https://pubs.rsc.org/en/content/articlepdf/2017/cp/c7cp03304d
11. Wang ZQ, Wu MS, Liu G, Lei XL, Xu B, Ouyang CY. Elastic properties of new solid state electrolyte material Li10GeP2S12: A study from first-principles calculations. Int J Electrochem Sci. 2014;9(2):562–8.
12. Karabelli D, Birke KP, Weeber M. A performance and cost overview of selected solid-state electrolytes: Race between polymer electrolytes and inorganic sulfide electrolytes. Batteries. 2021;7(1):18. Available from: https://www.mdpi.com/2313-0105/7/1/18/pdf
13. Yao X, Liu D, Wang C, Long P, Peng G, Hu Y-S, *et al.* High-energy all-solid-state lithium batteries with ultralong cycle life. Nano Lett. 2016;16(11):7148–54.
14. Zhang Q, Peng G, Mwizerwa JP, Wan H, Cai L, Xu X, *et al.* Nickel sulfide anchored carbon nanotubes for all-solid-state lithium batteries with enhanced rate capability and cycling stability. J Mater Chem A. 2018;6(25):12098–105. Available from: https://doi.org/10.1039/C8TA03449D
15. Chen A, Qu C, Shi Y, Shi F. Manufacturing strategies for solid electrolyte in batteries. Front Energy Res. 2020;8:226. Available from: https://www.frontiersin.org/articles/10.3389/fenrg.2020.571440/pdf
16. Zhao Y, Li X, Shen J, Gao C, van der Bruggen B. The potential of Kevlar aramid nanofiber composite membranes. J Mater Chem A. 2020;8(16):7548–68. Available from: https://doi.org/10.1039/D0TA01654C
17. Mao Y, Sun W, Qiao Y, Liu X, Xu C, Fang Li, *et al.* A high strength hybrid separator with fast ionic conductor for dendrite-free lithium metal batteries. Chem Eng J. 2021;416:129119. Available from: https://www.sciencedirect.com/science/article/pii/S1385894721007105
18. Ko J, Cho DH, Kim DJ, Yoon YS. Suppression of formation of lithium dendrite via surface modification by 2-D lithium phosphorous oxynitride as a highly stable anode for metal lithium batteries. J Alloys Compd. 2020;845:156280. Available from: https://www.sciencedirect.com/science/article/abs/pii/S092583882032644X
19. Kato Y, Hori S, Kanno R. Li10GeP2S12-type superionic conductors: Synthesis, structure, and ionic transportation. Adv Energy Mater. 2020;10(42). Available from: https://onlinelibrary.wiley.com/doi/pdf/10.1002/aenm.202002153
20. Ruiz AG, Sola PC, Palmerola NM. Advanced material and device applications with germanium: germanium: Current and novel recovery processes; 2018.

21. Kato Y, Saito R, Sakano M, Mitsui A, Hirayama M, Kanno R. Synthesis, structure and lithium ionic conductivity of solid solutions of Li10(Ge1−xMx)P2S12 (M = Si, Sn). J Power Sources. 2014;271:60–4. Available from: https://www.sciencedirect.com/science/article/abs/pii/S0378775314012130
22. Schnell J, Knörzer H, Imbsweiler AJ, Reinhart G. Solid versus liquid — a bottom-up calculation model to analyze the manufacturing cost of future high-energy batteries. Energy Technol. 2020;8(3):1901237.

https://doi.org/10.1142/9789811245626_bmatter

Index

www.ingramcontent.com/pod-product-compliance
Lightning Source LLC
LaVergne TN
LVHW060016210926
848481LV00013B/3725
9789819833337